Edited by
Katsuhiko Ariga

**Organized Organic
Ultrathin Films**

Related Titles

Decher, G., Schlenoff, J. B. (eds.)

Multilayer Thin Films

Sequential Assembly of Nanocomposite Materials
2 Volume Set

2012
ISBN: 978-3-527-31648-9

Knoll, W., Advincula, R. C. (eds.)

Functional Polymer Films

2 Volume Set

2011
ISBN: 978-3-527-32190-2

Friedbacher, G., Bubert, H. (eds.)

Surface and Thin Film Analysis

A Compendium of Principles, Instrumentation, and Applications

Second, Completely Revised and Enlarged Edition
2011
ISBN: 978-3-527-32047-9

Kumar, C. S. S. R. (ed.)

Nanostructured Thin Films and Surfaces

Series: Nanomaterials for the Life Sciences (Volume 5)

2010
ISBN: 978-3-527-32155-1

Martin, P.

Introduction to Surface Engineering and Functionally Engineered Materials

2011
ISBN: 978-0-470-63927-6

Edited by Katsuhiko Ariga

Organized Organic Ultrathin Films

Fundamentals and Applications

WILEY-VCH Verlag GmbH & Co. KGaA

The Editor

Dr. Katsuhiko Ariga
National Institute for Materials
Science (NIMS)
International Center for Materials
Nanoarchitectonics (MANA)
1-1 Namiki
Tsukuba-Shi
Ibaraki 305-0044
Japan

All books published by **Wiley-VCH** are carefully produced. Nevertheless, authors, editors, and publisher do not warrant the information contained in these books, including this book, to be free of errors. Readers are advised to keep in mind that statements, data, illustrations, procedural details or other items may inadvertently be inaccurate.

Library of Congress Card No.: applied for

British Library Cataloguing-in-Publication Data
A catalogue record for this book is available from the British Library.

Bibliographic information published by the Deutsche Nationalbibliothek
The Deutsche Nationalbibliothek lists this publication in the Deutsche Nationalbibliografie; detailed bibliographic data are available on the Internet at <http://dnb.d-nb.de>.

© 2013 Wiley-VCH Verlag & Co. KGaA, Boschstr. 12, 69469 Weinheim, Germany

All rights reserved (including those of translation into other languages). No part of this book may be reproduced in any form – by photoprinting, microfilm, or any other means – nor transmitted or translated into a machine language without written permission from the publishers. Registered names, trademarks, etc. used in this book, even when not specifically marked as such, are not to be considered unprotected by law.

Print ISBN: 978-3-527-32733-1
ePDF ISBN: 978-3-527-65469-7
ePub ISBN: 978-3-527-65468-0
mobi ISBN: 978-3-527-65467-3
oBook ISBN: 978-3-527-65466-6

Cover Design Formgeber, Eppelheim, Germany
Typesetting Toppan Best-set Premedia Limited, Hong Kong
Printing and Binding Markono Print Media Pte Ltd, Singapore

Printed in Singapore
Printed on acid-free paper

Contents

Preface *IX*
List of Contributors *XI*

1 **Introduction** *1*
  Katsuhiko Ariga

2 **Self-Assembled Monolayer (SAM)** *7*
  Toshihiro Kondo, Ryo Yamada, and Kohei Uosaki
2.1 Introduction *7*
2.2 Preparation and Characterization *8*
2.2.1 Organothiols on Au *8*
2.2.2 Organosilanes on SiO_x Surfaces *15*
2.2.3 SAMs on Si Surface via Si–C Bonding *17*
2.3 Functions and Applications *20*
2.3.1 Surface Coating and Patterning *21*
2.3.2 Sensor Applications *23*
2.3.3 Nanotribology *26*
2.3.4 Advanced Applications *28*
2.3.4.1 Electron Transfer *28*
2.3.4.2 Photoinduced Electron Transfer *29*
2.3.4.3 Luminescence *34*
2.3.4.4 Catalytic Activity *34*
2.4 Future Perspective *36*
   References *37*

3 **Langmuir–Blodgett (LB) Film** *43*
  Ken-ichi Iimura and Teiji Kato
3.1 Concept and Mechanism *43*
3.2 Preparation and Characterization *43*
3.2.1 Gibbs Monolayers *43*
3.2.2 Langmuir Monolayers *45*
3.2.2.1 Basic Measurements of Properties of Langmuir Monolayers *45*
3.2.2.2 A–T isobars *47*

3.2.2.3 π–A isotherms 50
3.2.2.4 Stability of Langmuir Monolayers 52
3.2.3 *In situ* Characterization of Monolayers at the Subphase Surface 56
3.2.3.1 Brewster-Angle Microscopy (BAM) 56
3.2.3.2 Fourier Transform Infrared (FTIR) Spectroscopy 57
3.2.3.3 X-ray Reflectometry and Grazing-Incidence X-ray Diffractometry 61
3.2.4 Transfer to Solid Supports 63
3.2.4.1 Instruments for LB Film Deposition 65
3.2.4.2 Turnover of Amphiphile Molecules during Deposition 67
3.2.4.3 Horizontal Lifting-Up Deposition 69
3.2.4.4 Horizontal Scooping-Up 71
3.3 Functions and Applications 73
3.3.1 Molecular Recognition 73
3.3.1.1 Molecular Recognition by Hydrogen Bonding and Electrostatic Interaction at the Air/Water Interface 73
3.3.1.2 Chiral Discrimination at the Air/Water Interface 77
3.3.1.3 Macrocyclic Hosts 79
3.3.1.4 Dynamic Host Cavity 80
3.3.2 Multilayer Films for Photoelectronic Functions 83
3.3.2.1 Molecular Photodiode 83
3.3.2.2 Fullerene C_{60} Containing LB Film 85
3.3.2.3 Optical Logic Gate/Photoswitch 87
3.3.3 Biomimetic Functions 88
3.3.3.1 Biomembrane Models – Langmuir Monolayers of Lipids 89
3.3.3.2 Lung Surfactants 93
3.3.3.3 Biomimetic Mineralization 94
3.3.4 Advanced Applications 95
3.3.4.1 Sensors 95
3.3.4.2 Photoresponsive Films 98
References 99

4 Layer-by-Layer (LbL) Assembly 107
Katsuhiko Ariga
4.1 Concept and Mechanism 107
4.2 Preparation and Characterization 109
4.2.1 Applicable Materials and Interactions 109
4.2.2 Thin-Film Preparation: Fundamental Procedure and Characterization 114
4.2.3 Various Driving Forces and Techniques 120
4.2.4 Three-Dimensional Assemblies 129
4.3 Functions and Applications 136
4.3.1 Physicochemical Applications of LbL Thin Films 137
4.3.2 Biomedical Applications of LbL Thin Films 143
4.4 Brief Summary and Perspectives 153
Further Reading 154

5	**Other Thin Films** *157*
	Mineo Hashizume, Takeshi Serizawa, and Norihiro Yamada
5.1	Bilayer Vesicle and Cast Film *157*
5.1.1	Definition of a Bilayer Structure, a Bilayer Membrane, and a Bilayer Vesicle *157*
5.1.2	Formation of a Bilayer Structure *159*
5.1.2.1	Bilayer Forming Amphiphiles *159*
5.1.2.2	Properties of Bilayer Membrane and Diagnostics of Bilayer Formation *162*
5.1.2.3	Mechanism and Preparation of Bilayer Formation *164*
5.1.2.4	Future of the Bilayer Vesicle *166*
5.1.3	Cast Films Containing a Bilayer Structure *166*
5.2	Self-Assembled Fibers, Tubes, and Ribbons *169*
5.2.1	Introduction *169*
5.2.2	Finding a Helical Superstructure *169*
5.2.3	Organogel *172*
5.2.4	Control of Aggregate Morphology *173*
5.2.4.1	Composite Structure with Two or More Different Parts *175*
5.2.4.2	Hydrogen Bonding to Immobilize Orderly Molecular Arrangement *175*
5.2.4.3	Hierarchic Interaction and Further Interaction *177*
5.3	Polymer Brush Layer *179*
5.3.1	Definition of Polymer Brushes *179*
5.3.2	Preparation of Polymer Brushes *180*
5.3.3	Properties and Applications of Concentrated Polymer Brushes *182*
5.4	Organic–Inorganic Hybrids *184*
5.5	Colloidal Layers *191*
5.6	Newly Appearing Techniques *195*
5.6.1	Material-Binding Peptide *195*
5.6.2	Block-Copolymer Films *197*
5.6.3	Nanoimprint Lithography *200*
	References *200*

Index *205*

Preface

Most important structures in the world are organized organic ultrathin films. They are much more important than the other well-known structures such as nanotubes, nanoparticles and graphene. Why do I (you) think so? Answers may be found within our body. Organic ultrathin films are universal structural units of life. Our body and inside mechanisms are made through assembly of functional units where various interfacial environments of thin films such as cell membranes provide medium for highly efficient molecular conversion, energy conversion, information conversion, and the other important life activities. These functions are surprisingly precise, specific and efficient as well as highly flexible, dynamic, and soft. Not limited to structures found inside of our bodies, organic ultrathin films are seen everywhere around us from soap bubbles to technologically important surface coatings and drug carriers. In addition, technique and science for organized organic ultrathin films can be regarded as one of the most practically advanced nanotechnologies where nanometer-level control of film structures and molecular functions are successfully realized. Organized organic ultrathin films are most important targets in our studies in many research fields including advanced biotechnology and nanotechnology.

This book describes fundamentals and frontiers of four major topics of organized organic ultrathin films: (i) self-assembled monolayer (SAM); (ii) Langmuir–Blodgett (LB) film; (iii) layer-by-layer (LbL) assembly; (iv) other thin films (bilayer vesicle and cast film). All the authors of these chapters are highly experienced in both basic education and advanced research, thus this book satisfies readers with research purposes and educational aims. Organized organic ultrathin films are most important structures in the world. Therefore, this book is a most valuable science guide for you.

August 2012 *Katsuhiko Ariga*

List of Contributors

Katsuhiko Ariga
National Institute for Materials Science (NIMS)
International Center for Materials Nanoarchitectonics (MANA)
1-1 Namiki
Tsukuba-Shi
Ibaraki 305-0044
Japan

Mineo Hashizume
Tokyo University of Science
Department of Industrial Chemistry
12-1 Ichigayafunagawara-machi
Shinjuku-ku
Tokyo 162-0826
Japan

Ken-ichi Iimura
Utsunomiya University
Graduate School of Engineering
Department of Advanced Interdisciplinary Science
7-1-2 Yoto
Utsunomiya
Tochigi 321-8585
Japan

Teiji Kato
Utsunomiya University
Graduate School of Engineering
Department of Advanced Interdisciplinary Science
7-1-2 Yoto
Utsunomiya
Tochigi 321-8585
Japan

Toshihiro Kondo
Ochanomizu University
Graduate School of Humanities and Sciences
Division of Chemistry
2-1-1 Ohtsuka
Bunkyo-ku
Tokyo 112-8610
Japan

Takeshi Serizawa
Tokyo Institute of Technology
Department of Organic and Polymeric Materials
2-12-1-H121 Ookayama
Meguro-ku
Tokyo 152-8550
Japan

Kohei Uosaki
National Institute for Materials
Science (NIMS)
International Center for Materials
Nanoarchitectonics (MANA)
1-1 Namiki
Tsukuba-Shi
Ibaraki 305-0044
Japan

Norihiro Yamada
Chiba University
Faculty of Education
1-33 Yayoi-cho
Inage-ku
Chiba 263-8522
Japan

Ryo Yamada
Osaka University
Graduate School of Engineering
Science
Division of Materials Physics
1-3 Machikaneyama
Toyonaka
Osaka 560-8531
Japan

1
Introduction

Katsuhiko Ariga

Developments of nanotechnology and microtechnology have been tremendous and are having huge social impact. Based on these technologies, various tools and machines are significantly miniaturized, leading to compact and efficient information processing and communication, as seen in mobile computers and cellular phones. Handy and wearable devices have been developed that enhance communication and reduce traffic congestion and overpopulation in certain areas, which may produce reductions in power consumption and environmentally unfriendly emissions. In order to obtain ultrasmall functional systems, advanced nanotechnology-based fabrication for highly precise small structures plays a central role. Most of them are called top-down nanofabrication methods. For example, photolithographic techniques have been widely used for miniaturization of structures especially in silicon-based technology. Unfortunately, these top-down lithographic approaches require a combination of instrumentation, clean-room environment, and materials that are accompanied by rapid cost increases. Industries moving along the current direction may encounter unavoidable limitations due to economical reasons and/or technical reasons.

Therefore, alternate methodology, bottom-up approaches, will become indispensable (Figure 1.1). In the bottom-up approaches, the principles of self-assembly are central to construct nanostructures through spontaneous processes. Self-assembled processes are sometimes capable of forming highly integrated and complicated three-dimensional structures in an energyless one-step process. However, such assemblies are not often predictable and designable. Therefore, nanostructure formation in three-dimensional ways remains as fundamental sciences rather than well-established methodologies. If the dimensions of objects are reduced from three to two, the situation drastically changes. We already have an established strategy to make well-organized two-dimensional films (ultrathin films) through molecular self-assembly with the aid of external processes such as substrate dipping and solution casting. Three representative methodologies for thin-film preparation would be (i) self-assembled monolayer (SAM) method, (ii) Langmuir–Blodgett (LB) technique, and (iii) layer-by-layer (LbL) assembly. In particular, these methods are good ways to provide organic ultrathin films. Therefore, studies on organic ultrathin films would be good starting points for bottom-up nanotechnology.

Organized Organic Ultrathin Films: Fundamentals and Applications, First Edition. Edited by Katsuhiko Ariga.
© 2013 Wiley-VCH Verlag GmbH & Co. KGaA. Published 2013 by Wiley-VCH Verlag GmbH & Co. KGaA.

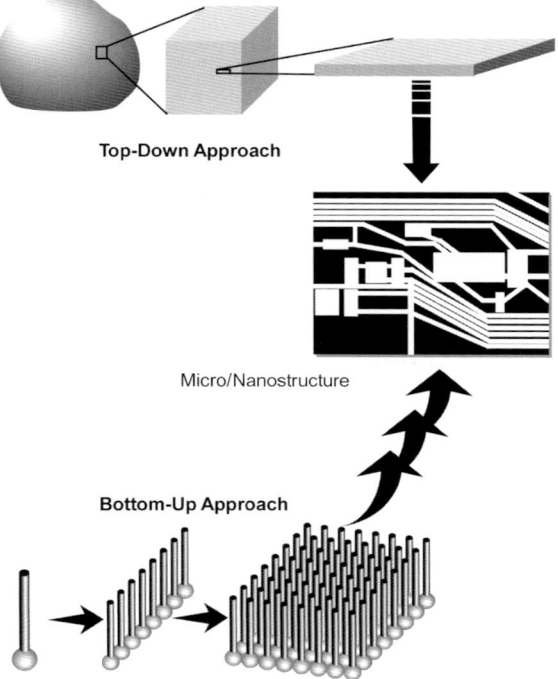

Figure 1.1 Top-down approach and bottom-up approach for fabrication of micro/nanostructures.

In this book, we describe the fundamentals and applications of organic ultrathin films upon classifications of fabrication strategies. Here, their outlines are summarized. Chapter 2 explains the self-assembled monolayer (SAM) method (Figure 1.2). The SAM method provides a monolayer strongly immobilized on a solid support. This method utilizes the strong interaction between the heads of the amphiphiles and the surface of the solid support, as seen in covalent linkages between silanol amphiphiles and a glass or metal oxide surface and strong interactions between thiol amphiphiles and a gold surface. These strong interactions with the solid surface sometimes allow molecules very different from those of typical amphiphiles to form monolayer structure on the surface. The formed SAM structure have great potential for a wide range of applications including sensors and various devices The formation of self-assembled monolayers is a powerful tool for surface modification.

In Chapter 3, the Langmuir–Blodgett (LB) technique is introduced (Figure 1.3). The LB technique is the most powerful method of achieving molecular assemblies with precisely layered structures. In this method, an insoluble monolayer of amphiphile molecules is first spread on the surface of a water phase. The monolayer can be highly compressed through lateral pressure application. The finally obtained highly condensed monolayer is transferred onto a solid support in a

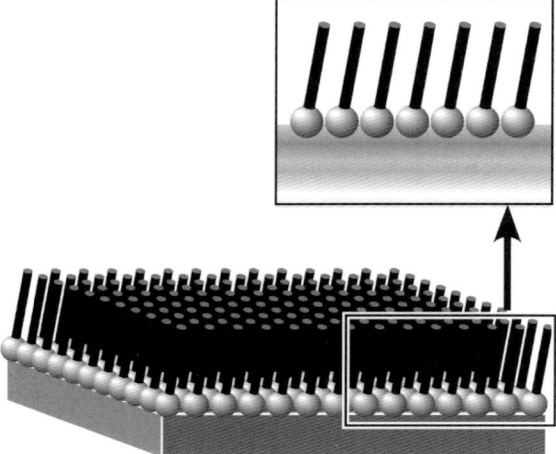

Figure 1.2 Self-assembled monolayer (SAM).

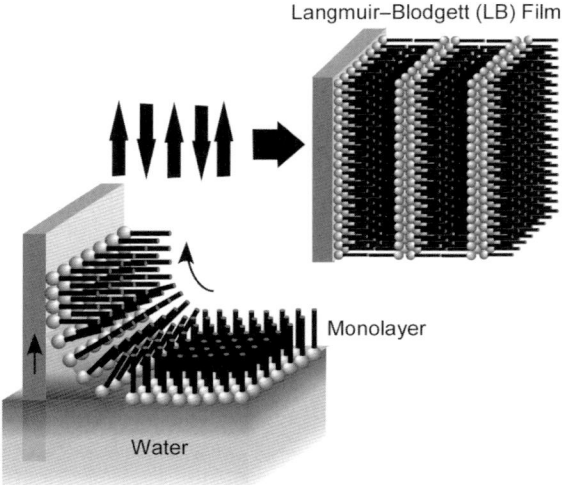

Figure 1.3 Langmuir–Blodgett (LB) film.

layer-by-layer manner by dipping the support through the monolayer. Film thickness (the number of the layers) is easily tuned in nanometer level just by controlling dipping cycles. The monolayer-forming amphiphile must have an appropriate hydrophilic–hydrophobic balance. The profile of monolayer compression can be interesting research subject of molecular assembly in two dimensions.

The LB method requires rather expensive apparatus, and water-soluble molecules are not usually appropriate targets. As compensation for these disadvantageous features, another type of technique for layered ultrathin films was developed.

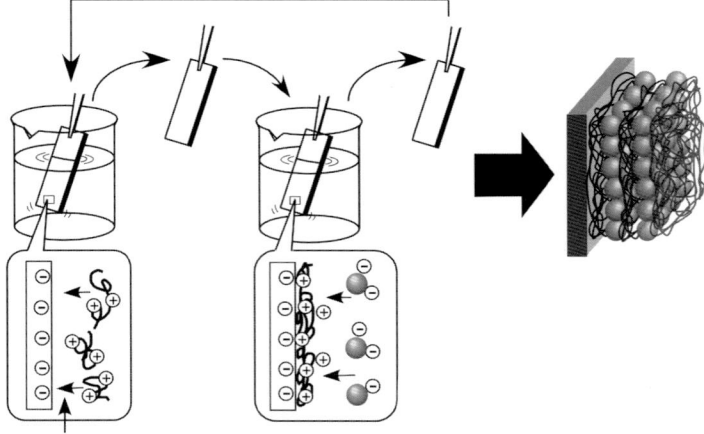

Figure 1.4 Layer-by-layer (LbL) assembly.

The so-called layer-by-layer (LbL) assembly is explained in Chapter 4 (Figure 1.4). Unlike the LB method, the LbL assembly can also be applicable to a wide range of water-soluble substances. In addition, this assembly method can be conducted using a very simple procedure with nonexpensive apparatuses such as beakers and tweezers. A typical LbL procedure is based on electrostatic adsorption. In the case of a solid support negative surface charge, adsorption of thin layer of cationic poelectrolyte neutralized surface change and subsequent overadsorption reverts surface charges. The subsequent process changes the surface charges alternately between positive and negative. Therefore, layered assembly can be continuously conducted to provide ultrathin films with desired thickness and layer sequence.

Chapter 5 describes the other types of organic ultrathin films and hybrid thin films some of which form assembling structures in solution (not on a solid surface) For example, formation of lipid bilayer structures in aqueous solution (Figure 1.5) and their transformation to thin films on a solid support by casting are exemplified. Lipids and related amphiphiles possess hydrophobic tails and a hydrophilic head. When they are dispersed in aqueous media, these molecules are usually assembled into bilayer films by avoiding unfavorable contact between hydrophobic parts of the molecules with external water media. Such organization often results in spherical assemblies having a water pool inside and lipid bilayer shell. These are called liposomes and/or vesicles. The formation mechanisms of these objects are basically identical to those for cell membranes. Casting of these dispersions onto a solid substrate leads to thin-film formation with multiple lipid layers. Upon appropriate designs of amphiphiles, their assemblies can extend to more complicated morphologies such as ribbons, sheets, and tubes.

In this book, various organic ultrathin films are described according to these categories, self-assembled monolayer (SAM), Langmuir–Blodgett (LB) films, layer-by-layer (LbL) assembly, and the other thin films such as lipid bilayers. Although

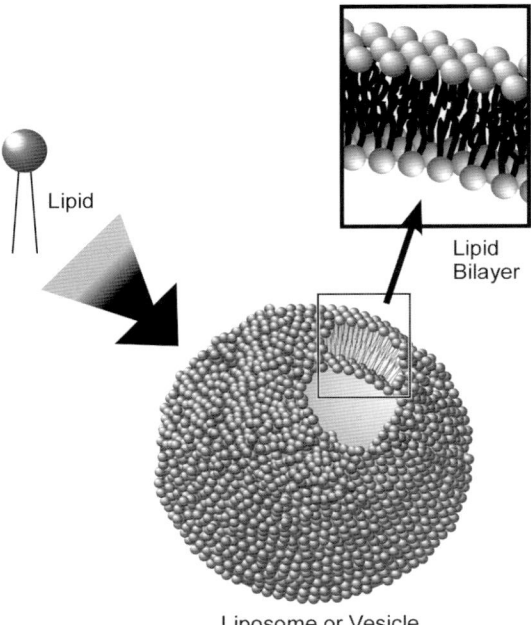

Figure 1.5 Formation of liposome (or vesicle).

the main aim of this book is to give an introduction to organic ultrathin films, some of the examples (especially in LbL assembly) are thin films of inorganic components. This means that a strategy useful for organic components can be applicable for inorganic nano-objects and their hybrids with organic components. We partially include this inorganic feature in a book entitled "Organic Ultrathin Films", because we want to demonstrate the wide versatility of the described methods and the availability of this typical bottom-up nanotechnology for all kinds of materials.

2
Self-Assembled Monolayer (SAM)

Toshihiro Kondo, Ryo Yamada, and Kohei Uosaki

2.1
Introduction

Construction of a molecular device with the desired functionality to fix and arrange molecules in order on a solid surface is one of the chemist's dreams. The attempts of fixing a molecular layer with various functionalities on a solid substrate and of controlling the surface properties have been carried out since Langmuir and Blodgett investigated Langmuir–Blodgett (LB) films, which are monolayers and multilayers transferred from the air/water interface onto a solid surface, in the early part of the last century [1–3]. Gaines [4] first summarized the details of LB films that were recently updated by Roberts [5]. Because the molecules are physisorbed onto the solid surface in LB films, their structures easily change and soon become random. In 1980, on the other hand, Sagiv found that molecules with a long alkyl chain can be fixed to the solid surface through the covalent linkage between the trimethoxysilyl (-Si(OCH$_3$)$_3$) or trichlorosilyl (-SiCl$_3$) group of the molecule and the surface hydroxyl (-OH) group of the solid substrate, and that an advanced orientation can be achieved by the interaction between the alkyl chains, and pointed out the similarity to the LB film [6]. Since the molecular layer can be spontaneously formed with a high orientation, this process is called self-assembly (SA) and the formed molecular film a self-assembled monolayer (SAM). Later, Nuzzo and Allara found in 1983 that an alkanethiol can react with a gold substrate to form Au–S bonds and that a highly oriented SAM with the hydrophobic interaction between the alkyl chains is formed on the gold surface [7], thus construction of a highly oriented molecular layer can be achieved on the electroconductive substrate, and the alkylthiol SAM on gold has been rapidly extended to both basic science and applications along with active research [8–13]. Recently, the modification of silicon surfaces of organic molecules through a covalent Si–C bonding has been realized by a wet reaction [14–18] and this research that makes the best use of the semiconductor property of the substrate has been actively carried out.

The molecule, which formed the SAM, consists of three parts as shown in Figure 2.1.

Organized Organic Ultrathin Films: Fundamentals and Applications, First Edition. Edited by Katsuhiko Ariga.
© 2013 Wiley-VCH Verlag GmbH & Co. KGaA. Published 2013 by Wiley-VCH Verlag GmbH & Co. KGaA.

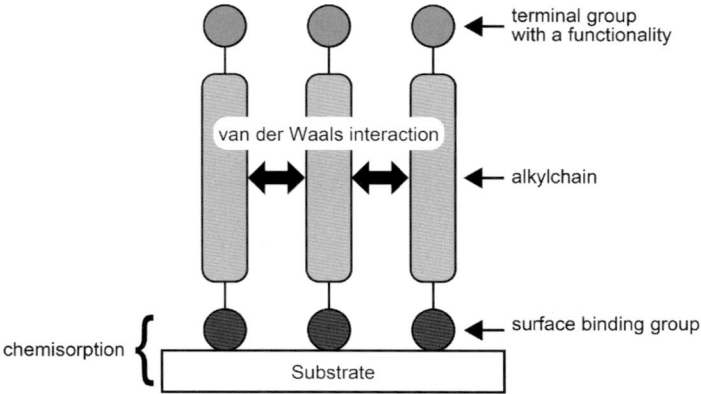

Figure 2.1 A schematic illustration of the SAM on a solid substrate.

The first part is the surface binding group (dark gray circle in Figure 2.1), which spontaneously binds to the substrate surface by a covalent linkage, so that this group depends on the substrate materials. In the case of using the oxide surface, which has a surface hydroxyl group, as the substrate, the trimethoxysilyl ($-Si(OCH_3)_3$) or trichlorosilyl ($-SiCl_3$) group is used as the surface binding group. When the thiol (-SH) group is used as a surface binding group, the semiconductor, such as GaAs [19], CdSe [20], and In_2O_3 [21], in addition to the metal, such as Pt, Ag, Cu, etc., besides Au, are used as a substrate. On these substrates, disulfide (-S–S-), selenol (-SeH), and isocyanide (-NC) can also be bonded [10]. When Si, as mentioned above, Ge, or diamond [22] are used as the substrate, the terminal olefin group can be used as a surface binding group [14–18]. The second part is the alkyl chain (rectangle in Figure 2.1), and the energy associated with its interchain van der Waals interaction is of the order of a few kcal/mol, which means that this interaction is exothermic [8]. The last part is the terminal group (light gray circle in Figure 2.1), which is generally a methyl group.

In this chapter, we introduce the fundamental science of the SAMs, such as molecular layer structures and formation processes of the SAMs, and applications of the SAMs to construct the surface with various functionalities.

2.2
Preparation and Characterization

2.2.1
Organothiols on Au

Organothiols adsorb on metal surfaces via metal–S bond formation, as shown in Figure 2.2. SAMs of alkanethiols can be prepared in solutions and vapors of molecules [7, 8, 10]. Self-assembly in solutions are commonly used because it is easy

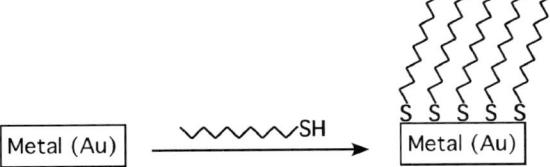

Figure 2.2 Schematic presentation of the SAMs of alkanethiols on metal substrates.

to prepare, control, and molecules that have small vapor pressure can be used. Typically, a several mM solution of alkanethiols in organic solvents, such as ethanol and hexane, is used to form SAMs.

A (111)-oriented gold thin film is the most widely used substrate since it is stable under ambient conditions and can be easily formed on various substrates, such as a glass slide, mica and silicon, by thermal vacuum evaporation [23]. A polished surface of a gold polycrystal is also used as a substrate [24]. Grains having atomically flat (111) terraces are grown when the substrate is heated around 300 °C during the deposition. Flame annealing is also employed to obtain flat and wide (111) terraces [23].

Most of the commercially available alkanethiols are usually used without further purification, though the residual impurity of sulfur is known to disturb the formation of an SAM of pyridinethiol on a Au(111) surface [25]. To form a monolayer, a clean substrate is immersed in the solution for 1–24 h. Sometimes, a much longer period, such as several days, is required for completion of the monolayer formation [26]. Both temperature and solvent strongly affect the density of the defects and domain size [27, 28]. These effects are discussed later.

The kinetics of the SA process in solution was monitored by a quartz crystal microbalance (QCM) [29], which can measure mass changes on surfaces. The SA process was found to be divided in an initial fast adsorption and subsequent slow steps [29–32]. The coverage reaches 80% during the initial first adsorption process (~10 min). The slow adsorption process continues for up to several hours.

Figure 2.3 shows the infrared spectrum of ferrocenyl-undecanethiol ($FcC_{11}SH$) monolayers formed on the Au(111) surface as a function of immersion time. The peak position assigned to CH vibration of the CH_2 group was found to shift from the frequency attributed to liquid-like phases of poly-methylene to that attributed to solid-like phases [32]. This result indicates that the alkyl chains in the monolayers become more solid-like during the slow adsorption process, thus, the slow adsorption process is interpreted as a defect-healing process.

Figure 2.4 shows schematic drawings of the top and side views of the monolayer. The alkyl chain is tilted from the surface normal about 30° with the all-*trans* conformation. This tilt angle comes from the conditions for closed packing of the alkyl chains. Close inspection of the IR data revealed that the plane defined by an all-*trans* carbon molecular skeleton alternatively changes its direction [33].

A high-resolution scanning tunneling microscopic (STM) image using large tunneling impedance revealed the molecular arrangement and local structures

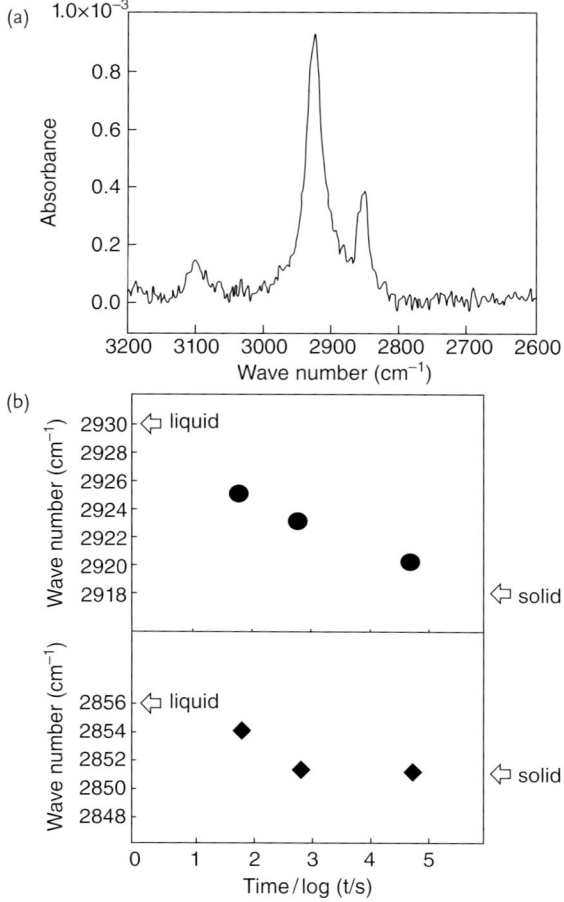

Figure 2.3 (a) IR absorption spectrum of a FcC$_{11}$SH monolayer on the Au (111) electrode. Modification time was 10 min. (b) Evolution of peak positions of the methylene asymmetric (upper panel) and symmetric (lower panel) stretching modes. Reprinted from [32].

of the SAMs [34]. Figure 2.5 shows a molecularly resolved STM image. The basic molecular arrangement is ($\sqrt{3} \times \sqrt{3}$)R30° with respect to the Au(111) surface. Close inspection of the structure revealed small differences in the height among the molecules [35–37]. The structure considering these modulations is called c(4 × 2) of ($\sqrt{3} \times \sqrt{3}$)R30°. The c(4 × 2) structure is attributed to the different orientations of the alkyl termination due to the different twist angles among the alkyl chains or small deviations of the sulfur atom positions from the hexagonal symmetry, indicating the existence of two kinds of sulfur positions [30, 38]. The different sulfur adsorption sites can result in a variable electronic structure and height in the monolayer. In fact, the position of the sulfur atom on a gold surface is still under debate.

(a) 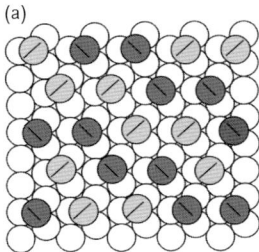 (b)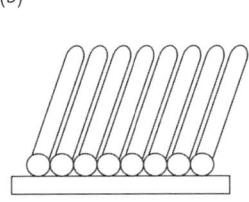

Figure 2.4 (a) Model of molecular arrangement with respect to Au(111) surface. Shadow circles and small open circles represent positions of molecules and gold atoms, respectively. A diagonal slash indicates the azimuthal orientation of the plane defined by the C–C–C backbone of an all *trans*-hydrocarbon chain. (b) Side view of the monolayer. Circles represent sulfur atoms.

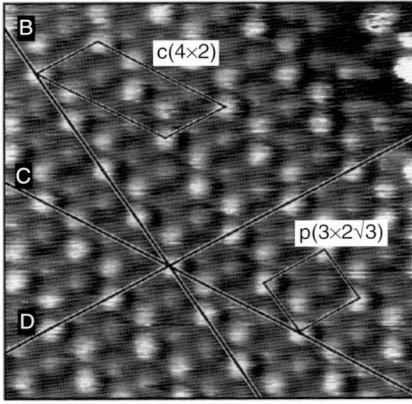

Figure 2.5 An STM image of a Au(111) surface covered with an octanethiol monolayer. The rectangular cell represents the unit cell for c(4 × 2) of ($\sqrt{3} \times \sqrt{3}$)R30°. Lines represent nearest neighbor (B) and two next nearest neighbor directions of the Au atoms of the substrate. Reprinted from [36].

An STM image revealed various defect structures as shown in Figure 2.6a. One significant feature is the pit-like structure. The holes are not pinholes in the monolayer but depressions of the Au surface created during the monolayer formation, as shown in Figure 2.6b. These depressions of the Au surface are called vacancy islands (VIs) of the gold surface. The VIs are known to be formed during the very initial stage of the SA and grow via an Ostwald ripening process [39, 40].

The other defect structure is the domain boundary. Typically, a domain boundary consists of void lines with a space of single or several molecules. These defects originate from the misfits in the tilt angles, stacking geometry and rotational direction of the c(4 × 2) geometry. Figures 2.6c and d show models of the typical domain boundaries caused by rotational and stacking misfits, respectively.

The defect density can be reduced by annealing after the SAM formation [41, 42], increasing the temperature of the solution during the SAM formation [27] and changing the solvent [28]. Figure 2.7 shows STM images of the Au(111) surface

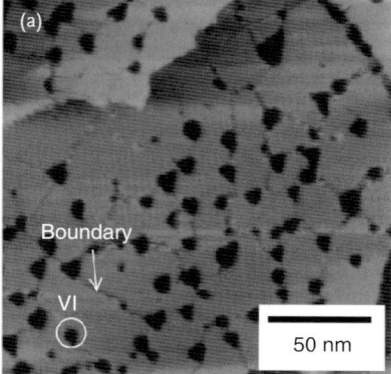
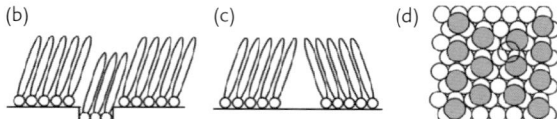

Figure 2.6 (a) An STM image of a Au(111) surface covered with decanethiol ($C_{10}SH$) SAM. A schematic model of VIs (b), domain boundary due to tilt angles (c) and stacking misfits (d).

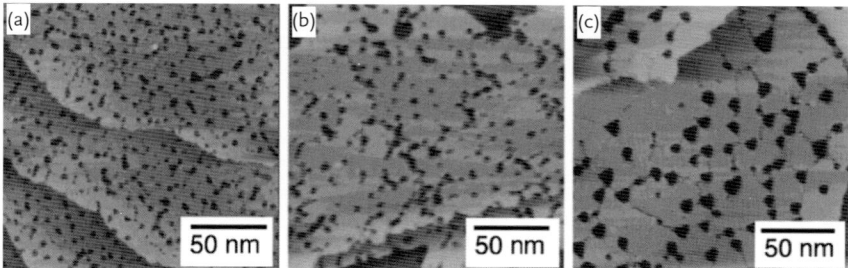

Figure 2.7 STM images of a Au(111) surface covered with a decanethiol monolayer modified in 1 mM solution in (a) ethanol (b) DMF and (c) toluene for 1 h. Reprinted from [28].

modified with decanethiol in ethanol, dimethylformamide (DMF) and toluene at room temperature [28]. The density of the VIs and size of the grain have changed.

Alkanethiols are known to change their orientations as a function of coverage [34, 43–46]. Figure 2.8 schematically shows the relationship between the coverage and orientation of the molecules. When the coverage is low, alkanethiol molecules do not form ordered structures (Figure 2.8a). As the coverage increased, the so-called pin-stripe patterns are formed (Figure 2.8b). In this structure, molecules are oriented parallel to the surface plane and arranged in a head-to-head configuration, that is, thiols are pointing to each other. After the pin-stripe phase, interdigit structures, in which the alkyl chains are stacked with those in the next rows, appeared (Figure 2.8c and d). As the coverage increased, the molecules stand up and form an island that consisted of a $\sqrt{3} \times \sqrt{3}$ molecular arrangement (Figure 2.8e).

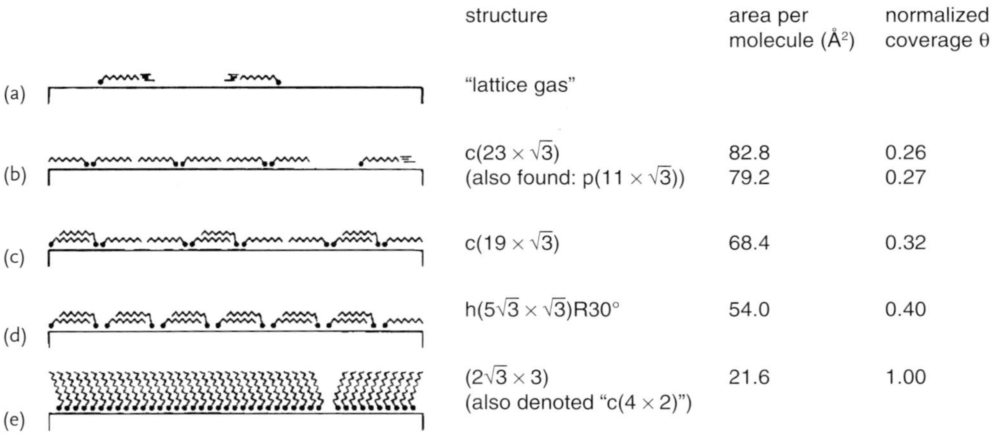

structure
area per molecule (Å²)
normalized coverage θ
(a)
"lattice gas"
(b)
c(23 × √3)
(also found: p(11 × √3))
82.8
79.2
0.26
0.27
(c)
c(19 × √3)
68.4
0.32
(d)
h(5√3 × √3)R30°
54.0
0.40
(e)
(2√3 × 3)
(also denoted "c(4 × 2)")
21.6
1.00

Figure 2.8 Schematic representation of evolution of structures of decanethiol on Au(111) as a function of coverage (a–e). See text for details. Reprinted from [30].

Current/μAcm⁻²
40
20
0
−20
−40
−60
−80
−100
−1200
−1000
−800
−600
−400
−200
0
Potential/mV

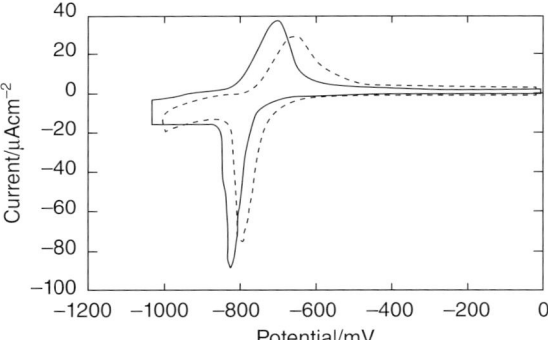

Figure 2.9 Cyclic voltammograms of a Au(111) electrode measured in 20 mM KOH ethanol solution containing 100 μM (dashed line) and 1 mM C_6SH (solid line). Reprinted from [52].

The SAMs of organothiols are known to be reductively desorbed by applying negative potential in alkaline aqueous solutions as shown in Eq. (2.1) [47].

$$\text{Au-SR} + e^- \rightarrow \text{Au} + \text{RS}^- \tag{2.1}$$

As a reverse process, thiol SAMs are expected to be oxidatively formed by applying a positive potential in solutions, containing thiolate molecules.

The desorption and readsorption processes of alkanethiol SAMs in electrochemical environment were investigated by various electrochemical techniques, scanning tunneling microscopy (STM), and surface X-ray diffraction (SXRD) [48–52]. Figure 2.9 shows cyclic voltammograms of a Au(111) electrode in 20 mM KOH ethanol solutions containing 100 μM and 1 mM hexane thiol (C_6SH) [52]. When the potential was swept in the negative direction, the cathodic peak (negative current peak) due to the reductive desorption of the monolayer was observed around −800 mV.

14 | *2 Self-Assembled Monolayer (SAM)*

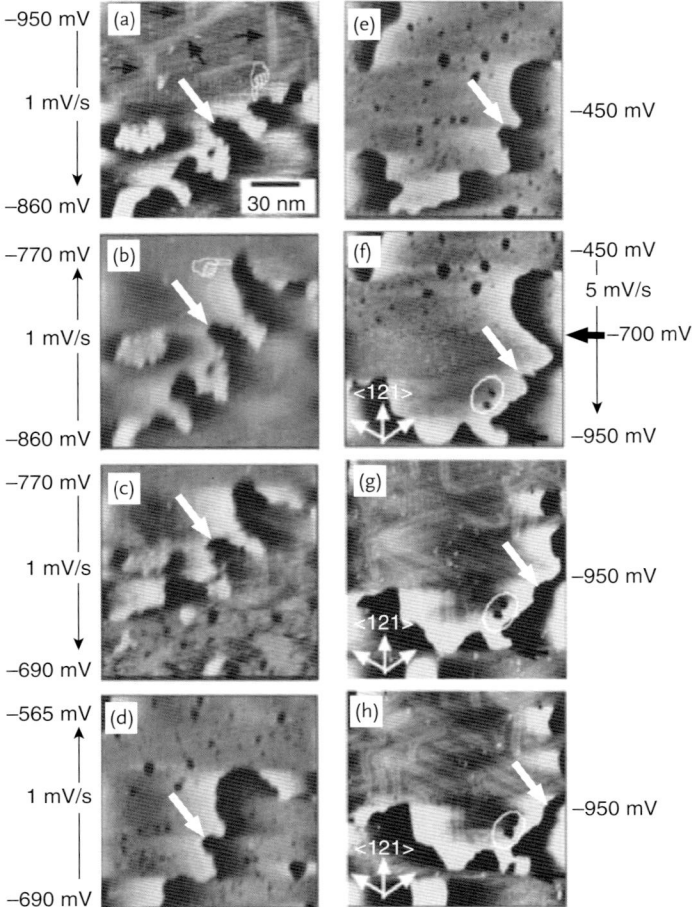

Figure 2.10 *In situ* STM images of a Au(111) electrode in 20 mM KOH ethanol solution containing 60 μM hexanethiol. Images (a–h) were sequentially taken. Arrows and numbers beside the image indicate the frame direction of STM image and electrochemical potential of the substrate during the scanning. A hexanethiol monolayer is known to desorb around −800 mV. Oxidative adsorption of desorbed molecules took place around −700 mV. Reprinted from [52].

The anodic peak (positive current peak) due to the oxidative adsorption of the molecules was observed at −700 mV during the positive potential scan.

Figure 2.10 shows sequential STM images during oxidative adsorption and reductive desorption of C_6SH taken in 20 mM KOH ethanol solution containing 60 μM of C_6SH [52]. The white arrow in the images is the marker indicating the same location of the surface. Initially, the potential of the Au electrode was −950 mV, at which C_6SH was desorbed. Double-stripe patterns attributed to herringbone structure of Au(111) surface were observed, as indicated by the black arrows in Figure 2.10a. This result indicates that the Au(111) surface is reconstructed and no molecules are chemisorbed on the surface, as expected.

When the potential was swept positively and reached the potential at which the anodic peak was observed, −700 mV, herringbone structure disappeared and the step structure of the gold changed as indicated by the pointing finger. No significant features assigned to the thiol molecules were clearly observed. This result indicates that the gold atoms became highly mobile. The thiol molecules were supposed to be adsorbed on the surface and make gold atoms mobile [53]. The VIs of the gold surface is observed in Figure 2.10c. When the potential of the electrode was more positive than −690 mV, typical features of the gold surface covered with SAMs of alkanethiols were observed. The desorption of the thiol molecules and formation of the double-stripe patterns due to the reconstruction of the Au(111) surface were again observed when the potential was swept negatively (Figures 2.10e–h).

The SAMs of organothiols on Au are becoming popular and conventional. This technique is used to functionalize not only single-crystalline surfaces but also nanoparticles and nanorods [54, 55]. Spontaneous adsorption of organodithiols to metal electrodes is used to fabricate single-molecular devices [56].

2.2.2
Organosilanes on SiO$_x$ Surfaces

Organosilane monolayers are formed via hydrolysis of the anchor group of molecules and hydroxyl group on the surfaces as shown in Figure 2.11. Whereas the most widely studied system is the alkyl-tri-chlorosilanes on SiO$_2$ including native oxide, glass and mica surfaces, the SAMs of organosilanes can be formed on various kinds of oxide surfaces [8]. Since the oxide surfaces play important roles in many electronic devices, such as field effect transistors, organosilane monolayers are used to control the interfacial properties of the devices [57].

One of the important characteristics of the SAMs of alkyl-tri-chlorosilanes is the lateral crosslinking of molecules. Due to this crosslinking, the SAMs become robust, but a long-range, two-dimensional order is not realized. The structural order, growth rate and growth mechanism are influenced by many environmental parameters, such as humidity, water content of the solution, wettability of the surface, pH and temperature [8], because the hydrolysis reaction is influenced by these factors. It was shown that most of OH group on the surface does not form

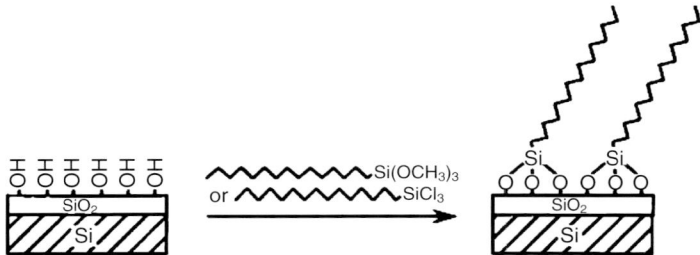

Figure 2.11 Schematic presentation of the SAMs of alkyl-tri-chlorosilanes and alkyl-tri-methoxysilanes on SiO$_2$ surfaces.

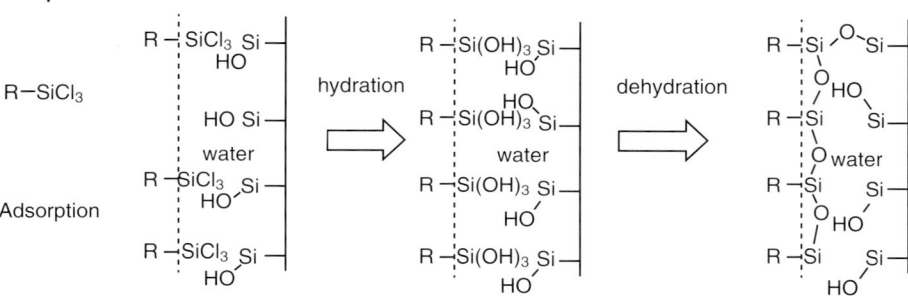

Figure 2.12 Schematic presentation of mechanism and structure of alkyl-try-chlorosilanes on SiO_2 surface. Reprinted from [58].

Figure 2.13 AFM images of the mica surface during octadecylsiloxane (ODS) monolayer growth in toluene solution. The concentrations of ODS and water were 0.5 mM and 4.5 mM, respectively. Image size: 1 μm × 1 μm. Reprinted from [59].

chemical bonds with molecules and physical adsorption through the water layer plays an important role, as shown in Figure 2.12 [58].

The structural evolution during the SAM formation was studied by *in situ* atomic force microscopy [59]. The solution containing 0.5 mM octadecyltrichlorosilane in toluene was injected into the atomic force microscope (AFM) cell to initiate the SAM formation. Figure 2.13 shows a series of AFM images taken after the injection of the solution. Initially, fractally shaped islands were observed to grow and cover the surface. This result indicates that the growth mechanism is a diffusion-limited aggregation. It is concluded that the polysiloxane oligomers formed in the solution adsorb onto the active centers of the surface.

2.2.3
SAMs on Si Surface via Si–C Bonding

The modification of the silicon surfaces by covalent attachment of organic molecules is realized by the chemical reaction between the olefin and hydrogen-terminated Si surfaces as shown in Figure 2.14. The H-terminated Si surface is prepared by chemical etching of the Si surface in NH_4F solution [60]. Si–H bonds on the surface are cleaved to generate radicals by heat [14], radical initiators [61], ultraviolet light [62], electrochemical [63, 64], and sonochemical [65] methods.

Figure 2.15 shows the proposed formation mechanism of the SAMs on Si [15–18, 66]. The initial reaction is removal of hydrogen on the H-terminated Si surface and formation of dangling bonds (Si radical) by UV light or thermal activation. The dangling bond reacts with the C=C bond, and this reaction results in the formation of the carbon radical. The carbon radical abstracts a hydrogen atom

Figure 2.14 Schematic presentation of the SAMs of alkanes directly connected to Si surface via a Si–C bond.

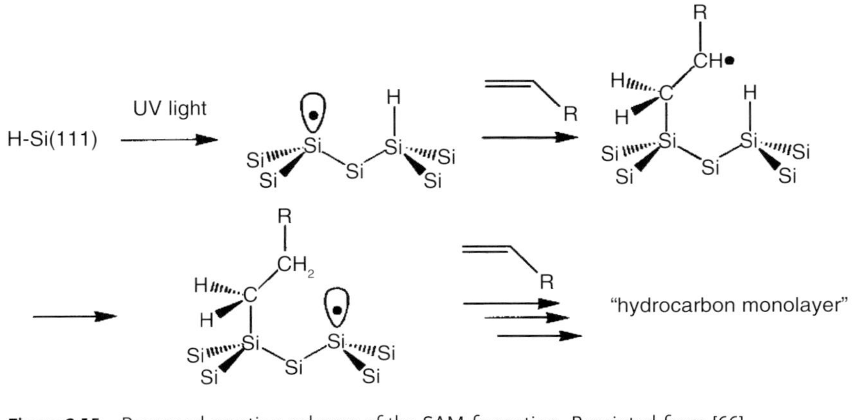

Figure 2.15 Proposed reaction scheme of the SAM formation. Reprinted from [66].

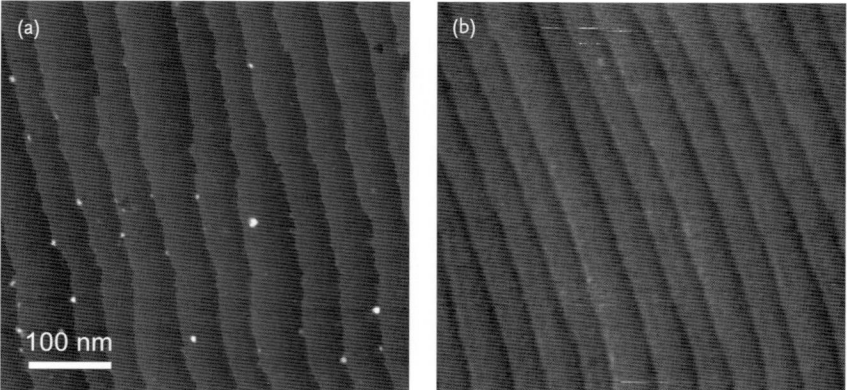

Figure 2.16 AFM images of H–Si(111) surface before (a) and after (b) modification of $CH_2=CH-(CH_2)_8-CH_3$ and $CH_2=CH-(CH_2)_8-SH$. Reprinted from [67].

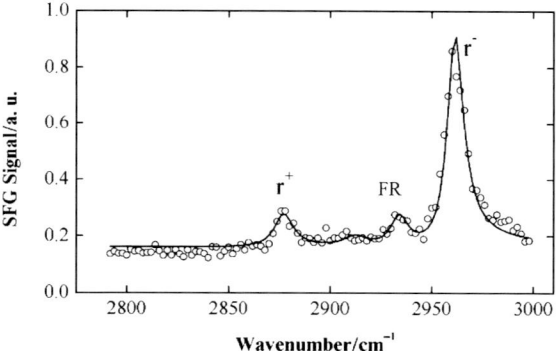

Figure 2.17 SFG spectrum of the Si(111)-C18 monolayer at the azimuthal angle of 0 in the CH stretching frequency region with the fitted curve. Reprinted from [70].

from the H-terminated surface and a new reactive dangling bond is formed. This chain reaction results in the formation of monolayers.

Figure 2.16 shows the H-terminated Si(111) surface before and after the monolayer formation, respectively [67]. The H-terminated Si(111) surface was prepared by chemical etching in 40% NH_4F. The monolayer was formed by immersing the Si substrate in a solution containing $CH_2=CH-(CH_2)_8-CH_3$ and $CH_2=CH-(CH_2)_8-SH$ (20:1) under UV irradiation for several hours. As shown in Figure 2.16, atomically flat terraces were observed before and after the SAM formation. Although the two-dimensional molecular order has not been confirmed, the SAM is uniformly formed on the surface.

The growth rate and conformational order were investigated by attenuated total reflection infrared spectroscopy and sum frequency generation (SFG) spectroscopy

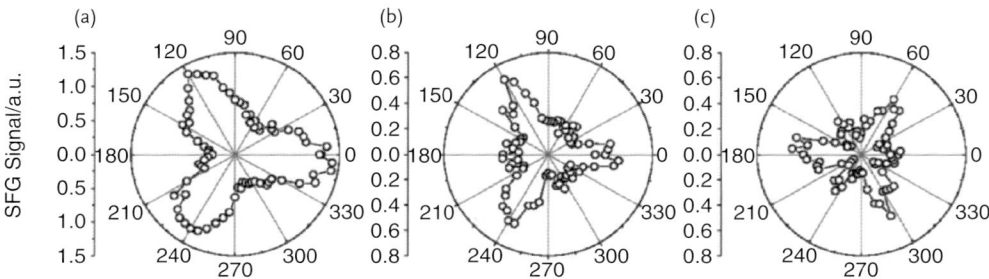

Figure 2.18 Rotational anisotropy of the SFG intensity of the Si(111)-C18 monolayer at (a) 2962 cm^{-1}, (b) 2878 cm^{-1}, and (c) 2800 cm^{-1} as a function of the azimuthal angle. Azimuthal angle was defined as the angle between the [211] direction of the Si(111) surface and the plane of the incidence angle. Reprinted from [70].

[68–71]. Figure 2.17 shows the SFG spectrum of the Si(111) surface modified with the octadecyl monolayer (Si(111)-C18) in the CH stretching frequency region [70]. The peaks observed at 2878, 2945 and 2962 cm^{-1} were attributed to the CH symmetric vibration (r^+), Fermi resonance between r^+ and the CH bending overtone, and the CH asymmetric vibration (r^-), respectively, of the terminal methyl (CH$_3$) group. The peaks due to the CH symmetric stretching (d^+) at 2850 cm^{-1} and asymmetric stretching (d^-) at 2917 cm^{-1} of the methylene (CH$_2$) group were very weak. Because a SFG signal is not generated in media with inversion symmetry, the weak signal from CH$_2$ indicates that the density of gauche defects, which break the inversion symmetry in alkyl chain layer, is small. Thus, it is concluded that the alkyl chains in the Si(111)-C18 are highly ordered, that is, in the all-*trans* conformation. The tilt angle of the alkyl chain can be determined from the ratio between the CH symmetric and asymmetric vibrations of the terminal methyl (-CH$_3$) group.

Rotation anisotropy of the SFG spectra is used to evaluate the lateral symmetry of the monolayer. Figure 2.18 shows the azimuthal angle dependencies of the SFG intensity at (a) 2962 cm^{-1} for r^-, (b) 2878 cm^{-1} for r^+, and (c) 2800 cm^{-1} where no peak was observed. All signals showed threefold patterns, although the pattern observed for 2800 cm^{-1} was rotated by 60° with respect to other two signals. The pattern observed for 2800 cm^{-1} originates in the symmetry of Si substrate since a similar pattern is observed for a H-terminated Si surface. Thus, the threefold pattern observed for r^- and r^+ is due to the symmetry of the two-dimensional arrangement of alkyl chains of the monolayer. Figure 2.19 shows the model of the octadecyl chain in the monolayer proposed from the quantitative analysis of the SFG spectrum [70].

The reaction mechanism was studied by UHV-STM [66]. Figure 2.20a shows an STM image of the H-terminated Si(111) surface with a small number of dangling bonds, which appear as bright spots. These dangling bonds were created by applying a pulse voltage from the STM tip. Figure 2.20b shows the STM image of the H-terminated Si(111) surface after exposure to styrene gas to 12 Langmuirs. The

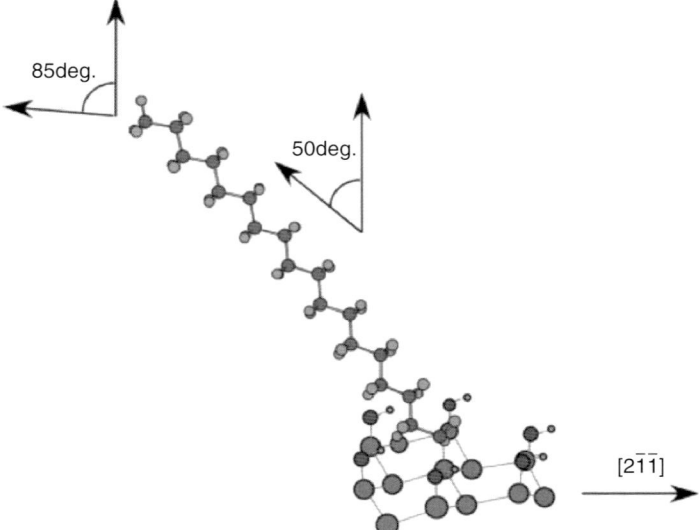

Figure 2.19 Model of the possible orientation of the octadecyl chain. The alkyl chain is in the all-*trans* conformation and epitaxially arranged to Si atoms on a Si(111) surface. The molecular axis is tilted by 50° toward the {211} direction. The tilt angle of the methyl group is about 85° with respect to the surface normal. Reprinted from [70].

black dots indicate the positions of the dangling bonds observed before the exposure. The islands consisted of styrene molecules that were observed around the location where the dangling bonds existed. This result supports the chain-reaction mechanism.

The SAMs formed on Si surface formed via Si–C bond is becoming popular because the monolayers are thermally, chemically and mechanically robust. It was shown that the monolayer is not decomposed up to 250 °C [72–74] and the corrosion and oxidation of Si surface by chemical reaction were inhibited [75] by the monolayer. A tip used for the atomic force microscopy modified with the monolayer via Si–C bond is shown to have longer lifetime compared to that modified by SAMs via Au–S bond [76]. The SAMs via Si–C bond are used to modify not only crystalline Si, but also the Si nanowires [77–79] and diamond [22] surfaces.

2.3
Functions and Applications

As described above, the characteristics of the solid surface can be controlled by the formation of the SAM on the solid substrate. For example, the gold electrode can be inert for most electrochemical species by modification by a long-chain alkanethiol SAM. Such a SAM can be used as a mask for lithography. The solid

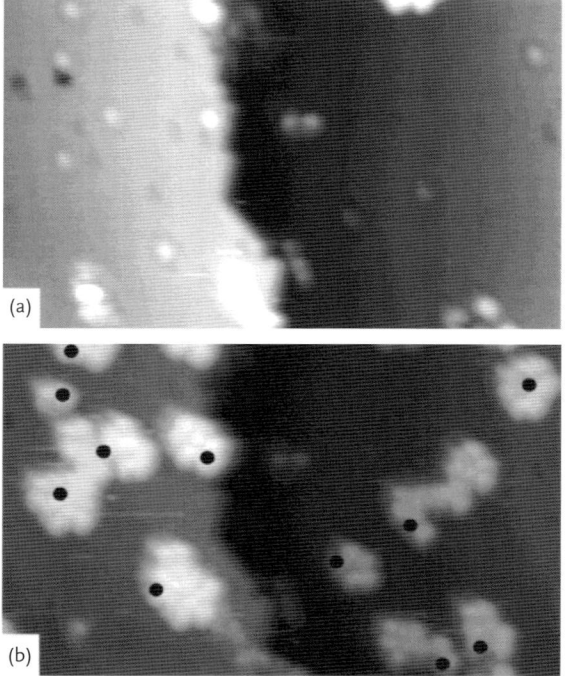

Figure 2.20 STM images of H-terminated Si(111) surface before (a) and after (b) modification of styrene under ultrahigh-vacuum conditions. Reprinted from [66].

surface is also changed from hydrophobic to hydrophilic only by replacing the methyl terminal group with the carboxyl or hydroxyl group of the SAM. Moreover, when a functional group is introduced into the SAM as a terminal group, it leads to the development of the fixation on the surface for biological samples, metal or semiconductor nanoparticles, and polymers, and with a multilayer construction. Such an approach is of great importance for considering applications such as sensors, surface coating and nanopatterning, tribology, and molecular devices.

In the next sections, the following topics are introduced: (i) surface coating and patterning, (ii) sensor applications, (iii) nanotribology, and (iv) advanced applications such as electron transfer, photoinduced electron transfer, luminescence, and catalytic activity.

2.3.1
Surface Coating and Patterning

In order to construct molecular devices with high functionalities, the two-dimensional structure of the molecular layers should be controlled and the nanofabrication techniques using SAMs have been widely investigated.

Xia and Whitesides proposed the "microcontact printing" (μCP) method in which an elastomeric poly(dimethylsiloxane) (PDMS) stamp was used to transfer molecules (various thiol molecules) of the ink to the substrate surface by contact (procedure is shown in Figure 2.21a), and aimed at various applications [80–82]. For example, when the substrate was dipped into a solution containing an alkanethiol with a terminal hydroxide or carboxylate group, after the pattern of the normal alkanethiol SAM using the PDMS stamp was formed, the other SAM formed part of the substrate surface where no SAMs originally existed. Namely, they can design the shape and space of the hydrophobic and hydrophilic parts on a molecular level. On this surface, the condensation of water according to the pattern takes place and it works as a diffraction grating at a suitable condensation

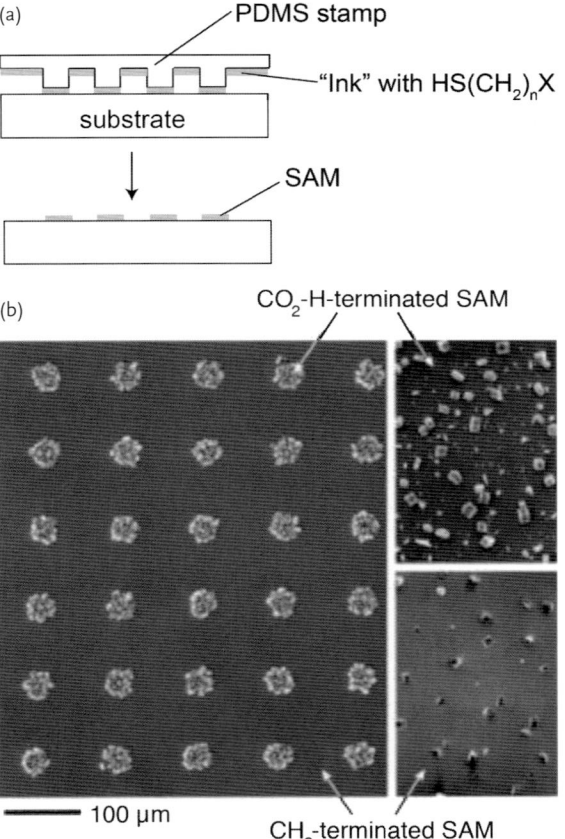

Figure 2.21 (a) Schematic presentation of the μCP method. (b) Scanning electron micrograph (SEM) images of the sample patterned surface-printed circles of $HS(CH_2)_{15}CO_2H$ in a background of $HS(CH_2)_{15}CH_3$ supported on Ag(111)-overgrown with calcite crystals. Reprinted from [80].

stage [82]. Moreover, they reported that the $CaCO_3$ crystals grew only on the hydrophilic part [80]. This result was proven to be an application not only for control at a monolayer level but also for the construction of a device that used ceramics, etc. (Figure 2.21b). Nanopatterning and nanofabrication have been actively studied using the μCP technique and alkanethiol [83–94] and/or alkylsilane [95–106] SAMs.

For combining the scanning probe microscopy (SPM) with the SA technique, it is possible to construct a surface in which the structure is controlled on the nm level. For example, using the probe of the AFM like a pen, the so-called "dip pen" nanolithography, where the molecule was fixed only in the part traced by the probe, was proposed [107]. The molecular fixation methods by mechanical [108] and electrochemical [109, 110] scratching of the other SAM modified surfaces were also proposed.

Shimazu et al. succeeded in constructing a nanopatterned SAM using an underpotentially deposited (UPD) lead submonolayer (Figure 2.22) [111, 112]. Alkanethiol SAMs of various chain lengths were constructed on an UPD Pb submonolayer on a Au(111) surface (steps 1 and 2 in Figure 2.22a). The UPD Pb submonolayer was then electrochemically stripped (step 3 in Figure 2.22a) and a 3-mercaptopropionic acid (MPA) SAM was next constructed on this surface (step 4 in Figure 2.22a). MPA was adsorbed only on the vacant site and therefore some patterns with an atomic/molecular dimension were formed (Figure 2.22b). The constructed nanopatterns were dependent on both the amount of the UPD Pb before the SAM formation and the chain length of the alkanethiol SAMs.

2.3.2
Sensor Applications

When a suitable terminal functional group is introduced in the SAM, the sensing systems, where the ions and/or the molecules are recognized, can be constructed. For example, when the electrochemical measurements in the electrolyte solution containing Cu^{2+} and Fe^{3+} were carried out at the gold electrode modified with the mixed SAM of n-octadecyl mercaptan (OM) and 2,2′-thiobisethyl acetoacetate (TBEA), as shown in Figure 2.23, the response of Fe^{3+} was completely suppressed and only the response of Cu^{2+} was selectively observed [113]. As a result that the two β-keto ester groups in TBEA became chelate centers and that only Cu^{2+} formed a complex with TBEA with a ratio of 1:1, Cu^{2+} can approach the electrode surface and an electrochemical reaction took place, although the monolayer became a barrier and then electron transfer did not take place from the electrode to Fe^{3+} in the solution.

It is also possible to sense a molecule by the SAM by the host–guest intermolecular interaction. When the functional group of cyclobis(paraquat-p-phenylene), which has a host property, as shown in Figure 2.24, was introduced in the SAM, the oxidation/reduction reaction of the bipyridyl group in the host molecule was easily observed in a buffer solution [114]. When small amounts of indole, catechol, benzonitrile, and nitrobenzene were added as a guest molecule to the solution, the

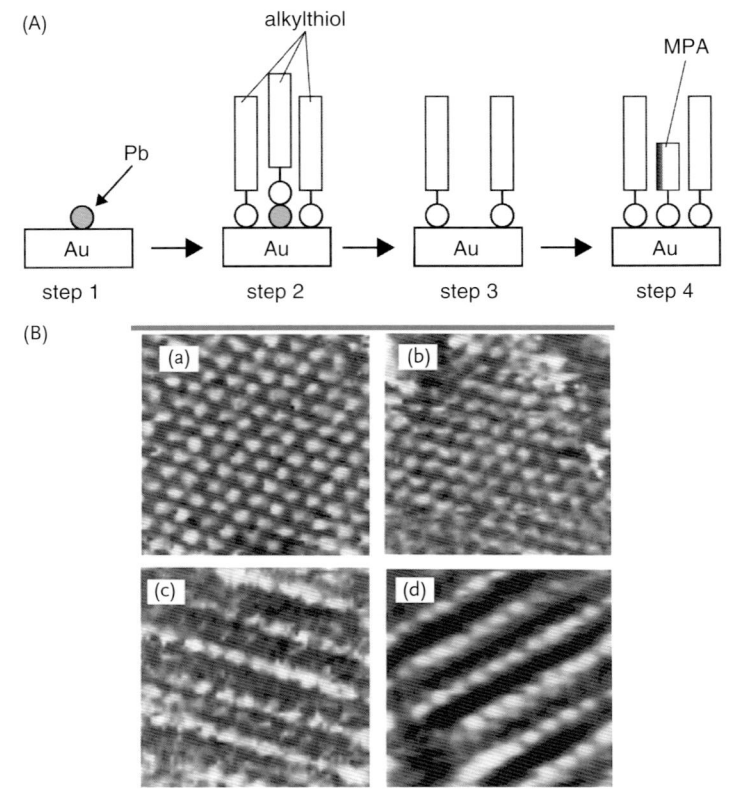

Figure 2.22 (A) Construction procedures of mixed monolayers of controlled compositions using Pb UPD submonolayer. (B) The 5 nm × 5 nm STM images of the (a) octanethiol SAM on Au(111), (b) octanethiol SAM on UPD Pb/Au(111) ($\theta_{Pb}^0 = 1.00$), (c) octanethiol SAM on UPD Pb/Au(111) ($\theta_{Pb}^0 = 0.60$), and (d) octanethiol adlayer that remained after the desorption of UPD Pb ($\theta_{Pb}^0 = 0.60$). Reprinted from [111].

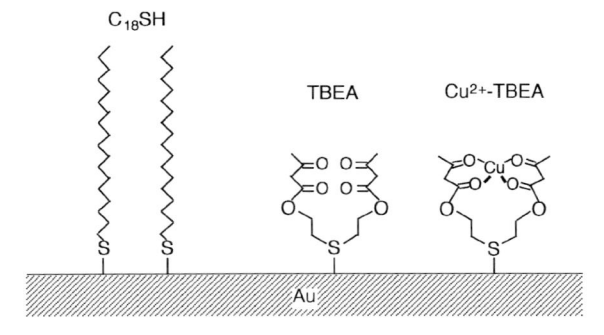

Figure 2.23 Schematic representation of TBEA, Cu^{2+}-TBEA complex, and OM, adsorbed on gold substrate. Note that the ligand binds Cu^{2+} in the enol form upon losing two protons, and thus the complex is neutral. TBEA was prepared by the 4-dimethylaminopyridine (DMAP) catalyzed reaction of 2,2′-thiobisethanol with diketene. Reprinted from [113].

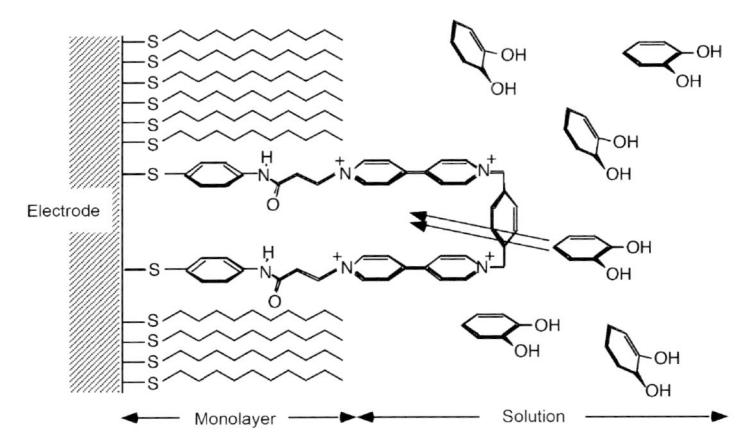

Figure 2.24 Idealized representation of an interfacial binding event at the mixed SAMs modified electrode. Reprinted from [114].

redox potential was negatively shifted as the concentration of the additives only in the case of indole and catechol, while the redox potential was not shifted for the benzonitrile and nitrobenzene. For indole and catechol, which are strong π-donors, they became the guest, and were fixed in the host molecule and then a negative shift of the reductive potential of the bipyridyl group in the host molecule occurred by the host–guest interaction due to the π-bond of the host–guest molecules.

In addition to this report, there were several applications as a Ca^{2+} sensor using the SAM of pyrroloquinoline quinone (PQQ) that worked as a coenzyme in the human body [115], of α-cyclodextrin (α-CD) as a host molecular receptor [116], and of 4-amino-6-hydroxy-2-mercaptopyrimidine as a ligand to form a metal complex [117].

Using the SAMs, biosensors can be easily constructed. Among them, Sato and coworkers have been aggressively executing the research of a biosensor using the SAM [118–125]. For example, they simultaneously detected glucose and ascorbic acid using the ferrocenylundecanethiol ($FcC_{11}SH$) SAM on gold by individually measuring the electrochemical response of $FcC_{11}SH$ and chemical luminescence intensity from luminol in the solution [121–123]. They also detected lectin with a high signal to noise ratio using a gold electrode modified with the mixed SAM of the short tri(ethylene glycol)-alkanethiol (EGC_2SH or EGC_6SH) and maltoside-terminated alkanethiol ($MalC_{12}SH$) for repelling proteins and capturing lectin (ConA), respectively (Figure 2.25) [124]. Moreover, they successfully detected Galectin, which was used as a diagnostic and prognostic marker, using the mixed SAM of a β-galactoside terminated alkanethiol and short tri(ethylene glycol)-alkanethiol for capturing Galectin and repelling proteins, respectively [125].

In addition to these studies, there are many recent research investigations of a biosensor that uses SAM, because it can easily construct a biosensor if the biomolecule is introduced into the SAM as a terminal group [126–135]. There is an

(a)

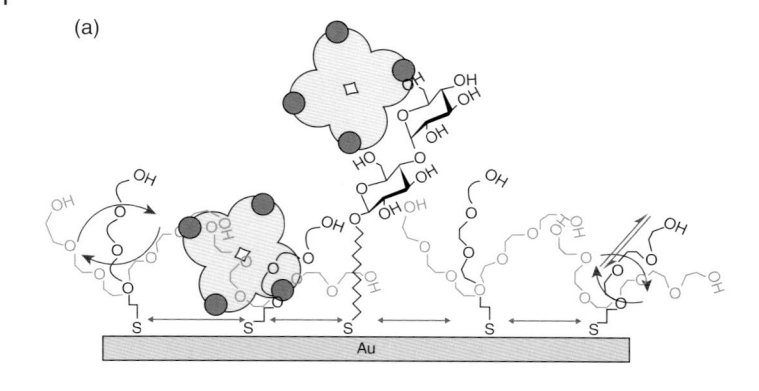

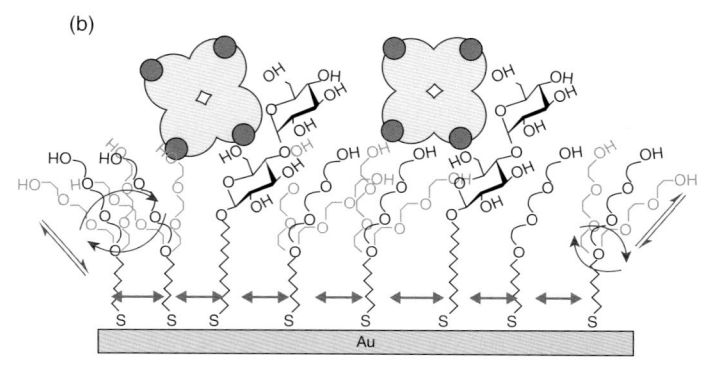

(b)

Figure 2.25 Models of lectin (ConA) recognition on MalC$_{12}$SH-EGC$_n$SH mixed SAMs. (a) MalC$_{12}$SH-EGC$_2$SH and (b) MalC$_{12}$SH-EGC$_6$SH mixed SAMs. Reprinted from [124].

example of using SAMs formed on a Si substrate via the Si–C bond, as described above (Section 2.3), for a pH sensor [136] and a biosensor [137].

2.3.3
Nanotribology

We need to understand the structure and stability of the monolayer to correlate the molecular properties of the SAM to its physicochemical characteristics. In order to do this, it is very important to examine the tribology of the SAM.

Prathima *et al.* investigated the thermal stability of the alkanethiol SAMs with various alkyl chain lengths on gold using grazing-angle reflection-absorption Fourier transform infrared spectroscopy (FT-IR RAS), cyclic voltammetry (CV), and a molecular dynamics simulation [138]. Disordering of the SAM by untilting and gauche-defect accumulation increased on increasing temperature in the 300–440 K range. For the relatively short-chain SAMs, with increasing temperature, disordering of those SAMs tended to saturate at temperatures below 360 K, as reflected in both untilting and gauche-defect accumulation, such that any further increase in temperature until desorption did not lead to any significant change in the conformational order. In contrast, for the relatively long-chain SAMs, the disorder monotonically increased with temperature beyond 360 K. Thus, they concluded that the ability of the SAMs to retain order with increasing thermal perturbations is governed by the state of disorder prior to the heat treatment.

Lee *et al.* proposed the pin-on-disk tribometry as a macroscopic tribological test for the SAMs [139]. This method employed an AFM and friction force microscopic (FFM) tip, which is coated with the other SAM (Figure 2.26). There are three advantages of this method, as follows: (i) While the contact pressure can be maintained low enough to retain the integrity of the SAM, as with nanotribological approaches, high-speed, macroscale tribological contacts can be achieved with an elastomeric counterface. (ii) The wide sliding tracks generated in this approach – even wider than those obtained in conventional tribometry using rigid sliders – allow

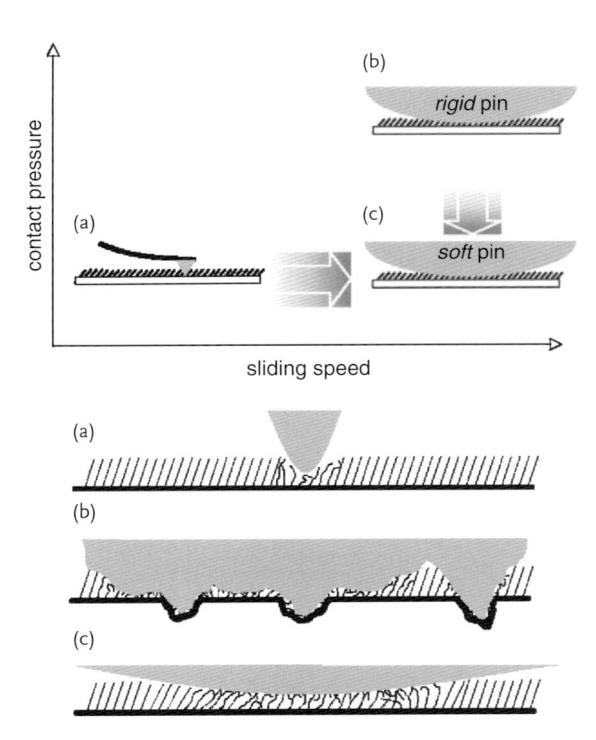

Figure 2.26 A pressure–sliding-speed diagram showing the contact configuration of an elastomeric slider on a SAM: schematic illustrations for (a) single-asperity contacts on a nanoscopic scale as in the AFM (low contact pressure, low speed), (b) multiasperity contacts on a macroscopic scale as in conventional pin-on-disk tribometry employing rigid sliders (high contact pressure, high speed), and (c) soft contacts on a macroscopic scale by employing an elastomeric slider in pin-on-disk tribometry (low contact pressure, high speed). Reprinted from [139].

for standard spectroscopic approaches to access the contact area and characterize the influence of the tribological contacts on the SAMs. (iii) By running the measurements in a liquid medium, it is possible to induce a range of lubrication regimes from boundary lubrication to fluid-film lubrication over the speed range available from an ordinary pin-on-disk tribometer.

By a method similar to that above, the frictional properties of n-alkane monolayers formed on the SAMs [140] and the SAMs on the Si surface through a Si–C bond [141] were investigated. Using these frictional properties of the SAMs, a nanopattern of the SAMs successfully formed on the solid substrates [109, 110].

2.3.4
Advanced Applications

2.3.4.1 Electron Transfer

The electrochemical behaviors of the gold electrodes modified with the SAMs of ferrocenylalkanethiols, which contain ferrocene (Fc) and thiol groups as electrochemically active and surface binding groups, respectively, have been extensively examined [21, 28, 31, 142–145] and they are known to show a reversible redox reaction (Figure 2.27) [142, 145]. The electron-transfer rate constant at the electrode/electrolyte interface can be precisely measured by analyzing these electrochemical

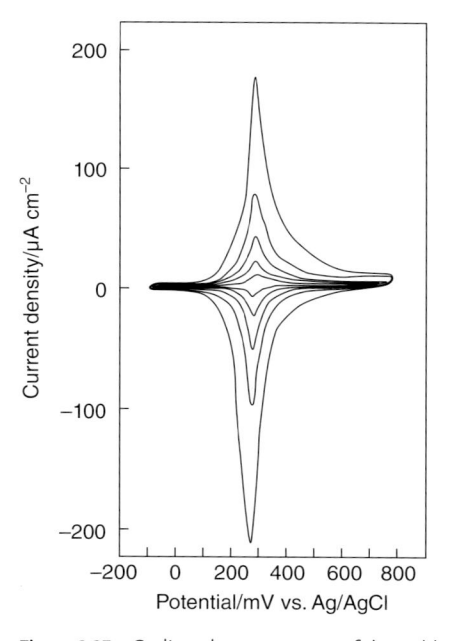

Figure 2.27 Cyclic voltammograms of the gold electrode modified with the SAM of ferrocenylhexanethiol measured in 0.1 M HPF_6 electrolyte solution at scan rates of 500, 200, 100, 50, and 20 mV s^{-1}. Reprinted from [145].

responses in detail. In order to understand the mechanism of the highly efficient energy conversion process, such as photosynthesis, and to achieve the molecular electronics like a transistor, etc., the electron-transfer rate constant at the electrode/electrolyte interfaces is an extremely important basic parameter.

Beside the Fc group as described above, the pH dependence of the redox reaction was studied at the alkanethiol SAM with the quinone/hydroquinone redox couple [146–149] and their application in a pH sensor was reported [150]. Moreover, control of the electron-transfer rate by photoisomerization of the azobenzene group, which was introduced into the SAM, was investigated [151].

There are examples of studies of electron transfer using SAMs on a Si substrate via Si–C bonding as described above (Section 2.3) [152–158]. Among them, viologen [152–155] and quinone [156, 157] moieties were introduced into the SAMs on Si as electrochemically active groups. Cai et al. introduced a single-strand DNA moiety into the SAM on Si and achieved DNA sensing by detecting an electrochemical response [158]. Because the Si substrate can be used as a semiconductor electrode, on the other hand, there are also examples of examining their photoelectrochemical properties [159–162]. These examples are described in the following section. Thus, a variety of electron-transfer controls by the SAM have been achieved.

2.3.4.2 Photoinduced Electron Transfer

When a photosensitizer (S), electron acceptor (A), and electron relay (R) are arranged in order on the electrode, the photoinduced electron-transfer systems that imitates photosynthesis can be constructed and a highly efficient up-hill electron transfer from the electrode to the electron acceptor, which has a higher energy, can be achieved (Figure 2.28a and b) [163–165].

As an example, a highly efficient visible-light-induced electron-transfer system using the SAM of the molecule ($PC_8FcC_{11}SH$, Figure 2.28c), which has ferrocene (Fc) and porphyrin (Por) as an electron relay and photosensitizer, respectively, in the solution containing methylviologen (MV^{2+}) as an electron acceptor is introduced here [165]. When the visible light was irradiated on the gold electrode modified with the $PC_8FcC_{11}SH$ SAM in the solution containing 5 mM MV^{2+} at a constant potential, a cathodic current flowed just after irradiation and returned to zero when the irradiation was stopped (inset of Figure 2.29). Figure 2.29 shows the potential dependence of the photocurrent density. A cathodic photocurrent was observed when the potential was more negative than 600 mV (vs. Ag/AgCl), which is in very good agreement with the redox potential of Fc (610 mV), indicating that the Fc group acted as an electron-relay group. This result revealed that an up-hill electron transfer with ca. 1.2 eV was achieved because the redox potential of MV^{2+}, used as an electron acceptor, is −630 mV. Moreover, the shape of the photocurrent action spectrum is well matched with the absorption spectrum of the Por group, showing that the Por group acted as a photosensitizer. The quantum efficiency of this system was ca. 11% measured at −200 mV. This value of 11% was the best among those that were observed at metal electrodes modified with organic molecular

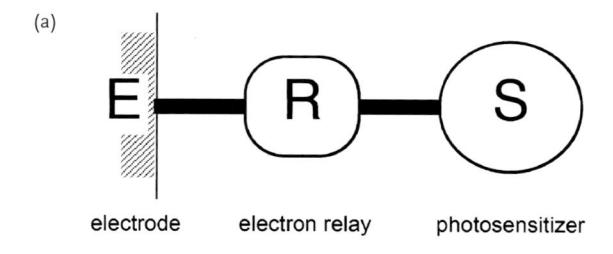

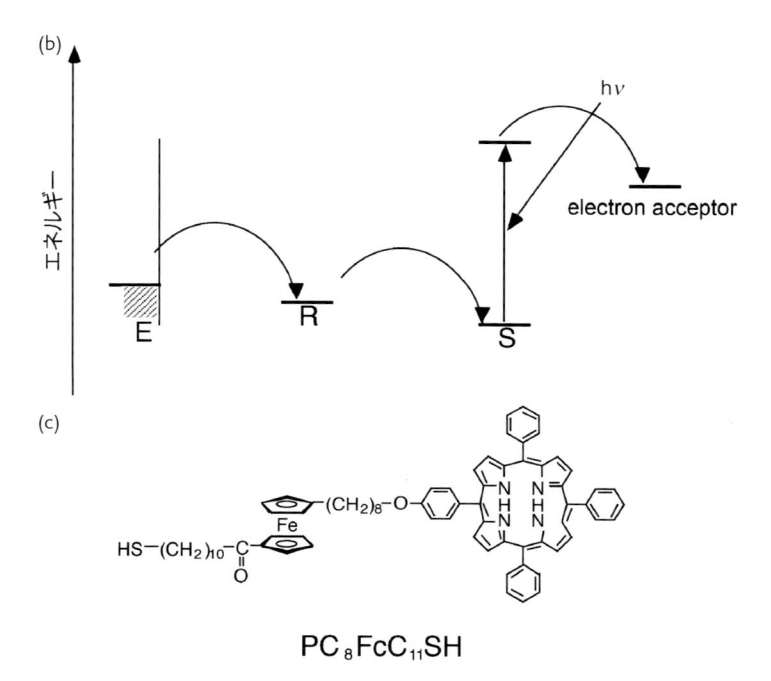

Figure 2.28 (a) Schematic illustration of photoinduced electron-transfer system at the SAM-modified electrode. (b) Energy diagram of (a). (c) $PC_8FcC_{11}SH$ molecule. Reprinted from [165].

layers and therefore, this result indicated that a very highly efficient visible-light-induced up-hill electron transfer was achieved using the SAM-modified metal electrode. There were the following three possibly reasons why this system has such a high efficiency: (i) the high orientation of the SAM due to the interaction between alkyl chains, which were introduced between the functional groups [166], (ii) reduction of the backward electron transfer due to the relatively long distance between the functional groups [167], and (iii) relatively higher electron-transfer rate of the Fc group, which was introduced as an electron-relay group.

In addition to these studies, the research that aims at the mutual conversion of photoenergy and electrical energy when considering the development of an artificial photosynthesis has been actively carried out. For examples, Imahori *et al.*

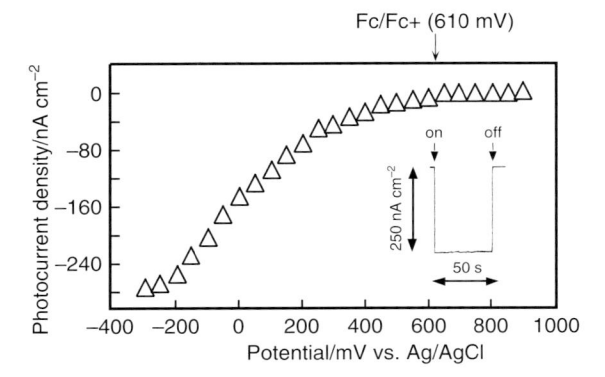

Figure 2.29 Electrode potential dependence of cathodic photocurrent when the gold electrode modified with the SAM of $PC_8FcC_{11}SH$ was irradiated by monochromatized light (430 nm, 40 µW cm^{-2}) in the electrolyte solution containing 5 mM MV^{2+} as an electron acceptor. Inset: Photocurrent response when the irradiation was switched on and off measured at −200 mV (vs. Ag/AgCl). Reprinted from [165].

investigated the photocurrent generation using the SAM of the triad molecule, which has the Por, Fc, and fullerene groups as a photosensitizer, electron relay, and electron acceptor, respectively [168]. They also constructed a light-harvesting system using the mixed SAM of the antenna and these three mentioned molecules [169]. Moreover, the intermolecular photoinduced transfer using the electrode modified with the alternating Por and MV^{2+} multilayers based on the SAM was investigated by Katz and coworkers [170].

Yamada et al. [171] used the $Ru(bpy)_3^{2+}$ complex as a sensitizer and Takimiya and coworkers [172] and Fujihira and coworkers [173] used the origothiopnene and phenylene oligomer, respectively, as a linking part in order to increase the quantum efficiency.

Recently, Ikeda et al. achieved a very highly efficient photoinduced up-hill electron transfer at a gold electrode modified with the SAM of $PC_8FcC_{11}SH$ using plasmonic enhancement by the Au nanoparticle modification (Figure 2.30) [174].

As the semiconductor electrode property of the silicon substrate has been optimized, there are studies of the photoelectrochemistry using SAMs on the Si substrate via Si–C bonding as already described (Section 2.3) [159–162]. Nakato and coworkers reported that a modification of the SAM stabilizes the Si electrode and it can control the size of the coating metal nanoparticles [159–161]. Fabre et al. photoelectrochemically constructed conducting polymer films on the SAMs on the Si substrate [162].

Since Brust and coworkers reported that the gold nanoclusters, whose surfaces were covered with the alkanethiol SAMs, are stable and easy to introduced to functional groups by a place-exchange method [55, 175, 176], many studies on photoinduced electron transfer using the alkanethiol SAM-modified metal or semiconductor nanoclusters were reported [177–188]. Yamada et al. observed the photocurrent at the ITO electrodes modified with the multilayer of gold

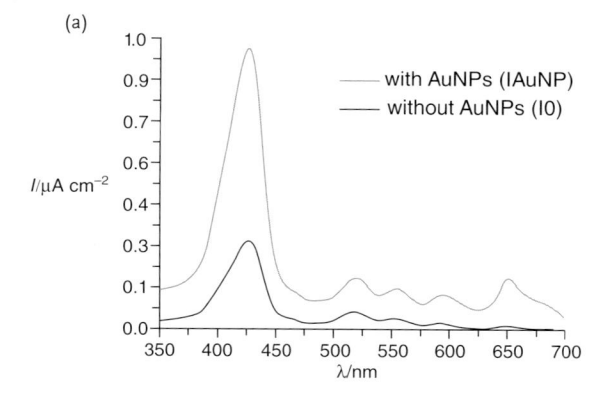

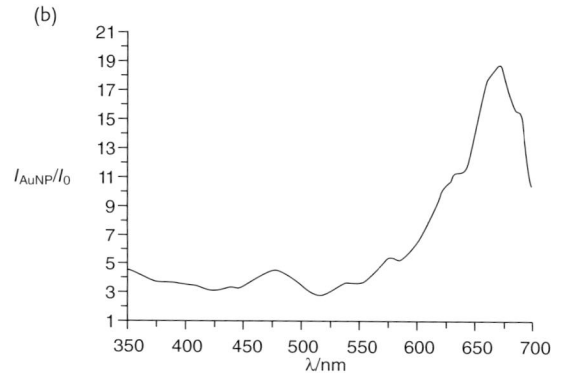

Figure 2.30 (a) Action spectra of $PC_8FcC_{11}SH$ SAM on an Au electrode without gold nanoparticles (AuNPs) (I_0) and with AuNPs (I_{AuNP}), measured in 0.1 M $NaClO_4$ electrolyte solution under an applied potential of −200 mV (vs. Ag/AgCl). (b) Excitation wavelength dependence of the efficiency enhancement. Reprinted from [174].

nanoclusters and porphyrin-tetraalkylthiol molecules [177]. Imahori et al. investigated the photophysical properties of gold nanoclusters modified with the SAM of a porphyrin-thiol coupling molecule [178, 179] and observed a photocurrent at the SnO_2 electrode modified with electrophoretically deposited layers of gold nanoclusters, whose surface is covered with the mixed SAMs of porphyrin-thiol and fullerene-thiol coupling molecules [180]. Li et al. also observed a photocurrent at the gold electrode modified with electrostatically deposited layers of gold nanoclusters, whose surface is covered with the SAM of a porphyrin-viologen coupling molecule [181].

For the gold electrode modified with the semiconductor nanocluster layers, a unique preparation procedure is employed. First, semiconductor nanoclusters, which are covered with the surfactant, sodium bis(2-ethylhexyl) sufrosuccinate (Aerosol OT, AOT), were prepared in reverse micelles [182]. After the dithiol SAMs were prepared on the gold surface, the layers of the semiconductor nanoclusters were prepared by dipping the dithiol SAM-modified gold in a nanocluster disper-

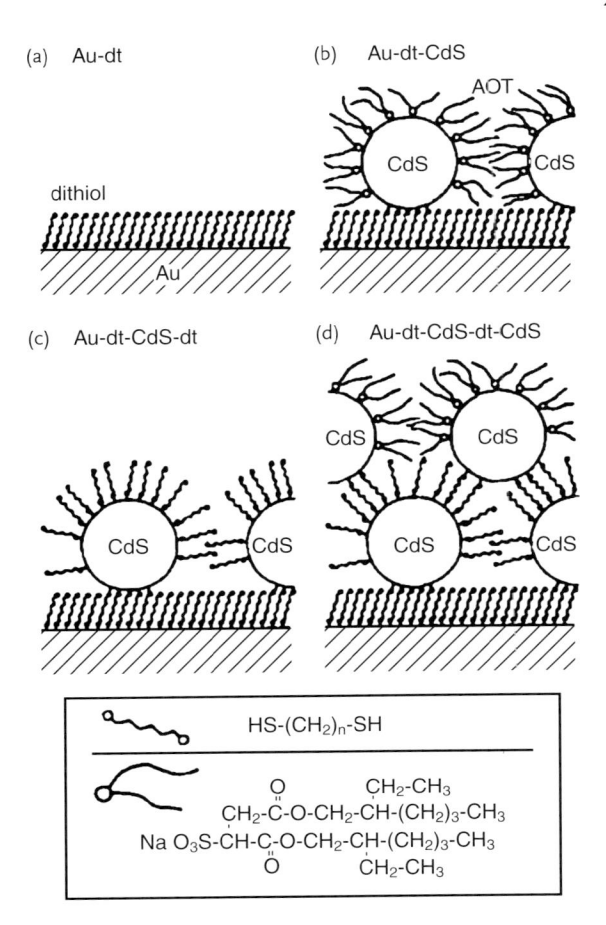

Figure 2.31 Schematic illustration of binding of the CdS nanoclusters from reverse micelles onto gold via dithiol and the formation of alternating layer-by-layer structure: (a) dithiol SAM on a gold substrate (Au-dt); (b) CdS nanoclusters attached on the SAM (Au-dt-CdS); (c) adsorption of dithiol layers on CdS nanoclusters (Au-dt-CdS-dt); (d) formation of a second CdS-nanocluster layer (Au-dt-CdS-dt-CdS). Each component is drawn in size according to the estimation from experimental results. Reprinted from [183].

sion (Figure 2.31) [183–190]. It was confirmed by X-ray photoelectron spectroscopy (XPS) that the terminated thiol group, which is not connected with gold, in the dithiol SAMs on gold is covalently bonded to the surface atoms of the semiconductor nanoclusters [183] and, as a result, the SAM forms on the semiconductor nanocluster surface. Relatively large photocurrents were observed at the gold electrode modified with many kinds of semiconductor nanoclusters, such as CdS [184–187], ZnS [186], PbS [188], and CdSe [189], which were prepared by the above procedures shown in Figure 2.31. There is an interesting preparation method by which the tellurium nanoclusters were electrochemically deposited on the gold electrodes modified with the SAMs of molecular templates, whose terminal group is β-cyclodextrin [190]. Woo et al. observed a photocurrent at this modified gold electrode.

2.3.4.3 Luminescence

Luminescence from the SAM-modified electrodes has been extensively studied. Fox and Wooten constructed the SAM of an anthracene-thiol linked molecule, measured the luminescent intensity and FT-IR spectrum of the SAMs, and investigated dimer formation of the anthracene moiety in this SAM [191]. Guo et al. constructed the photoactive and electrochemical active myoglobin protein layer on the gold electrodes modified with the SAMs of metalloporphyrin-thiol linked molecules by reconstitution of apomyoglobin in solution with the corresponding metalloporphyrin and investigated their fluorescence spectra [192]. Fluorescence from the mixed SAMs of the ferrocene-thiol derivative and Zn tetraarylporphyrin-thiol derivative were measured under open-circuit conditions and the amounts of the photostoraged charge in the SAM were quantitatively examined by Roth and coworkers [193]. Bohn and coworkers constructed the protein-connected SAMs and measured the fluorescence intensity from the polystyrene nanosphere doped as a fluorescent label in the SAMs, and investigated cellular adhesion and motility by measuring the surface composition gradients of extracellular matrix proteins such as fibronectin [194, 195].

When the tris-bipyridyl ruthenium complex ($Ru(bpy)_3^{2+}$) is anodically polarized in an oxalate solution, it is known that an electrochemical luminescence (ECL) takes place [196, 197]. Just after the oxidation of $Ru(bpy)_3^{2+}$ to $Ru(bpy)_3^{3+}$, the oxalate anion ($C_2O_4^{2-}$) is resolved and then the carbon dioxide anion radical ($CO_2^{\cdot-}$) is produced. This anion radical reacts with $Ru(bpy)_3^{3+}$ and then the excited state of $Ru(bpy)_3^{2+}$ ($Ru(bpy)_3^{2+*}$) is produced. When the excited state ($Ru(bpy)_3^{2+*}$) is back to the ground state ($Ru(bpy)_3^{2+}$), the ECL occurs. The same phenomenon was reported using the SAM of $Ru(bpy)_3^{2+}$ derivative, in which the alkyl chain was introduced between the $Ru(bpy)_3^{2+}$ and the thiol groups (Figure 2.32) [198, 199].

Sato et al. simultaneously detected glucose and ascorbic acid using $FcC_{11}SH$ SAM on gold by individually measuring the electrochemical response of $FcC_{11}SH$ and the chemical luminescence intensity from luminol in the solution [121–123], as already described.

Passivation of the photoluminescence from the semiconductor substrate by the formation of the SAM was investigated [200, 201]. On the other hand, there is a paper that discusses the formation process of the SAM on the semiconductor surface in detail by measuring the photoluminescence from the semiconductor [202]. A polymer light-emitting diode using the SAM as an interfacial spacer was also investigated [203].

2.3.4.4 Catalytic Activity

Much attention has focused on fixing the metal complex by oxygen reduction on the electrode surface in relation to the catalyst development for the fuel cell. Oxygen reduction of the SAM of the alkanethiol with Por that imitates heme, on

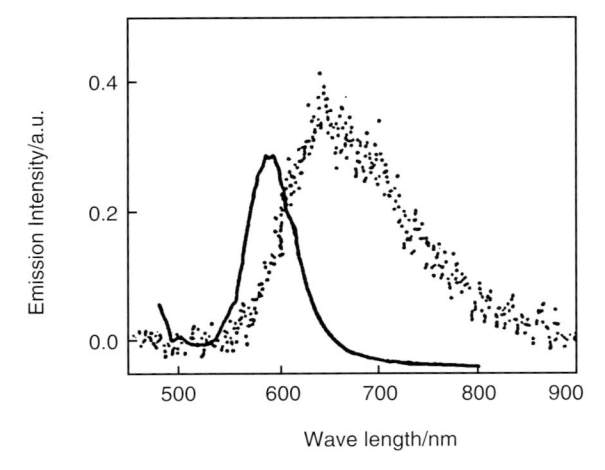

Figure 2.32 ECL spectrum of the Ru(bpy)$_3^{2+}$-thiol SAM-modified ITO electrode in a solution containing 0.4 M Na$_2$SO$_4$ + 0.1 M Na$_2$C$_2$O$_4$ at 1.15 V (vs. Ag/AgCl) (dots) and the emission spectrum of Ru(bpy)$_3^{2+}$-thiol in CH$_2$Cl$_2$ solution (solid curve). Reprinted from [198].

which the oxygen reduction takes place with the biologic system, has been significantly evaluated [204, 205]. At the bare gold electrode, a cathodic current due to the oxygen reduction reaction (ORR) was observed at a potential more negative than 0 V, and it was observed at a potential more negative than 0.3 V at the gold electrode modified with the SAM by the cobalt-Por (CoPor) derivative, clearly indicating that the modification of the SAM by the CoPor accelerates the ORR. Moreover, based on the comparison with the cathodic current for the ORR at the gold electrodes modified with the CoPor derivatives, which have one (CoPor1) and two (CoPor2) alkanethiol chains, the porphyrin ring of CoPor2 SAM seemed to be oriented more parallel to the surface than that of the CoPor1 one, showing that the metal center of Por is very important for the ORR. When the metal center of cobalt was substituted by zinc, the ORR activity was completely lost. There is also an example of using the Por SAM for the oxidation of tryptophan [206]. There are other examples of using cobalt phthalocyanine [207] and iron, cobalt, and manganese phthalocyanine [208–210] SAMs for the detection of $_L$-cysteine.

There are examples of using the SAM as a catalyst for organic reactions [211–213]. For example, Morrin et al. constructed the SAM of cyclopentadienylnickel(II) thiolate on gold and used it as a Schiff-base compound on the surface [211]. Besler et al. introduced the chiral rhodium-diphosphine complex into the alkanethiol SAM covered with the gold nanoparticle surface, used it as a catalyst for the hydrogenation reaction, and achieved high enantioselectivities (Figure 2.33) [212]. Hara et al. constructed an N-heterocyclic carbine-rhodium(I) complex terminated SAM on gold as a heterogeneous metal catalyst [213].

When enzyme or coenzyme is introduced into the SAM, the biomimetic catalytic reaction can be easily achieved. Carvalhal et al. immobilized glucose oxidase, which is the coenzyme for the glucose oxidation, onto the mercaptocarboxylic acid SAM on gold and investigated the effect of the redox reaction on the chemical

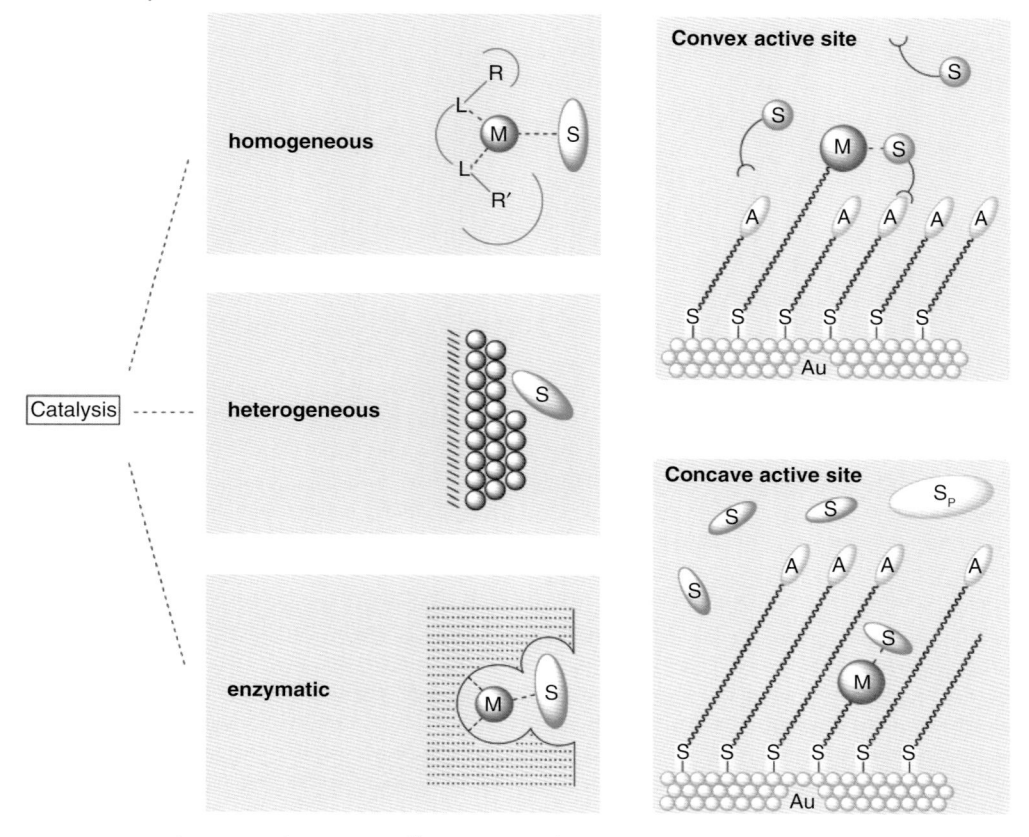

Figure 2.33 Comparison of homogeneous, heterogeneous, and enzymatic catalysts (left) and schematic representations of metal catalysts attached to the SAM on gold. Reprinted from [212].

environment around riboflavin [214]. Song *et al.* estimated the catalytic mechanism of the arsenite methylation reaction using the arsenite/S-adenosyl-$_L$-methionine SAM [215]. A protein structure-sensitive electrocatalytic reaction was achieved by Ostetná *et al.* using the gold electrode modified with the SAM of dithiothreitol [216]. The electrocatalytic properties of a synthetic heme peptide, mimochrome VI, which was designed to reproduce the catalytic activity of heme oxidases, were studied by Ranieri *et al.* [217].

2.4
Future Perspective

In this chapter, we have briefly described the self-assembled monolayer (SAM), including the formation mechanism, characterization, and its applications, by providing several examples. However, we think that the readers have the possibility of introducing functionalities to the solid surface and constructing molecular devices using the SAM although we showed only a few examples of this huge research effort. It is expected that more upgrades, such as the introduction of various molecules, three-dimensional accumulation, and the combining with metal and/or semiconductor nanoclusters, will occur in the future.

References

1 Langmuir, I. (1917) *J. Am. Chem. Soc.*, **39**, 1848.
2 Blodgett, K.A. (1935) *J. Am. Chem. Soc.*, **57**, 1007.
3 Blodgett, K.A. (1937) *Phys. Rev.*, **51**, 964.
4 Gaines, G.L. (1966) *Insoluble Monolayers Liquid–Gas Interfaces*, Interscience, New York.
5 Roberts, G. (1990) *Langmuir-Blodgett Films*, Plenum Press, New York.
6 Sagiv, J. (1980) *J. Am. Chem. Soc.*, **102**, 92.
7 Nuzzo, R.G., and Allara, D.L. (1983) *J. Am. Chem. Soc.*, **105**, 4481.
8 Ulman, A. (1991) *An Introduction to Ultrathin Organic Films from Langmuir-Blodgett to Self-Assembly*, Academic Press, New York.
9 Finklea, H.O. (1996) in *Electroanalytical Chemistry*, vol. 19 (eds A.J. Bard and I. Rubinstein), Marcel Dekker, New York, p. 109.
10 Love, J.C., Estroff, L.A., Kriebel, J.K., Nuzzo, R.G., and Whitesides, G.M. (2005) *Chem. Rev.*, **105**, 1103.
11 Kondo, T., and Uosaki, K. (2007) in *Encyclopedia of Electrochemistry*, vol. 10 (eds A.J. Bard, M. Stratmann, M. Fujihira, I. Rubinstein, and J.F. Rusling), Wiley-VCH Verlag GmbH, Weinheim, p. 80.
12 Kondo, T., and Uosaki, K. (2007) *J. Photochem. Photobiol. C, Photochem. Rev.*, **8**, 1.
13 Kondo, T., and Uosaki, K. (2009) in *Bottom-up Nanofabrication: Supramolecules, Self-Assemblies, and Organized Films*, vol. 4 (ed. K. Ariga), American Scientific Publishers, p. 409.
14 Linford, M.R., and Chidsey, C.E.D. (1993) *J. Am. Chem. Soc.*, **115**, 12631.
15 Buriak, J.M. (2002) *Chem. Rev.*, **102**, 1271.
16 Stutzmann, M., Garrido, J. A., Eickhoff, M., and Brandt, M. S. (2006), *Phys. Status Solidi*, **203**, 3424.
17 Ciampi, S., Harper, J.B., and Gooding, J.J. (2010) *Chem. Soc. Rev.*, **39**, 2158.
18 Cummings, S.P., Savchenko, J., and Ren, T. (2011) *Coord. Chem. Rev.*, **255**, 1587.
19 Baum, T., Ye, S., and Uosaki, K. (1999) *Langmuir*, **15**, 8577.
20 Hong, R., Fischer, N.O., Yerma, A., Goodman, D.M., Emrick, T., and Rotello, V.M. (2004) *J. Am. Chem. Soc.*, **126**, 739.
21 Kondo, T., Takechi, M., Sato, Y., and Uosaki, K. (1995) *J. Electroanal. Chem.*, **381**, 203.
22 Nichols, B.M., Butler, J.E., Russell, J.N., Jr., and Hamers, R.J. (2005) *J. Phys. Chem. B*, **109**, 20938.
23 Gewirth, A., and Siegenthaler, H. (1995) *Nanoscale Probes of the Solid/Liquid Interface*, Springer, New York.
24 Uosaki, K., Shen, Y., and Kondo, T. (1995) *J. Phys. Chem.*, **99**, 14117.
25 Taniguchi, I., Yoshimoto, S., Yoshida, M., Kobayashi, S., Miyawaki, T., Aono, Y., Sunatsuki, Y., and Taira, H. (2000) *Electrochem. Acta*, **45**, 2843.
26 Abe, M., Michi, T., Sato, A., Kondo, T., Zhou, W., Ye, S., Uosaki, K., and Sasaki, Y. (2003) *Angew. Chem. Int. Ed.*, **42**, 2912.
27 Yamada, R., Wano, H., and Uosaki, K. (2000) *Langmuir*, **16**, 5523.
28 Yamada, R., Sakai, H., and Uosaki, K. (1999) *Chem. Lett.*, **28**, 667.
29 Shimazu, K., Yagi, I., Sato, U., and Uosaki, K. (1992) *Langmuir*, **8**, 1385.
30 Schreiber, F. (2000) *Prog. Surf. Sci.*, **65**, 151.
31 Schwartz, D.K. (2001) *Annu. Rev. Phys. Chem.*, **52**, 107.

32. Sato, Y., Fray, B.L., Corn, R.M., and Uosaki, K. (1994) *Bull. Chem. Soc. Jpn.*, **67**, 21.
33. Nuzzo, R.G., Korenic, E.M., and Dubois, L.H. (1990) *J. Chem. Phys.*, **93**, 767.
34. Poirier, G.E. (1997) *Chem. Rev.*, **97**, 1117.
35. Delamarche, E., Michel, B., Gerber, C., Anselmetti, D., Güntherodt, H.J., Wolf, H., and Ringsdorf, H. (1994) *Langmuir*, **10**, 2869.
36. Poirier, G.E., and Tarlov, M.J. (1994) *Langmuir*, **10**, 2853.
37. Bucher, J.P., Santesson, L., and Kern, K. (1994) *Appl. Phys. A*, **59**, 135.
38. Vericat, C., Vela, M.E., and Salvarezza, R.C. (2005) *Phys. Chem. Chem. Phys.*, **7**, 3258.
39. Poirier, G.E., and Tarlov, M.J. (1995) *J. Phys. Chem.*, **99**, 10966.
40. Cavalleri, O., Hirstein, A., and Kern, K. (1995) *Surf. Sci.*, **340**, L960.
41. Cavalleri, O., Hirstein, A., Bucher, J.P., and Kern, K. (1996) *Thin Solid Films*, **284/285**, 392.
42. Cavalleri, O., Gilbert, S.E., and Kern, K. (1997) *Chem. Phys. Lett.*, **269**, 479.
43. Poirier, G.E., and Tarlov, M.J. (1996) *Science*, **272**, 1145.
44. Poirier, G.E., Fitts, W.P., and White, J.M. (2001) *Langmuir*, **17**, 1176.
45. Yamada, R., and Uosaki, K. (1998) *Langmuir*, **14**, 855.
46. Yamada, R., and Uosaki, K. (1997) *Langmuir*, **13**, 5218.
47. Walczak, M.M., Popenoe, D.D., Deinhammer, R.S., Lamp, B.D., Chung, C., and Porter, M.D. (1991) *Langmuir*, **7**, 2687.
48. Uosaki, K. (2009) *Chem. Record*, **9**, 199.
49. Sumi, T., and Uosaki, K. (2004) *J. Phys. Chem. B*, **108**, 6422.
50. Kondo, T., Sumi, T., and Uosaki, K. (2002) *J. Electroanal. Chem.*, **538/539**, 59.
51. Wano, H., and Uosaki, K. (2001) *Langmuir*, **17**, 8224.
52. Wano, H., and Uosaki, K. (2005) *Langmuir*, **21**, 4024.
53. Poirier, G.E. (1997) *Langmuir*, **13**, 2019.
54. Reed, M.A., and Lee, T. (2003) *Molecular Nanoelectronics*, American Scientific Publishers, Los Angeles.
55. Brust, M., Walker, D., Bethell, D., Schiffrin, D.J., and Whyman, R. (1994) *J. Chem. Soc. Chem. Commun.*, 801.
56. Haick, H., and Cahen, D. (2008) *Prog. Surf. Sci.*, **83**, 217.
57. Miozzo, L., Yassar, A., and Horowitz, G. (2010) *J. Mater. Chem.*, **20**, 2513.
58. Zhao, X., and Kopelman, R. (1996) *J. Phys. Chem.*, **100**, 11014.
59. Resch, R., Grasserbauer, M., Friedbacher, G., Vallant, T., Brunner, H., Mayer, U., and Hoffmann, H. (1999) *Appl. Surf. Sci.*, **140**, 168.
60. Higashi, G.S., Chabal, Y.J., Trucks, G.W., and Raghavachari, K. (1990) *Appl. Phys. Lett.*, **56**, 656.
61. Linford, M.R., Fenter, P., Eisenberger, P.M., and Chidsey, C.E.D. (1995) *J. Am. Chem. Soc.*, **117**, 3145.
62. Barrelet, C.J., Robinson, D.B., Cheng, J., Hunt, T.P., Quate, C.F., and Chidsey, C.E.D. (2001) *Langmuir*, **17**, 3460.
63. Fidélis, A., Ozanam, F., and Chazalviel, J.-N. (2000) *Surf. Sci.*, **444**, L7.
64. Allongue, P., de Villeneuve, C.H., and Pinson, J. (2000) *Electrochim. Acta*, **45**, 3241.
65. Zhong, Y.L., and Bernasek, S.L. (2011) *J. Am. Chem. Soc.*, **133**, 8118.
66. Cicero, R.L., and Chidsey, C.E.D. (2002) *Langmuir*, **18**, 305.
67. Fukumitsu, H., Masuda, T., Qu, D., Waki, Y., Noguchi, H., Takakusagi, S., Chun, W.-J., Kondo, T., Shimazu, K., Asakura, K., Taketsugu, T., and Uosaki, K. submitted, (2012).
68. Quayum, M.E., Kondo, T., Nihonyanagi, S., Miyamoto, D., and Uosaki, K. (2002) *Chem. Lett.*, **31**, 208.
69. Uosaki, K., Quayum, M.E., Nihonyanagi, S., and Kondo, T. (2004) *Langmuir*, **20**, 1207.
70. Nihonyanagi, S., Miyamoto, D., Idojiri, S., and Uosaki, K. (2004) *J. Am. Chem. Soc.*, **126**, 7034.
71. Ishibashi, T., Ara, M., Tada, H., and Onishi, H. (2003) *Chem. Phys. Lett.*, **367**, 376.
72. Faucheux, A., Yang, F., Allongue, P., de Villeneuve, C.H., Ozanam, F., and Chazalviel, J.-N. (2006) *Appl. Phys. Lett.*, **88**, 193123.
73. Faucheux, A., Gouget-Laemmel, A.C., Allongue, P., de Villeneuve, C.H., Ozanam, F., and Chazalviel, J.-N. (2007) *Langmuir*, **23**, 1236.

References

74 Yamada, R., Ara, M., and Tada, H. (2004) *Chem. Lett.*, **33**, 492.
75 Gorostiza, P., de Villeneuve, C.H., Sun, Q.Y., Sanz, F., Wallart, X., Boukherroub, R., and Allongue, P. (2006) *J. Phys. Chem. B*, **110**, 5576.
76 Ara, M., and Tada, H. (2003) *Appl. Phys. Lett.*, **83**, 578.
77 Haick, H., Hurley, P.T., Hochbaum, A.I., Yang, P., and Lewis, N.S. (2006) *J. Am. Chem. Soc.*, **128**, 8990.
78 Streifer, J.A., Kim, H., Nichols, B.M., and Hamers, R.J. (2005) *Nanotechnology*, **16**, 1868.
79 Bunimovich, Y.L., Shin, Y.S., Yeo, W.-S., Amori, M., Kwong, G., and Heath, J.R. (2006) *J. Am. Chem. Soc.*, **128**, 16323.
80 Aizenberg, J., Black, A.J., and Whitesides, G.M. (1999) *Nature*, **398**, 495.
81 Xia, Y., and Whitesides, G.M. (1998) *Angew. Chem. Int. Ed. Engl.*, **37**, 550.
82 Kumar, A., and Whitesides, G.M. (1994) *Science*, **263**, 60.
83 Love, J.C., Wolfe, D.B., Chabinyc, M.L., Paul, K.E., and Whitesides, G.M. (2002) *J. Am. Chem. Soc.*, **124**, 1576.
84 Pardo, L., Wilson, W.C., Jr., and Boland, T. (2003) *Langmuir*, **19**, 1462.
85 Zhou, Y., Valiokas, R., and Liedberg, B. (2004) *Langmuir*, **20**, 6206.
86 Datwani, S.S., Vijayendran, R.A., Johnson, E., and Biondi, S.A. (2004) *Langmuir*, **20**, 4970.
87 Saalmink, M., van der Marel, C., Stapert, H.R., and Burkinski, D. (2006) *Langmuir*, **22**, 1016.
88 O'Brien, B., Stebe, K.J., and Searson, P.C. (2007) *J. Phys. Chem. C*, **111**, 8686.
89 Soolaman, D.M., and Yu, H.-Z. (2007) *J. Phys. Chem. C*, **111**, 14157.
90 Fan, X., Rogow, D.L., Swanson, C.H., Tripathi, A., and Oliver, S.R.J. (2007) *Appl. Phys. Lett.*, **90**, 163114.
91 Mullen, T.J., Zhang, P., Srinivasan, C., Horn, M.W., and Weiss, P.S. (2008) *J. Electroanal. Chem.*, **621**, 229.
92 Andersson, O., Ulrich, C., Björefors, F., and Liedberg, B. (2008) *Sens. Actuators B*, **134**, 545.
93 Benor, A., Gburek, B., Wagner, V., and Knipp, D. (2010) *Org. Electron.*, **11**, 831.
94 Wendeln, C., Rinnen, S., Schulz, C., Arlinghaus, H.F., and Ravoo, B.J. (2010) *Langmuir*, **26**, 15966.
95 Nakagawa, T., and Hiwatashi, T. (2002) *Jpn. J. Appl. Phys.*, **41**, 3896.
96 Almanze-Workman, A.M., Raghavan, S., Petrovic, S., Gogoi, B., Deymier, P., Monk, D.J., and Roop, R. (2003) *Thin Solid Films*, **423**, 77.
97 Harada, Y., Girolami, G.S., and Nuzzo, R.G. (2003) *Langmuir*, **19**, 5104.
98 Harada, Y., Girolami, G.S., and Nuzzo, R.G. (2004) *Langmuir*, **20**, 10878.
99 Hale, P.S., Kappen, P., Brack, N., Prissanaroon, W., Pigram, P.J., and Liesengang, J. (2006) *Appl. Surf. Sci.*, **252**, 2217.
100 Kalyankar, K.D., Sharma, M.K., Vaidya, S.V., Calhoun, D., Maldarelli, C., Couzis, A., and Gilchrist, L. (2006) *Langmuir*, **22**, 5403.
101 Duan, X., Sadhu, V.B., Perl, A., Péter, M., Reinhoudt, D.N., and Huskens, J. (2008) *Langmuir*, **24**, 3621.
102 Cau, J.-C., Cerf, A., Thibault, C., Geneviéve, M., Séverac, C., Peyrade, J.-P., and Vieu, C. (2008) *Microelectron. Eng.*, **85**, 1143.
103 Zheng, Z., Jang, J.-W., Zheng, G., and Mirkin, C.A. (2008) *Angew. Chem. Int. Ed.*, **47**, 9951.
104 Jones, C.N., Tuleuova, N., Lee, J.Y., Ramanculov, E., Reddi, A.H., Zern, M.A., and Revzin, A. (2009) *Biomater*, **30**, 3733.
105 Packard, C.E., Murata, A., Lam, E.W., Schmidt, M.A., and Bulović, M.A. (2010) *Adv. Mater.*, **22**, 1840.
106 Mizuno, H., and Buriak, J.M. (2010) *ACS Appl. Mater. Interface*, **2**, 2301.
107 Piner, R.D., Zhu, J., Xu, F., Hong, S., and Mirkin, C.A. (1999) *Science*, **283**, 661.
108 Liu, G.-Y., Xu, S., and Qian, Y. (2000) *Acc. Chem. Res.*, **33**, 457.
109 Zhao, J., and Uosaki, K. (2002) *Nano Lett.*, **2**, 137.
110 Zhao, J., and Uosaki, K. (2001) *Langmuir*, **17**, 7784.
111 Shimazu, K., Kawaguchi, T., and Isomura, T. (2002) *J. Am. Chem. Soc.*, **124**, 652.
112 Shimazu, K., Hashimoto, Y., Kawaguchi, T., and Tada, K. (2002) *J. Electroanal. Chem.*, **534**, 163.

113 Rubinstein, I., Steinberg, S., Tor, Y., Shanzer, A., and Sagiv, J. (1988) *Nature*, **332**, 426.
114 Rojas, M.T., and Kaifer, A.E. (1995) *J. Am. Chem. Soc.*, **117**, 5883.
115 Katz, E., Lötzbeyer, T., Schlereth, D.D., Schuhmann, W., and Schmidt, H.-L. (1994) *J. Electroanal. Chem.*, **373**, 189.
116 Tredici, I., Merli, D., Zavarise, F., and Profumo, A. (2010) *J. Electroanal. Chem.*, **645**, 22.
117 Oztekin, Y., Ramanaviciene, A., and Ramanavisius, A. (2011) *Sens. Actuators B*, **155**, 612.
118 Mizutani, F., Sato, Y., Yabuki, S., Sawaguchi, T., and Iijima, S. (1999) *Electrochim. Acta*, **44**, 3833.
119 Sato, Y., and Mizutani, F. (2000) *Electrochim. Acta*, **45**, 2869.
120 Mizutani, F., Yabuki, S., Sato, Y., Sawaguchi, T., and Iijima, S. (2000) *Electrochim. Acta*, **45**, 2945.
121 Sato, Y., Kato, D., Niwa, O., and Mizutani, F. (2005) *Sens. Actuators B*, **108**, 617.
122 Sato, Y., Sawaguchi, T., and Mizutani, F. (2001) *Electrochem. Commun.*, **3**, 131.
123 Sato, Y., Yabuki, S., and Mizutani, F. (2000) *Chem. Lett.*, **29**, 1330.
124 Sato, Y., Yoshioka, K., Tanaka, M., Murakami, T., Ishida, M.N., and Niwa, O. (2008) *Chem. Commun.*, 4909.
125 Yoshioka, K., Sato, Y., Murakami, T., Tanaka, M., and Niwa, O. (2010) *Anal. Chem.*, **82**, 1175.
126 Vergheese, T.M., and Berchmans, S. (2004) *J. Electroanal. Chem.*, **570**, 35.
127 Gobi, K.V., Matsumoto, K., Toko, K., Ikezaki, H., and Miura, N. (2007) *Anal. Bioanal. Chem.*, **387**, 2727.
128 Tencer, M., Nie, H.-Y., and Berini, P. (2009) *J. Electrochem. Soc.*, **156**, J386.
129 Uzawa, T., Cheng, R.R., White, R.J., Makarov, D.E., and Plaxco, K.W. (2010) *J. Am. Chem. Soc.*, **132**, 16120.
130 Loaiza, O.A., Lamas-Ardisana, P.J., Jubete, E., Ochoteco, E., Loinaz, I., Cabañero, G., García, I., and Penadés, S. (2011) *Anal. Chem.*, **83**, 2987.
131 Walter, A., Wu, J., Flechsig, G.-U., Haake, D.A., and Wang, J. (2011) *Anal. Chim. Acta*, **689**, 29.
132 Rezaei, B., Majidi, N., Rahmani, H., and Khayamian, T. (2011) *Biosens. Bioelectron.*, **26**, 2130.
133 Behpour, M., Ghoreishi, S.M., Honarmand, E., and Salavati-Niasari, M. (2011) *J. Electroanal. Chem.*, **653**, 75.
134 Akin, M., Prediger, A., Yuksel, M., Höpfiner, T., Demirkol, D.O., Beutel, S., Timur, S., and Scheper, T. (2011) *Biosens. Bioelectron.*, **26**, 4532.
135 Bian, C., Tong, J., Sun, J., Zhang, H., Xue, Q., and Xia, S. (2011) *Biomed. Microdevices*, **13**, 345.
136 Faber, E.J., Sparreboom, W., Groeneveld, W., de Smet, L.C.P., Bomer, J., Olthuis, W., Zuihof, H., Sudhölter, E.J.R., Bergveld, P., and van den Berg, A. (2007) *ChemPhysChem*, **8**, 101.
137 Huang, Y., and Suni, I. (2008) *J. Electrochem. Soc.*, **155**, J350.
138 Prathima, N., Harini, M., Rai, N., Chandrashekara, R.H., Ayappa, K.G., Sampath, S., and Biswas, S.K. (2005) *Langmuir*, **21**, 2364.
139 Lee, S., Heeb, R., Venkataraman, N.V., and Spencer, N.D. (2007) *Tribol. Lett.*, **28**, 229.
140 Lu, L., and Cai, Y. (2011) *Langmuir*, **27**, 5953.
141 Booth, B.D., Vilt, S.G., McCabe, C., and Jennings, G.K. (2009) *Langmuir*, **25**, 9995.
142 Uosaki, K., Sato, Y., and Kita, H. (1991) *Langmuir*, **7**, 1510.
143 Uosaki, K., Sato, Y., and Kita, H. (1991) *Electrochim. Acta*, **36**, 1799.
144 Sato, Y., Itoigawa, H., and Uosaki, K. (1993) *Bull. Chem. Soc. Jpn.*, **66**, 1032.
145 Kondo, T., Okamura, M., and Uosaki, K. (2001) *J. Organometal. Chem.*, **637/639**, 841.
146 Sato, Y., Fujita, M., Mizutani, F., and Uosaki, K. (1996) *J. Electroanal. Chem.*, **409**, 145.
147 Ye, S., Yashiro, A., Sato, Y., and Uosaki, K. (1996) *J. Chem. Soc. Faraday Trans.*, **92**, 3813.
148 Nann, T., and Urban, G.A. (2001) *J. Electroanal. Chem.*, **505**, 125.
149 Hong, H.-G., and Park, W. (2005) *Electrochim. Acta*, **51**, 579.
150 Hickman, J.J., Ofer, D., Laibinis, P.E., Whitesides, G.M., and Wrighton, M.S. (1991) *Science*, **252**, 688.
151 Kondo, T., Kanai, T., and Uosaki, K. (2001) *Langmuir*, **17**, 6317.

152 Masuda, T., and Uosaki, K. (2004) *Chem. Lett.*, **33**, 768.
153 Masuda, T., Shimazu, K., and Uosaki, K. (2008) *J. Phys. Chem. C*, **112**, 10923.
154 Masuda, T., Irie, M., and Uosaki, K. (2009) *Thin Solid Films*, **518**, 591.
155 Fukumitsu, H., Masuda, T., Qu, D., Waki, Y., Noguchi, H., Shimazu, K., and Uosaki, K. (2010) *Chem. Lett.*, **39**, 768.
156 Bunimovich, Y.L., Ge, G., Beverly, K.C., Ries, R.S., Hood, L., and Heath, J.R. (2004) *Langmuir*, **20**, 10630.
157 Rohde, R.D., Agnew, H.D., Yeo, W.-S., Bailey, R.C., and Heath, J.R. (2006) *J. Am. Chem. Soc.*, **128**, 9518.
158 Cai, W., Peck, J.R., van der Weide, D.W., and Hamers, R.J. (2004) *Biosens. Bioelectron.*, **19**, 1013.
159 Nakato, K., Takabayashi, S., Imanishi, A., Murakoshi, K., and Nakato, Y. (2004) *Sol. Energy Mater. Sol. Cells*, **83**, 323.
160 Takabayashi, S., Ohashi, M., Mashima, K., Liu, Y., Yamazaki, S., and Nakato, Y. (2005) *Langmuir*, **21**, 8832.
161 Ohashi, M., Takabayashi, S., Mashima, K., and Nakato, Y. (2006) *Chem. Lett.*, **35**, 956.
162 Fabre, B., Lopinski, G.P., and Wayner, D.D.M. (2003) *J. Phys. Chem. B*, **107**, 14326.
163 Kondo, T., Ito, T., Nomura, S., and Uosaki, K. (1996) *Thin Solid Films*, **284/285**, 652.
164 Kondo, T., Yanagida, M., Nomura, S., Ito, T., and Uosaki, K. (1997) *J. Electroanal. Chem.*, **438**, 121.
165 Uosaki, K., Kondo, T., Zhang, X.-Q., and Yanagida, M. (1997) *J. Am. Chem. Soc.*, **119**, 8367.
166 Yanagida, M., Kanai, T., Zhang, X.-Q., Kondo, T., and Uosaki, K. (1998) *Bull. Chem. Soc. Jpn.*, **71**, 2555.
167 Kondo, T., Kanai, T., Iso-o, K., and Uosaki, K.Z. (1999) *Phys. Chem.*, **212**, 23.
168 Imahori, H., Yamada, H., Nishimura, Y., Yamazaki, I., and Sakata, Y. (2000) *J. Phys. Chem. B*, **104**, 2099.
169 Imahori, H., Norieda, H., Yamada, H., Nishimura, Y., Yamazaki, I., Sakata, Y., and Fukuzumi, S. (2001) *J. Am. Chem. Soc.*, **123**, 100.
170 Ungashe, S.B., Wilson, W.L., Katz, H.E., Scheller, G.R., and Putvinski, T.M. (1992) *J. Am. Chem. Soc.*, **114**, 8717.
171 Terasaki, N., Akiyama, T., and Yamada, S. (2002) *Langmuir*, **18**, 8666.
172 Otsubo, T., Aso, Y., and Takimiya, K. (2002) *J. Mater. Chem.*, **12**, 2565.
173 Wakamatsu, S., Akiba, U., and Fujihira, M. (2002) *Jpn. J. Appl. Phys.*, **41**, 4998.
174 Ikeda, K., Takahashi, K., Masuda, T., and Uosaki, K. (2011) *Angew. Chem. Int. Ed.*, **50**, 1280.
175 Brust, M., Fink, J., Bethell, D., Schiffrin, D.J., and Kiely, C. (1995) *J. Chem. Soc. Chem. Commun.*, 1655.
176 Hostetler, M.J., Templeton, A.C., and Murray, R.W. (1999) *Langmuir*, **15**, 3782.
177 Yamada, S., Tasaki, T., Akiyama, T., Terasaki, N., and Nitahara, S. (2003) *Thin Solid Films*, **438/439**, 70.
178 Imahori, H., Kashiwagi, Y., Endo, Y., Hanada, T., Nishimura, Y., Yamazaki, I., Araki, Y., Ito, O., and Fukuzumi, S. (2004) *Langmuir*, **20**, 73.
179 Imahori, H., and Fukuzumi, S. (2001) *Adv. Mater.*, **13**, 1197.
180 Hasobe, T., Imahori, H., Kamat, P.V., and Fukuzumi, S. (2003) *J. Am. Chem. Soc.*, **125**, 14962.
181 Li, G., Fudickar, W., Skupin, M., Klyszcz, A., Draeger, C., Lauer, M., and Fuhrhop, J.-H. (2002) *Angew. Chem. Int. Ed.*, **41**, 1828.
182 Steigerwald, M.L., Alivisatos, A.P., Gibson, J.M., Harris, T.D., Kortan, R., Muller, A.J., Thayer, A.M., Duncan, T.M., Douglass, D.C., and Brus, L.E. (1988) *J. Am. Chem. Soc.*, **110**, 3046.
183 Nakanishi, T., Ohtani, B., and Uosaki, K. (1998) *J. Phys. Chem. B*, **102**, 1571.
184 Ogawa, S., Fan, F.-R.F., and Bard, A.J. (1995) *J. Phys. Chem.*, **99**, 11182.
185 Nakanishi, T., Ohtani, B., and Uosaki, K. (1998) *J. Electroanal. Chem.*, **455**, 229.
186 Nakanishi, T., Ohtani, B., and Uosaki, K. (1999) *Jpn. J. Appl. Phys.*, **38**, 518.
187 Miyake, M., Torimoto, T., Nishizawa, M., Sakata, T., Mori, H., and Yoneyama, H. (1999) *Langmuir*, **15**, 2714.
188 Ogawa, S., Hu, K., Fan, F.-R.F., and Bard, A.J. (1997) *J. Phys. Chem. B*, **101**, 5707.
189 Bakkers, E.P.A.M., Roest, A.L., Marsman, A.W., Jenneskens, L.W., de Jong-van Steensel, L.I., Kelly, J.J., and Vanmaekelbergh, D. (2000) *J. Phys. Chem. B*, **104**, 7266.

190 Woo, D.-H., Choi, S.-J., Han, D.-H., Kang, H., and Park, S.-M. (2001) *Phys. Chem. Chem. Phys.*, **3**, 3382.
191 Fox, M.A., and Wooten, M.D. (1997) *Langmuir*, **13**, 7099.
192 Guo, L.-H., McLendon, G., Razafitrimo, H., and Gao, Y. (1996) *J. Mater. Chem.*, **6**, 369.
193 Roth, K.M., Lindsey, J.S., Bocian, D.F., and Kuhr, W.G. (2002) *Langmuir*, **18**, 4030.
194 Plummer, S.T., and Bohn, P.W. (2002) *Langmuir*, **18**, 4142.
195 Plummer, S.T., Wang, Q., and Bohn, P.W. (2003) *Langmuir*, **19**, 7528.
196 Rubinstein, I., and Bard, A.J. (1981) *J. Am. Chem. Soc.*, **103**, 512.
197 Zhang, X., and Bard, A.J. (1988) *J. Phys. Chem.*, **92**, 5566.
198 Sato, Y., and Uosaki, K. (1995) *J. Electroanal. Chem.*, **384**, 57.
199 Zanarini, S., Rampazzo, E., Bich, D., Canteri, R., Ciana, L.D., Marcaccio, M., Marzocchi, E., Montalti, M., Panciatichi, C., Pederzolli, C., Paolucci, F., Prodi, L., and Vanzetti, L. (2008) *J. Phys. Chem. C*, **112**, 2949.
200 Budz, H.A., Biesinger, M.C., and La Pierre, R.R. (2009) *J. Vac. Sci. Technol. B*, **27**, 637.
201 Schvartzman, M., Sidorov, V., Ritter, D., and Paz, Y. (2001) *Semicon. Sci. Technol.*, **16**, L68.
202 Kim, C.-K., Marshall, G.M., Martin, M., Bisson-Viens, M., Wasilewski, Z., and Dubowski, J.J. (2009) *J. Appl. Phys.*, **109**, 083518.
203 Lee, T.-W. (2007) *Adv. Funct. Mater.*, **17**, 3128.
204 Nishimura, N., Ooi, M., Shimazu, K., Fujii, H., and Uosaki, K. (1999) *J. Electroanal. Chem.*, **473**, 75.
205 Shimazu, K., Takechi, M., Fujii, H., Suzuki, M., Saiki, H., Yoshimura, T., and Uosaki, K. (1996) *Thin Solid Films*, **273**, 250.
206 Chambrier, I., Russell, D.A., Brundish, D.E., Love, W.G., Jori, G., Magaraggia, M., and Cook, M.J. (2010) *J. Porphyrins Phthalocyanines*, **14**, 81.
207 Nombona, N., Geraldo, D.A., Hakuzimana, J., Schwarz, A., Westbroek, P., and Nyokong, T. (2009) *J. Appl. Electrochem.*, **39**, 727.
208 Mashazi, P.N., Westbroek, P., Ozoemena, K.I., and Nyokong, T. (2007) *Electrochim. Acta*, **53**, 1858.
209 Ozoemena, K.I., and Nyokong, T. (2006) *Electrochim. Acta*, **51**, 2669.
210 Ozoemena, K.I., Nyokong, T., and Westbroek, P. (2003) *Electroanalysis*, **14**, 1762.
211 Morrin, A., Moulaoli, R.M., Killard, A.J., Smyth, M.R., Darkwa, J., and Iwuoha, E.I. (2004) *Talanta*, **64**, 30.
212 Belser, T., Stöhr, M., and Pfaltz, A. (2005) *J. Am. Chem. Soc.*, **127**, 8720.
213 Hara, K., Iwahashi, K., Takakusagi, S., Uosaki, K., and Sawamura, M. (2007) *Surf. Sci.*, **601**, 5127.
214 Carvalhal, R.F., Mendes, R.K., and Kubota, L.T. (2007) *Int. J. Electrochem. Sci.*, **2**, 973.
215 Song, X., Geng, Z., Li, X., Hu, X., Bian, N., Zhang, X., and Wang, Z. (2010) *Biochimie*, **92**, 1397.
216 Ostatná, V., Černocká, H., and Paleček, E. (2010) *J. Am. Chem. Soc.*, **132**, 9408.
217 Ranieri, A., Monari, S., Sola, M., Borsari, M., Battistuzzi, G., Ringhieri, P., Nastri, F., Pavone, V., and Lombardi, A. (2010) *Langmuir*, **26**, 17831.

3
Langmuir–Blodgett (LB) Film
Ken-ichi Iimura and Teiji Kato

3.1
Concept and Mechanism

There are two monolayer systems of amphiphilic substances at the aqueous solution surfaces, that is, Gibbs monolayers and Langmuir monolayers. Gibbs monolayers are formed by spontaneous adsorption of water-soluble amphiphiles such as surfactants from the bulk of aqueous solutions to the solution surfaces. Gibbs monolayers are also called adsorbed monolayers or soluble monolayers. Langmuir monolayers are formed by spreading water-insoluble amphiphiles from their dilute solutions of volatile organic solvents on the surface of aqueous subphase. Langmuir monolayers can be compressed with movable Teflon barriers to increase the two-dimensional densities of amphiphile molecules. Langmuir–Blodgett (LB) multilayers are prepared by successive transfer of Langmuir monolayers from subphase surface to smooth solid substrates. They are molecularly organized multilayer systems. Because of some advantages of LB films such as high homogeneity, ordered molecular orientation and/or arrangement, controllable nanometer-scale thickness, a wide variety of useable film molecules and their easy chemical modification, etc., the films have been applied to the development of diverse functional assemblies in molecular engineering.

3.2
Preparation and Characterization

3.2.1
Gibbs Monolayers

Surfactant molecules dissolved in water spontaneously adsorb at the air/water interface to form the Gibbs monolayer. The surface tension of an aqueous surfactant solution reduces from that of pure water with an increase of the adsorbed amount (or with time) to attain a constant value. This constant surface tension is the thermodynamic equilibrium value of the surfactant solution system, and depends on the bulk concentration of surfactant and on temperature. When the equilibrium surface tension is plotted against the bulk concentration at a constant temperature, it decreases with increasing bulk concentration, but becomes constant above the critical micelle concentration (cmc) of the surfactant solution. This is a representative procedure to determine the cmc of a surfactant by the break point of this plot.

The surface pressure (π, two-dimensional pressure) of a monolayer is defined as follows,

$$\pi = \gamma_w - \gamma_f \tag{3.1}$$

where γ_f is the surface tension of the monolayer-covered water surface, and γ_w is that of pure water at the same temperature. After removal of the already adsorbed Gibbs monolayer by sucking the surface with a suction pump or by sweeping the surface with movable Teflon barriers, one can measure the surface pressure (π)–time (t) adsorption isotherm of the Gibbs monolayer by following the change of surface pressure with time. However, it is very difficult to measure the exact adsorption isotherm of common surfactants because the adsorption process is very rapid and it is difficult to assure $\pi = 0$ at the starting point ($t = 0$). The surface pressure π increases rapidly and monotonously, and reaches the equilibrium value within several minutes for most common surfactants. It was believed for a long time that the equilibrated Gibbs monolayers of common surfactants are uniform with no characteristic higher-order structures, corresponding to an expanded phase. Recently, however, it was reported that some kinds of soluble amphiphiles formed Gibbs monolayers of condensed phases exhibiting highly ordered structures [1–5].

The common features of these amphiphiles are as follows; (i) the solubilities of these amphiphiles against water are considerably lower in comparison with those of normal surfactants, (ii) there are strong attractive interactions among the hydrophilic groups of the amphiphiles such as formation of hydrogen bonding, (iii) these amphiphiles have longer alkyl groups than common surfactants, and therefore, there is rather strong van der Waals attractive interaction among the hydrophobic alkyl groups.

The π–t adsorption isotherms of these amphiphiles do not increase monotonously with time, but exhibit cusp points followed by plateaus and increase again to reach the equilibrium surface pressures. By observing the Gibbs monolayers during the adsorption processes with Brewster-angle microscopy (BAM, described later), one can see the formation of condensed-phase domains in the expanded uniform phase after the cusp points, and growing of the condensed domains with time. Finally, at equilibrium, the whole Gibbs monolayers become a condensed phase of highly ordered structures. Figure 3.1 shows π–t adsorption isotherms and BAM images of Gibbs monolayers of bis(ethylene glycol) mono-n-tetradecyl ether (C14E2) at different temperatures [5c]. All π–t adsorption isotherms show cusp points followed by plateaus. After the cusp points, the formation and growth of condensed domains of various shapes are observed at different temperatures by BAM.

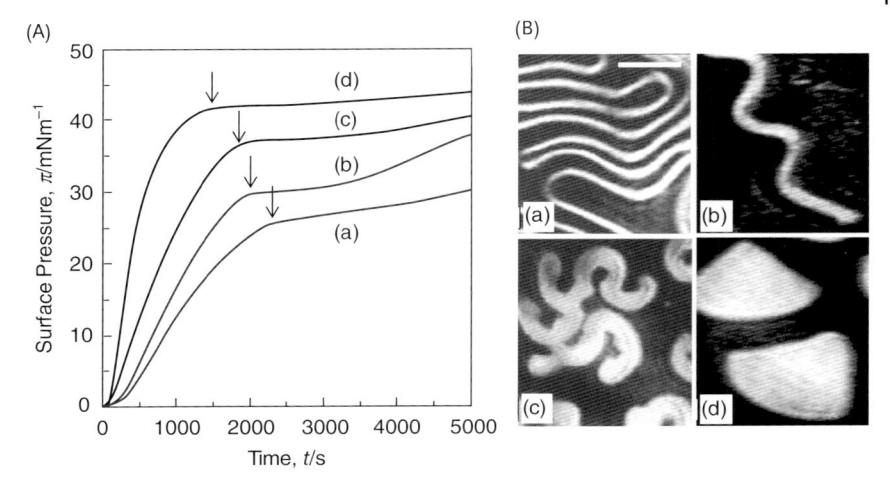

Figure 3.1 (A) π–t adsorption isotherms of a 2.0×10^{-5} M aqueous solution of C14E2 at different temperatures: (a) 5 °C, (b) 10 °C, (c) 15 °C, (d) 18 °C. (B) BAM images after the phase transition: (a) 5 °C, (b) 10 °C, (c) 15 °C, (d) 18 °C. The bar in image (A) indicates 100 µm. Reprinted with permission from Ref. [5c]. Copyright 2004 The American Chemical Society.

Melzer and his collaborators have shown by simulation that these types of π–t adsorption isotherms can be converted to the π–A isotherms (described later) of Langmuir monolayers [3b]. Figure 3.2 (top) shows π–A isotherms of N-(γ-hydroxypropyl)-tridecanoic acid amide (HTRAA) and simulated π–A isotherms converted from the π–t adsorption isotherms of the same material at two different temperatures (4 °C and 10 °C). Figure 3.2 (bottom) shows BAM images of the HTRAA monolayers at point (a) during measurement of the π–A isotherm, and at points (b) and (c) during measurement of the π–t adsorption isotherm.

There was a long history that Gibbs monolayers and Langmuir monolayers were the subjects of studies in the different fields. The facts described above confirmed that there is no clear boundary between them. Furthermore, when we observe condensed phases of Gibbs monolayers and Langmuir monolayers of the same amphiphiles by BAM, the shapes of the condensed domain in Gibbs monolayers are more roundish, whereas those in Langmuir monolayers are dendritic, as shown in Figure 3.2 (bottom). This means that formation of Gibbs monolayers proceeds maintaining near equilibration, whereas compression of Langmuir monolayers causes deviation from maintaining equilibrium.

3.2.2
Langmuir Monolayers

3.2.2.1 Basic Measurements of Properties of Langmuir Monolayers
Langmuir monolayers are prepared by spreading of solutions of water-insoluble amphiphiles in volatile organic solvents, such as benzene, diethyl ether,

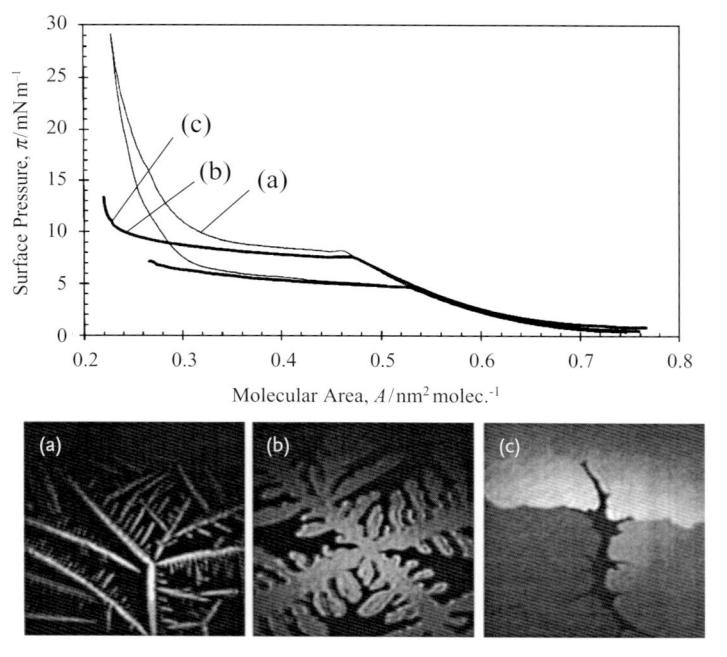

Figure 3.2 π–A isotherms of HTRAA at 4 °C and 10 °C, and two simulated π–A isotherms converted from π–t adsorption isotherms. BAM images were observed at point (a) during π–A measurement and observed at points (b) and (c) during π–A measurement. Reprinted with permission from Ref. [3b]. Copyright 1998 The American Chemical Society.

chloroform, etc., on the surface of aqueous subphase in a shallow, rectangular trough, called a Langmuir trough. The trough is equipped with several Teflon barriers to sweep the subphase surface to remove surface active pollutants, and to compress monolayers to change the occupied molecular area of the amphiphile. A surface balance should also be equipped to detect the surface pressure of the monolayer with compression. There are some different types of surface balance, but the Wilhelmy-type (a hanging-plate type) surface balance is the most common and very easy to use: a rectangular plate of thin glass or filter paper of the size of 1 cm × 2 cm is hung from a force sensor, and the bottom half of the plate is immersed in the subphase.

According to Gains [6], the number of degrees of freedom of a system of a Langmuir monolayer on the water surface is given by the following equation,

$$F = C^b + C^s - P^b - q + 3 \tag{3.2}$$

where F is the number of degree of freedom, C^b is the number of components in bulk and equilibrated throughout the system, C^s is the number of components confined completely to the surface, P^b is the number of bulk phases, and q is the number of surface phases in equilibrium with one another.

In this derivation of the phase rule, it is assumed that the monolayer material should be confined to the surface phase, and neither dissolution into subphase nor vaporization into gas phase should occur. In the case of a single-component Langmuir monolayer, as $C^s = 1$, $C^b = 2$ (water and gas), $P^b = 2$ (water and a gas phase), F of the system is given as follows;

$$F = 4 - q \qquad (3.3)$$

There are three degrees of freedom when the monolayer consists of one surface phase. They are external pressure (three-dimensional pressure), temperature, and surface pressure (two-dimensional). Since experiments on Langmuir monolayers are usually performed under atmospheric pressure, bulk pressure is dealt with not as a variable but rather as a constant (fixed to the atmospheric pressure). There is another experimental variable used usually, that is, occupied molecular area (A) that can be calculated from the geometrical area (S) of the surface occupied by the Langmuir monolayer and the number of amphiphile molecules (N) in the monolayer,

$$A = S/N \qquad (3.4)$$

A can be changed by compression with moving Teflon barriers confining the Langmuir monolayer. There are three ways to take two variables from three, and there are three types of basic experiments on Langmuir monolayers. The first one is the surface pressure–occupied molecular area (π–A) isotherm, where surface pressure (π) is measured as a function of molecular area (A), keeping temperature constant. The second is the molecular area–temperature (A–T) isobar where molecular area is measured as a function of temperature, keeping surface pressure constant. The third is the surface pressure–temperature (π–T) isochore or isoarea where surface pressure is measured as a function of temperature, keeping molecular area constant. But it is very difficult to measure precise A–T isobars or especially π–T isochores because dissolution of film material into the subphase water or leakage of the film material through corners of barriers, even in a very small amount, will introduce fatal errors to the results. There were only several papers dealing with A–T isobars and its relating subjects of Langmuir monolayers [7, 8], but no report on π–T isochores yet.

3.2.2.2 A–T isobars

To measure A–T isobars, the temperature of the subphase surface must be controlled linearly with time in both directions of heating and cooling at various rates. To suppress convection in the bulk of subphase water, the depth of the subphase should be less than 2 mm. The bottom of the Langmuir trough is made of gold-plated copper plate to assure good thermal conductivity. Rim of the trough is made of Teflon block of 2 mm thickness. On the backside of the copper bottom plate, a number of integrated Peltier elements are fixed. At the surface of the subphase water, a platinum wire temperature sensor of very small heat capacity is set. A microcomputer feedback controls the direction and amount of direct current runs through the series of Peltier elements based on the surface temperature detected

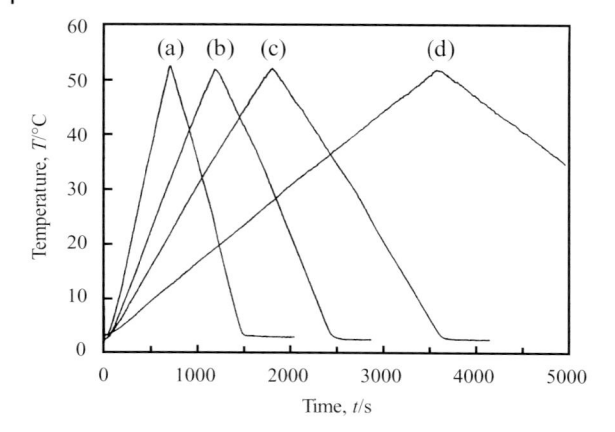

Figure 3.3 Linear heating and linear cooling at various rates of the subphase surface: (a) ±5 °C/min, (b) ±3 °C/min, (c) ±2 °C/min, (d) ±1 °C/min.

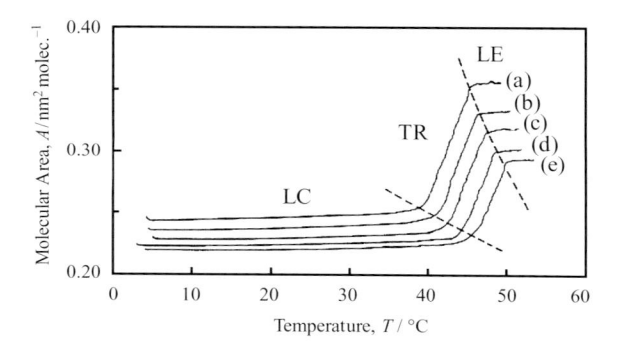

Figure 3.4 A–t isobars of palmitic acid monolayers at different surface pressures: (a) 5 mN/m, (b) 7.5 mN/m, (c) 10 mN/m, (d) 12.5 mN/m, (e) 15 mN/m.

[9]. Figure 3.3 shows results of linear heating and cooling at various rates of the surface temperature of the subphase in the trough.

It is not easy to keep the surface pressure of a Langmuir monolayer constant during temperature rising or lowering because the surface pressure is defined by Eq. (3.1) at the same temperature. Both the surface pressure of the monolayer and the surface tension of the clear water surface change on changing temperature. Therefore, the instrument is equipped with two Wilhelmy-type surface balances, one is situated in the monolayer-covered area and the other is situated in the clean water surface area in the same trough. The computer feedback controls the surface pressure to be constant based on the detected change of surface tension of clean water surface during the temperature change. Figure 3.4 shows the surface pressure dependency of A–T isobars of Langmuir monolayers of palmitic acid at a heating rate of 1 °C/min. The monolayers show a phase transition from a liquid condensed (LC) phase to a liquid expanded (LE) phase at around 30 to 40 °C by increasing the temperature through a transition region (TR). The transition tem-

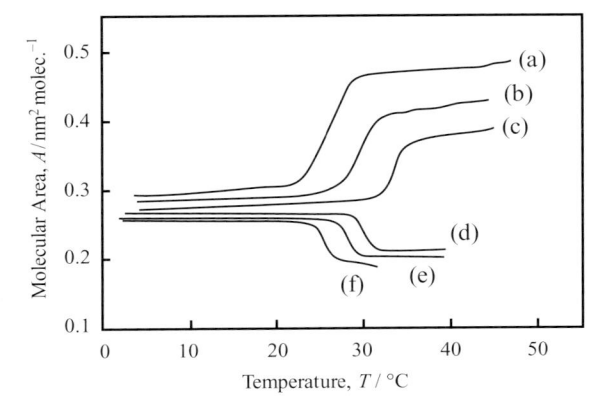

Figure 3.5 A–t isobars of N-methyl OU at (a) 5 mN/m, (b) 10 mN/m, (c) 15 mN/m, and those of OU at (d) 5 mN/m, (e) 10 mN/m, (f) 15 mN/m. Reproduced with permission from Ref. [10]. Copyright 1995 Elsevier.

perature increases with increasing surface pressure. Figure 3.5 shows the surface pressure dependency of A–T isobars of Langmuir monolayers of two amphiphiles, octadecyl urea (OU) and N-methyl octadecyl urea (N-methyl OU) [10].

N-methyl OU monolayers exhibit a transition from a condensed phase to an expanded phase with increasing temperature, keeping surface pressure constant. The transition temperature shifts to higher values with increase of the constant surface pressure, as shown by the case of palmitic acid in Figure 3.4. A–T isobars of OU exhibit a contraction-type phase transition from one condensed phase (β-form) to another condensed phase (α-form) with increasing temperature. In contrast to N-methyl OU and heptadecanoic acid monolayers, phase transition temperatures of OU shift to lower values with the increase of the constant surface pressure. This abnormal phase transition of OU monolayers is explained by melting of the hydrogen-bond network among the urea head groups of the amphiphile. This corresponds to the two-dimensional version of melting of ice. For N-methyl OU, the methyl group connected to the nitrogen atom in the head group disturbed the formation of the hydrogen-bond network. The reliability of A–T isobars is confirmed by the coincidence of isobars measured under temperature rising and lowering. Figure 3.6 shows two isobars of OU, successively measured at the heating and cooling rate of 0.5 °C/min. Two isobars coincide well except in the phase-transition region that shows supercooling of the phase transition under cooling.

The tangential slopes of A–T isobars give two-dimensional expansivities (β^s), which are important thermal properties of Langmuir monolayers, defined by the following equation.

$$\beta^s = (1/A)(\partial A/\partial T)_\pi \tag{3.5}$$

The thermal expansivity of OU monolayers of α-form is 2×10^{-3}/K, while that of the β-form is a negative value of -2×10^{-4}/K. This negative expansivity of OU monolayers in β phase is not due to the partial dissolution or partial collapse of

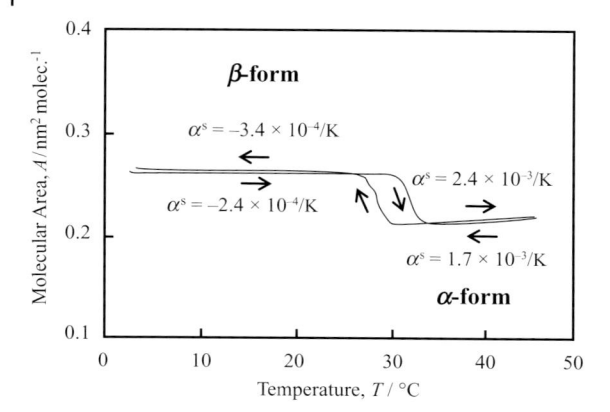

Figure 3.6 Two isobars of OU successively measured at a heating and cooling rate of 0.5 °C/min. Reprinted with permission from Ref. [8c]. Copyright 1992 The Chemical Society of Japan.

the film material with increasing temperature because the molecular areas at the start point and end point coincided well. This means that this small but negative thermal expansivity is an inherent property of OU monolayers in the β-form. In general, β^s of Langmuir monolayers in the liquid expanded state is around 0.03/K, that in the liquid condensed state is around 0.002/K, and that in the solid state is around 0.001/K.

3.2.2.3 π–A isotherms

When an exact amount of an amphiphile solution of known concentration is spread with a microsyringe on the subphase surface of known area in a Langmuir trough, one can easily calculate the occupied molecular area (A) of the amphiphile on the subphase surface by Eq. (3.4). Then, the Langmuir monolayer is compressed by movable Teflon barriers to change A, and the surface pressure change is followed as a function of A with a Wilhelmy surface balance during compression at constant temperature. This procedure gives the π–A isotherms of Langmuir monolayers. Figure 3.7 shows schematically π–A isotherms of Langmuir monolayers of long-chain acids. At molecular areas larger than 20–30 nm^2/molecule, the monolayers are in the two-dimensional gaseous (G) phase. Upon compression, the monolayer undergoes a phase transition to the liquid-expanded (LE) phase through the G–LE coexistence region, or to the liquid-condensed (LC) phase through the G–LC coexistence region, depending on temperature or on the alkyl chain length of the amphiphile used. At 25 °C, the surface pressures of the G–LE and the G–LC phase transitions are 95 μN/m for myristic acid and about 3 μN/m for stearic acid, respectively. Upon further compression, the LE phase changes to the LC phase, and then finally reaches the two-dimensional solid (S) phase. At the S phase, the surface pressure increases rapidly on compression and then abruptly drops. The maximum surface pressure is called the collapse pressure of the monolayer and this phenomenon corresponds to collapse of the Langmuir monolayer by overcompression to form three-dimensional microcrystals of the amphiphile. It is well known that the collapse pressure markedly depends on the compression speed.

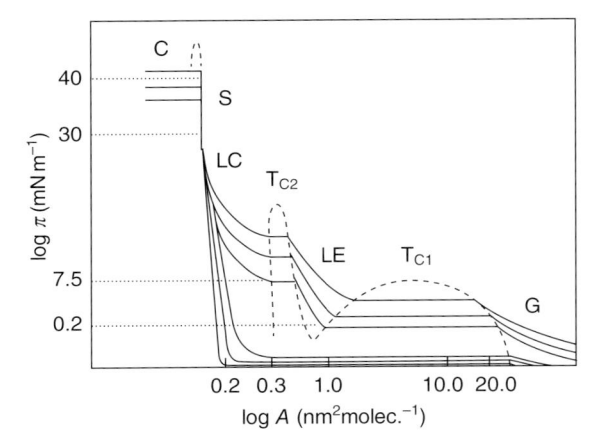

Figure 3.7 Schematic π–A isotherms of Langmuir monolayers of long-chain acids at different temperature, modified from the figure in R. Defay, I. Prigogine and A. Sanfeld, *J. Colloid Interface Sci.*, 58, 498 (1977) (see text).

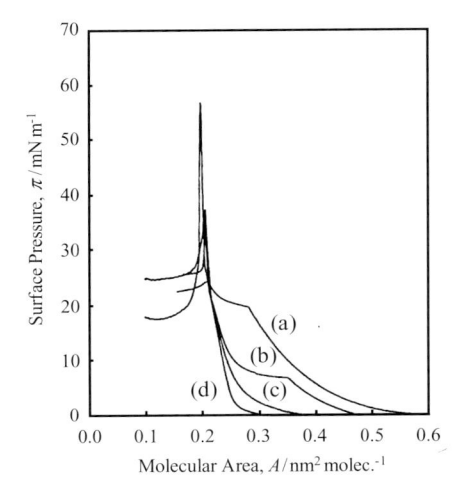

Figure 3.8 π–A isotherms of four long chain acids at 25 °C on the pH 2 subphase: (a) myristic acid, (b) pentadecanoic acid, (c) palmitic acid, (d) arachidic acid.

The G–LE phase transition is first order accompanying the constant surface pressure region. However, the phase-transition pressure is commonly too low to be measured in practice, and very sensitive surface balances are necessary. The LE–LC phase transition is also first order with a constant surface pressure region, but at the latter parts of the transition regions, there is a tendency to a gradual increase of the surface pressures. The LC–S transition is of the second order and is shown only by a break of slope in the isotherm.

Figure 3.8 shows π–A isotherms of four long-chain acids at 25 °C on the pH 2 subphase. Arachidic (C20) acid and palmitic (C16) acid monolayers only exhibit the LC phase and the S phase at this temperature, but myristic (C14) acid and pentadecanoic (C15) acid monolyers exhibit the LE–LC phase transition and the myristic acid monolayer collapses before a transition to the S phase. Temperature lowering and increase of the carbon number in hydrophobic alkyl chains of simple long-chain amphiphiles give the same effect on the behavior of π–A isotherms.

The reciprocal of tangential slope of π–A isotherm gives surface compressibility (κ^s), which is an important mechanical property of Langmuir monolayer, defined by the following equation,

$$\kappa^s = -(1/A)(\partial A/\partial \pi)_T \tag{3.6}$$

The reciprocal of κ^s gives the two-dimensional area modulus (k^s) of a Langmuir monolayer, which corresponds to the three-dimensional bulk modulus.

$$k^s = (\kappa^s)^{-1} \tag{3.7}$$

In two-dimensional monolayer systems, the first derivative of Gibbs free energy with respect to π gives area (A). Therefore, κ^s and β^s correspond to the second derivative of Gibbs free energy. Phase transitions accompanying discontinuous jumps of the first derivatives of Gibbs free energy are of first order, and those accompanying discontinuous jumps of the second derivatives of Gibbs free energy are of second order. Therefore, κ^s (k^s) and β^s become good indicators of phase transitions in Langmuir monolayers.

π–A isotherms are the most common and most frequently measured properties of Langmuir monolayers. Mangotaud and his colleagues have collected a few thousand π–A isotherms of various amphiphiles including long-chain acids, alcohols, chromophore derivative amphiphiles, natural occurring amphiphiles such as cholesterols, chlorophylls, etc., from the literature [11].

3.2.2.4 Stability of Langmuir Monolayers

As described in a later section, Langmuir–Blodgett films are fabricated by successive transfer of Langmuir monolayers to solid substrates, and it takes rather a long time to deposit multilayers of say, several tens of layers at constant surface pressures. Therefore, the long-term stability of the object Langmuir monolayers under constant high surface pressures must be taken into consideration. As described above, it is trivial that the shape of π–A isotherms changes with the change of compression speed even at the same temperature especially in the condensed-phase regions. Collapse pressures of S phase monolayers are changed considerably by changes of compression speed.

When a Langmuir monolayer of condensed phase is kept at a certain high surface pressure, the occupied molecular area is relaxed with time. When compression of a Langmuir monolayer is stopped at a certain high surface pressure, the surface pressure at a constant area is also relaxed with time. In both cases, the relaxations are caused by some reasons such as partial collapse of the monolayer to three-dimensional structures, partial dissolution of the film material to subphase, or structural relaxations, etc. An example of structural relaxation in a Langmuir monolayer is shown in Figure 3.9 [12]. An erucic acid monolayer under-

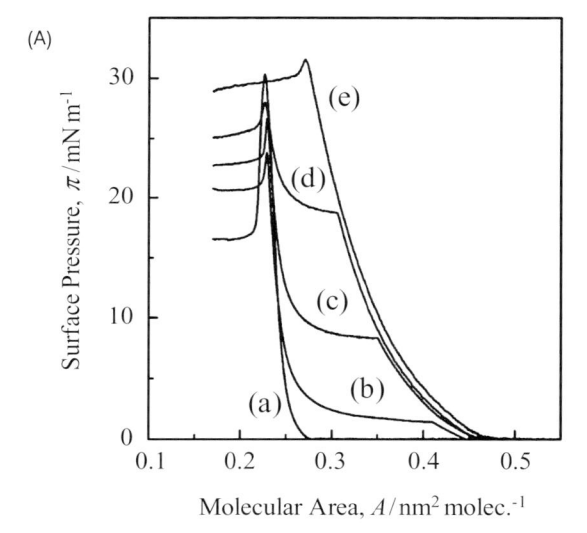

Figure 3.9 (A) π–A isotherms of erucic acid on the pH 3 subphase at (a) 1.8 °C, (b) 5.0 °C, (c) 10.0 °C, (d) 15.0 °C, and (e) 20.0 °C. (B) BAM images showing shape relaxation of the condensed-phase domains in the erucic acid monolayer at 10 °C with time (t) after stop of compression at 0.34 nm^2/molec: (a) t = 0, (b) 6 min, (c) 350 min, (d) 107 min, (e) 128 min, and (f) 143 min. Reprinted with permission from Ref. [12]. Copyright 2001 The American Chemical Society.

goes the LE–LC phase transition at 10 °C (Figure 3.9A), and during the transition two-dimensional dendritic crystals of condensed phase are formed (image (a) in Figure 3.9). However, when monolayer compression is stopped at 0.34 nm^2/molecule, the dendrites relax to a round shape with time. This observation suggests that the dendritic structure is the growing shape and the round shape is the thermodynamically stable one. The dendrite formation is due to high supersaturation in the monolayer, originating from the low incorporation rate of the film molecules to the condensed-phase domains compared to the rate of molecular area reduction upon compression. The *cis*-double bond prevents rotation of the hydrocarbon chain, so that the molecules are packed with the bend shape in the dendrites. This means that the molecules can be incorporated into the condensed-phase domains only when they fit into a "*cis*-template". This directivity decreases the incorporation rate, and then makes the surrounding phase highly supersaturated. In monolayers of normal fatty acids, once incorporated molecules can easily move and/or reorient at the domain circumference to attain a thermodynamically stable state.

Kato *et al.* has proposed a concept of the characteristic time of measurements of π–A isotherms, "time of observation" (t_{ob}), defined as follows [13];

$$t_{ob} = 1/(\text{strain rate of compression})$$

To keep t_{ob} constant during measurement of π–A isotherms, the strain rate of compression should be kept constant. This means that the compression speed of Teflon barriers should follow an exponential function of time. Figure 3.10 shows the t_{ob} dependency of π–A isotherms of arachidic acid monolayers at 10 °C on the pH 2 subphase. As is seen clearly, the collapse pressure decreases considerably with increase of t_{ob}. This fact shows that a Langmuir monolayer of arachidic acid is in a metastable existence under the experimental conditions. The shape of π–A isotherms is governed by the balance between the relaxation time of phenomena occurring in a monolayer and t_{ob}. If the relaxation time is shorter than t_{ob}, surface pressure does not increase even under compression. Many of the amphiphiles containing bulky functional groups such as chromophore tend to form unstable monolayers, in such cases mixing with other amphiphiles forming stable monolay-

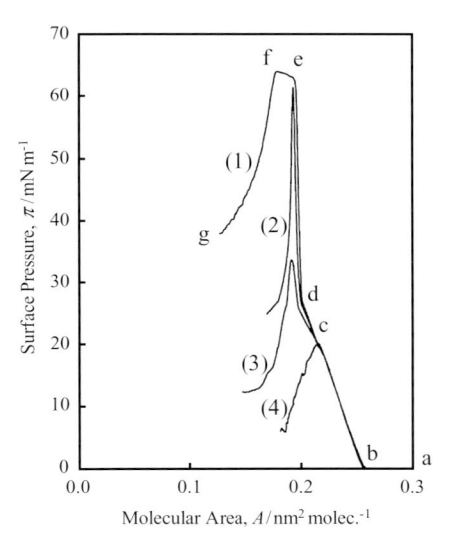

Figure 3.10 π–A isotherms of arachidic acid monolayers at 10 °C on the pH 2 subphase, measured under t_{ob} of (1) 6×10^2 s, (2) 6×10^3 s, (3) 7×10^4 s, (4) 3×10^5 s. Reprinted with permission from Ref. [13a]. Copyright 1990 The American Chemical Society.

ers, such as long-chain acids and amides, provides the monolayer stability. If LB transfer of mutilayers needs 2 h, it is better to measure π–A isotherms of the mixed amphiphiles monolayer with a t_{ob} of 2 h to determine the control surface pressure for the LB deposition. If not, collapse of the monolayer or squeeze out of the functional amphiphile may occur at the latter stage of the successive LB deposition, and accordingly the regular structure of the LB film may be disturbed.

When long-chain acids, such as arachidic acid or behenic acid are spread on the alkaline subphase containing di- or trivalent metal ions, such as Ba^{2+}, Ca^{2+}, La^{3+}, etc., very rigid monolayers of S phase are formed, and one can see bright images of irregular shape fragments of various sizes by BAM (described in the next section), even at zero surface pressure before compression. π–A isotherms of these monolayers are almost straight lines of very steep slopes at around $0.20\,nm^2/$ molec. Figure 3.11 shows area relaxation of a calcium arachidate monolayer at 15 °C under a constant surface pressure of 15 mN/m. As is seen from Figure 3.11, about 2% area reduction was observed after 300 min. Figure 3.12 shows A–T

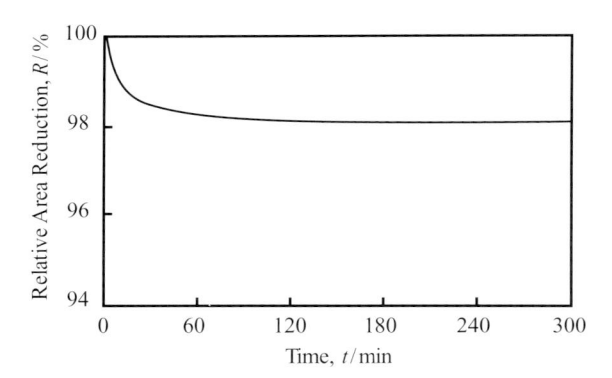

Figure 3.11 Isothermal area relaxation of a calcium arachidate monolayers at 15 mN/m and 15 °C.

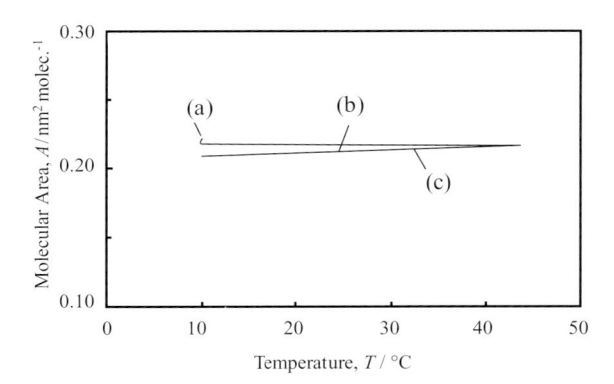

Figure 3.12 A–T isobars of a calcium arachidate monolayer at 15 °C under a constant surface pressure of 15 mN/m. See text for details. Reprinted with permission from Ref. [8b]. Copyright 1989 Elsevier.

isobars of a calcium arachidate monolayer under a constant surface pressure of 10 mN/m. Isobar (a) was measured by the first temperature rising from 10 °C to 43 °C, isobar (b) was measured by the first temperature lowering, and isobar (c) was measured by the second temperature rising [8b]. The isobars (b) and (c) completely coincide with each other. Hence, these are the true isobars of this monolayer, and it was assured that there was neither dissolution nor slow collapse of the monolayer during measurements even at higher temperature. However, the isobar (a) exhibited area reduction and the thermal expansivity calculated from the isobar (a) was negative. The area discrepancy at 10 °C of the isobars (a) and (b) was 4.2%, and this discrepancy becomes large with increase of constant surface pressure; 4.4% at 15 mN/m, 5.2% at 25 mN/m, and 5.8% at 35 mN/m. Dipalmitoyl phosphatidylcholine (DPPC) monolayers and trioctadecylglyceride (TOG) monolayers have also shown a few % of thermally stimulated area reduction at first temperature rising up to around 45 °C. These results mean that the isobars of the first temperature rising of such rigid monolayers are spurious ones that include thermally stimulated or thermally accelerated area relaxation information. The rigid monolayers of S phase or LC phase spread at low temperatures and compressed without thermal treatment contain a large number of pin holes or defects introduced by the lack of fluidity of the irregular shaped rigid fragments of condensed phases already exist before compression starts. The process of temperature rising up to some high temperatures under constant surface pressures, which was named an "isobaric thermal treatment", can reduce pin holes or irregular defects in rigid monolayers by the two-dimensional sintering mechanism, even below the melting temperatures of the amphiphiles in monolayers [14].

3.2.3
In situ **Characterization of Monolayers at the Subphase Surface**

3.2.3.1 Brewster-Angle Microscopy (BAM)
BAM is an epochal instrument to visualize the monolayer morphology directly without using any probe impurities as used in the cases of fluorescence microscopy. BAM was first reported by two groups in 1991 [15]. The principle of BAM is rather simple. When a p-polarized laser light beam is incident on the clean water surface at the Brewster angle (θ_B, about 53.1° from the surface normal), no light is reflected. However, if a monolayer exists on the water surface, a little reflection occurs because the different refractive index of the monolayer from that of water breaks the Brewster angle condition for the air/water interface. However, the strength of the reflected light from the monolayers is very weak, and it is necessary to use a highly sensitive CCD camera through a microscope of rather long focal distance. Then, one can observe microscopic structures of Langmuir monolayers or Gibbs monolayers, especially in the transition regions between expanded and condensed phases. BAM can detect the internal anisotropy of condensed-phase domains, caused by the tilt azimuthal angle of the hydrophobic alkyl groups of the amphiphiles. Two problems occur by oblique glancing of the monolayer at the Brewster angle of water. The first is the distortion of the surface structures; a

circular structure looks elliptic, for example. The second is that only a band shape zone of the BAM image is in focus when a microscope of high resolution and small depth of field is used. The first problem of the image distortion is easily corrected by uniform extension of the distorted BAM images by the factor $1/\sin(90-\theta_B) = 1.82$ in the shrunken direction. To solve the second problem, the microscope is moved stepwise along the optical axis direction rapidly, say 10 μm for each step, and an image is taken at each step. Then, by collecting the focused band shape zones of every image, a wholly focused BAM image can be synthesized with image processing software. Recent advanced BAM instruments are equipped with functions of ellipsometry and can get information about local refractive indices or information about local thicknesses of Langmuir monolayers at the water surface as well as wholly focused BAM images without distortion by oblique glancing by image processing as described above.

3.2.3.2 Fourier Transform Infrared (FTIR) Spectroscopy

FTIR spectroscopy is particularly useful for characterization of ultrathin films on an aqueous subphase as well as those on solid supports. This technique nondestructively provides a lot of valuable information on the chemical species existing, their interactions (hydrogen bonding, ion binding, etc.), conformation, and orientation in the monolayers.

In many cases of organic monolayer studies, it is very fruitful to know the orientation of functional groups in films. For this purpose polarization-dependent IR spectroscopy has been performed, probing the transition moment distribution in films. Orientation evaluation by FTIR spectroscopy is based on the fact that the infrared absorption intensity of a vibration mode is proportional to the square of the dot product of the transition dipole moment of the vibration and the electric field of infrared light, and varies with the square of $\cos\theta$ where θ is the angle between the transition dipole and the electric field [16]: the intensity reaches a maximum when the transition dipole and the infrared electric field are parallel, but decreases with increasing θ. Before moving to the FTIR spectroscopy for the liquid surface, it is better to introduce that on solid supports for easy understanding. The combination of transmission and RA measurements are often used for analyzing molecular orientation in LB films since the electric vector in the transmission measurement of LB films on infrared-transparent substrates is parallel to the film surface, while that in the RA measurement with metal substrates is perpendicular to the surface. An example is shown in Figure 3.13, where the RAS and transmission spectra for a 7-monolayer LB film of cadmium stearate deposited on a Ag-coated glass slide and ZnSe plate, respectively, are compared [17]. The symmetric COO^- stretching band ($v_s COO^-$) and the band progression due to the CH_2 wagging modes (ωCH_2) are exclusively observed in the RA spectrum. Consequently, it is evident that the transition moments of these bands are nearly perpendicular to the film surface. On the other hand, the antisymmetric and symmetric CH_2 stretching bands ($v_a CH_2$, $v_s CH_2$) and the antisymmetric COO^- stretching band ($v_a COO^-$) appear strongly in the transmission spectrum but weakly in the RA spectrum, indicating that the transition moments of these bands lie almost

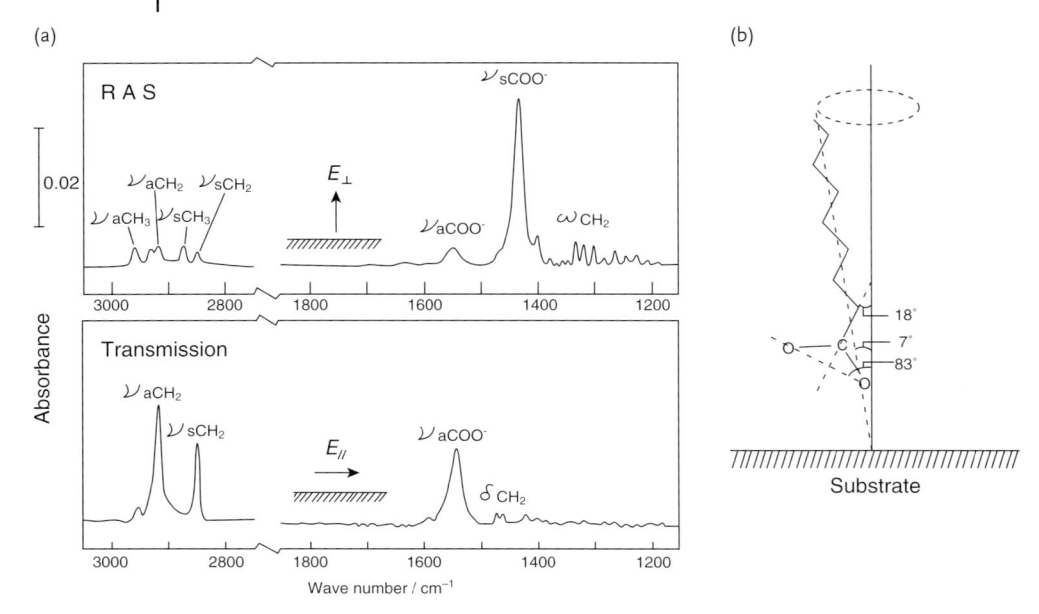

Figure 3.13 (a) Infrared RA and transmission spectra of a 7-monolayer LB film of cadmium stearate. (b) Schematic illustration of molecular orientation in a 7-monolayer LB film of cadmium stearate. Reprinted with permission from Ref. [17]. Copyright 1990 The American Chemical Society.

parallel to the film surface. From the differences between the spectra, one can expect the almost perpendicular orientation of the long axis of the film molecules in their LB film. By using theoretically calculated enhancement factors and experimentally determined intensity ratios of the RA to transmission spectra, the orientation angles of the transition moments of major infrared bands have been quantitatively estimated (Figure 3.13b).

The above-mentioned combination is useful as a molecular-orientation evaluation technique. However, one needs careful consideration in data interpretation because different solid substrates may lead to different orientation of film molecules. In this regard, the ordinary infrared external reflection (ER) spectroscopy for LB films on nonmetallic substrates has some advantages. This spectroscopy gives negative or positive reflection absorbance depending on the direction of the transition moment, the angle of incidence, and the polarization of the incident beam [18a]. The quantitative analysis approach for evaluating uniaxial molecular orientation in LB films by ER spectroscopy has also been established, which could be adapted to many kinds of substrate. Furthermore, the infrared multiple-angle incidence resolution spectroscopy (MAIRS) has been developed to deduce simultaneously two spectra corresponding to the conventional transmission and RA spectra from an identical thin film deposited on a nonmetallic substrate [18b–d].

The pioneering works of Dluhy and coworkers from the mid-1980s demonstrated that it was possible to acquire molecular structure information from

Langmuir monolayers at the air/water interface using the RA spectroscopy [19]. However, the main problem of application of the conventional RA method to the air/water interface lies in the difficulty of extracting a very weak monolayer signal from very strong absorptions of the surrounding water vapor. Nowadays, the following two ways are typically applied in order to overcome this problem. One is the use of polarization modulation infrared reflection absorption spectroscopy (PM–IRRAS) [20]. This technique combines FTIR spectroscopy with fast polarization modulation of the incident infrared beam between parallel (p) and perpendicular (s) polarized states. The spectra obtained are in principle insensitive to all the polarization-independent signals such as strong absorption of water vapor in environment. In the PM-IRRAS, contrast to the normal RAS for metal substrate surfaces, as the electric field at the air/water interface has an inplane component, absorptions with a transition moment parallel to the surface are also detected, and absorption bands appear upward or downward, depending on the orientation of their transition moment and the incident angle. Figure 3.14a shows theoretically calculated PM-IRRA spectra for different orientations of the transition moment [20c]. The PM-IRRAS signal intensity changes from the negative minimum to the

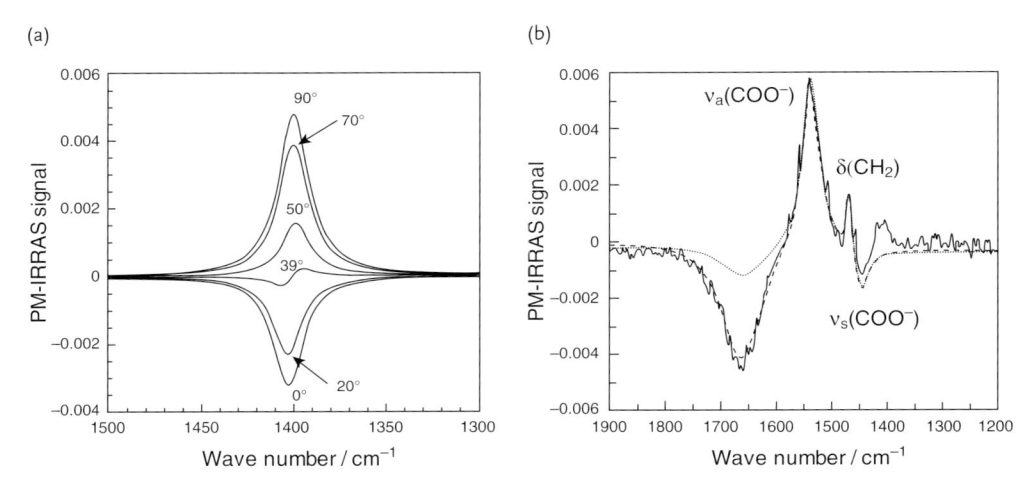

Figure 3.14 (a) Calculated PM-IRRAS spectra for an orientation of the transition moment varying from parallel ($\theta = 0°$) to perpendicular ($\theta = 90°$) to the surface normal at an incident angle of 71°. (b) PM-IRRAS spectrum measured for a cadmium arachidate monolayer on the $CdCl_2$ aqueous solution (pH = 6.7), and corresponding spectrum calculated on the assumption of preferred orientation of transition moments (see text) with (dashed line) and without (dotted line) an anisotropic water film between the monolayer and the bulk water subphase. PM-IRRAS signal is presented as $(S(d) - S(0))/S(0)$, where $S(d)$ and $S(0)$ stand for the monolayer-covered and uncovered subphase, respectively. Absorption vibration mode assignments in (b); 1540 cm^{-1}: asymmetric carboxylate stretching $\nu_a(COO^-)$, 1445 cm^{-1}: symmetric carboxylate stretching $\nu_s(COO^-)$, 1469 cm^{-1}: CH_2 deformation (scissoring) $\delta(CH_2)$. Negative dip at 1160 cm^{-1} and positive band at 1410 cm^{-1} are due to deformation vibration mode of liquid water $\delta(OH_2)$ and to carbonate ions, respectively, in the subphase. Reprinted with permission from Ref. [20c]. Copyright 1996 The Royal Society of Chemistry.

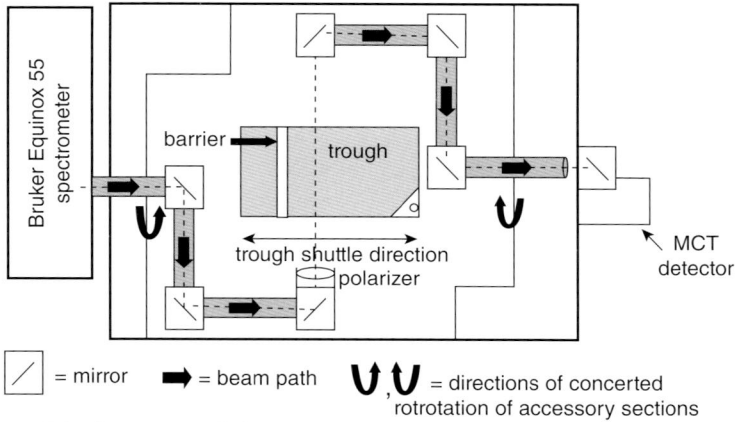

Figure 3.15 IRRAS accessory. The Wilhemy plate is placed on the right-hand side of the trough. The reference channel is to the left of the barrier. The entire trough shuttles as indicated to sample the desired channel. The incident light is guided to the surface via three mirrors. The beam path is represented as a dashed line. The reflected light follows an equivalent optical path. The angles of incidence and reflection as well as the state of polarization are under computer control. Reprinted with permission from Ref [21e]. Copyright 2010 Elsevier.

positive maximum when the tilt of the transition moment varies from parallel to perpendicular to the surface normal. An experimentally obtained spectrum of a cadmium arachidate monolayer at the air/water interface is shown in Figure 3.14b, together with corresponding spectra calculated on the assumption of parallel orientation of transition moments of asymmetric carboxylate stretching ($v_a(COO^-)$) and CH_2 scissoring ($\delta(CH_2)$) bands and perpendicular orientation of symmetric carboxylate stretching band ($v_s(COO^-)$) with respect to the water surface.

Another means is the conventional infrared RA spectroscopy with a trough shuttle system, where a computer-controlled sample shuttle allows alternative acquisition of the reference (or background) and the sample spectrum from the subphase surface and from the monolayer-covered surface, respectively, in a sample box (Figure 3.15) [21]. The theoretical treatment of RA spectra of organic monolayers on the water surface is essentially similar to that of ER spectroscopy. The molecular orientation can be estimated by comparison of the dichroic ratios of intensities measured with p- and s-polarized infrared beams and those simulated under the assumption of a certain molecular orientation, at different incidence angles (Figure 3.16).

IR vibrational bands are very sensitive to the conformation, packing state, and intra- and intermolecular interactions, and thus it is convenient to estimate them qualitatively. It is well known that the CH_2 asymmetric and symmetric stretching bands are indicators of the alkyl-chain conformation: they typically appear in the ranges of 2916–2920 and 2848–2850 cm^{-1} for all-*trans* extended chains but shifted to higher wave numbers of ~2928 and 2858 cm^{-1} for disordered chains, respec-

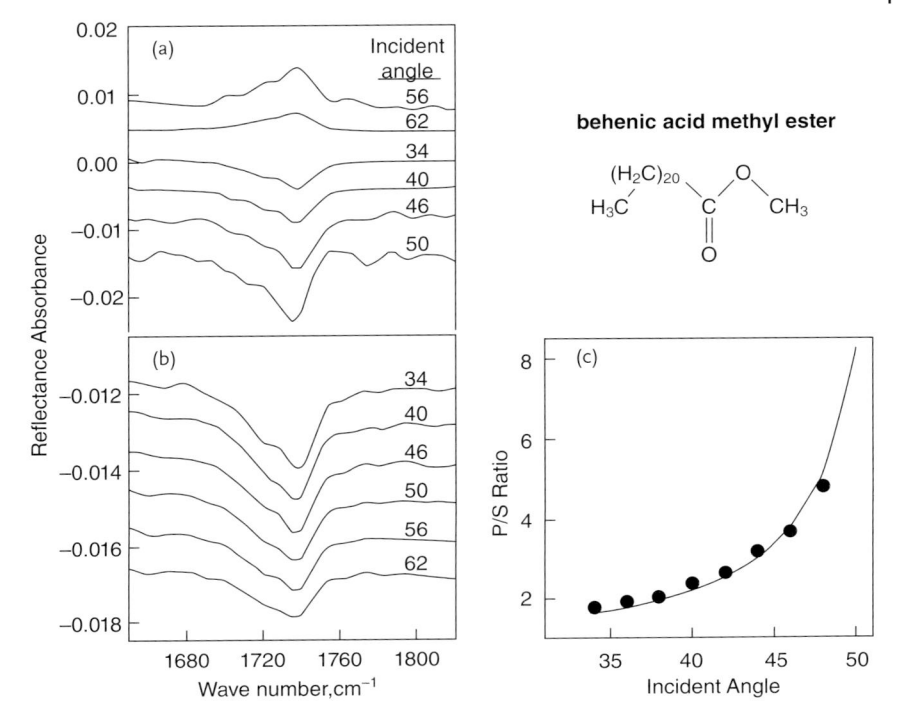

Figure 3.16 RA spectra measured by (a) p- and (b) s-polarized radiation for behenic acid methyl ester monolayer on the aqueous subphase. The orientation of the 1737 cm^{-1} band arising from unhydrated C=O groups was determined. The experimental p/s band intensity ratios are plotted as a function of angle of incidence in (c) along with the theoretical simulations for a 90° tilt angle. Reprinted with permissions from Refs. [21e]. Copyright 2010 Elsevier.

tively. The wave number of the CH$_2$ scissoring band is characteristic of alkyl chain packing. When the chains are in an orthorhombic or monoclinic unit cell, doublet bands emerge at 1463 and 1473 cm^{-1}, whereas in hexagonal and triclinic unit cells, a singlet band at 1468 cm^{-1} and 1473 cm^{-1}, respectively [22, 23]. On the other hand, absorption peaks assigned to functional groups in the head groups are largely influenced by the chemical interactions with the neighbors such as the hydrogen bonding, ion binding, hydration, and so on. From the frequencies of the so-called amide bands, one can estimate the secondary structure of corresponding portions in proteins [20e, 21d–e].

3.2.3.3 X-ray Reflectometry and Grazing-Incidence X-ray Diffractometry

X-ray reflectivity (XR) and grazing-incidence X-ray diffraction (GIXD) are the most powerful tools to investigate the monolayer structures on liquid surfaces. In Figure 3.17, general geometries for specular XR and GIXD measurements are displayed [24]. In the specular XR experiment, a monochromatic X-ray beam is irradiated on the water surface while keeping the angular relation of α_i (incidence angle) = α_r

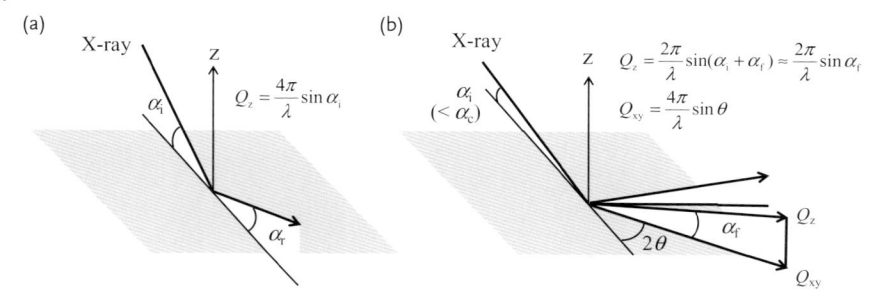

Figure 3.17 (a) Specular XR and (b) GIXD geometries. For the specular XR measurement, the detection angle α_r always equals the incident angle α_i. In GIXD experiments, the incident angle is fixed at α_i ($<\alpha_c$), and the $2q$ scan determines Q_{xy} and Q_z components simultaneously when a vertically mounted position-sensitive detector is used.

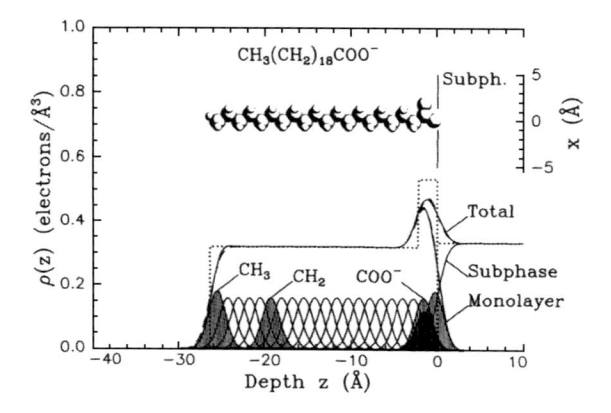

Figure 3.18 Electron-density profiles across a fatty acid monolayer at the air/water interface. The total electron density at the interface is the sum of contributions from the subphase and the monolayer. The monolayer density, in turn, is the sum of contributions from individual atoms of the monolayer molecules. Due to thermally excited capillary waves, all electron densities are smeared. Reprinted with permission from Ref. [25]. Copyright 1994 Elsevier.

(reflection angle), and then the reflected X-ray intensity is detected as a function of α_i. The reflection has a wavevector transfer Q perpendicular to the liquid surface, and thus measures the electron-density distribution across the surface. Figure 3.18 shows the electron-density profile of a closed-packed monolayer of arachidic acid on an aqueous subphase [25]. The profile is constructed as the sum of electron densities of individual atoms in the monolayer molecules and subphase (full lines), and also regarded as being composed of two slabs at constant electron densities (i.e., hydrocarbon and head group parts with lower and higher electron densities, respectively). In actual XR data evaluation, the XR curve observed is analyzed most typically by a so-called box-model fitting with parameters of electron density for each slab and bulk phases, thickness for each slab, and roughness at boundaries between slabs and monolayer-bulk phases.

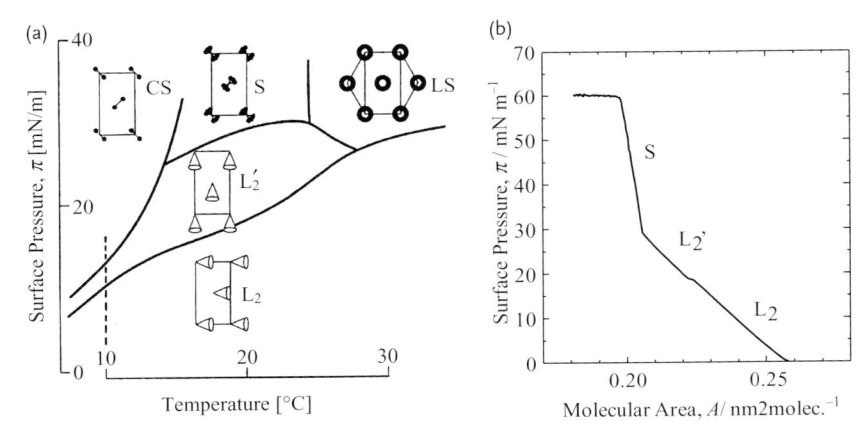

Figure 3.19 (a) A phase diagram of behenic acid monolayer on the water surface. Reprinted with permisiion from Ref. [25]. Copyright 1994 Elsevier. (b) A π–A isotherm of a behenic acid monolayer at 20 °C.

For GIXD measurements, a monochromatic X-ray beam is incident on the water surface at a very shallow angle below the critical angle for total reflection at the air/water interface. In such a condition, since the refractive index for X-rays is slightly less than 1, the beam undergoes total reflection and only the resulting evanescent wave penetrates into the subphase less than 10 nm, leading to reduced background scattering from water and accordingly enhanced surface sensitivity of X-rays. The surface is probed by scanning the horizontal angle 2θ, corresponding to the horizontal wavevector transfer Q_{xy}, and the vertical angle α_f, or Q_z. In Figure 3.19, a phase diagram of behenic acid monolayer on the water surface [26] is shown together with a π–A isotherm measured at 20 °C. The isotherm reveals that the behenic monolayer undergoes two phase transitions, which corresponds to transitions from L_2 to L_2' phase and further to S phase according to the phase diagram.

3.2.4
Transfer to Solid Supports

Most common and traditional method to fabricate Langmuir–Blodgett (LB) films is the so-called Langmuir–Blodgett technique or vertical dipping technique, in which a solid substrate passes through a surface-pressure-controlled Langmuir monolayer up and down vertically to transfer the monolayer from the subphase surface to the solid substrate, successively. In this method, there are three types of deposition, that is, X-type deposition in which the monolayer transfers only at downstroke onto the hydrophobic surface, Y-type deposition in which the monolayer transfers at both up and down strokes, and Z-type deposition in which the monolayer transfers only at upstroke onto the hydrophilic surface. Figure 3.20 shows these deposition processes, schematically. There are also three types of LB film structures; X-type, Y-type and Z-type structure. In X-type structure, all amphiphile molecules direct their hydrophobic groups to the solid substrate and

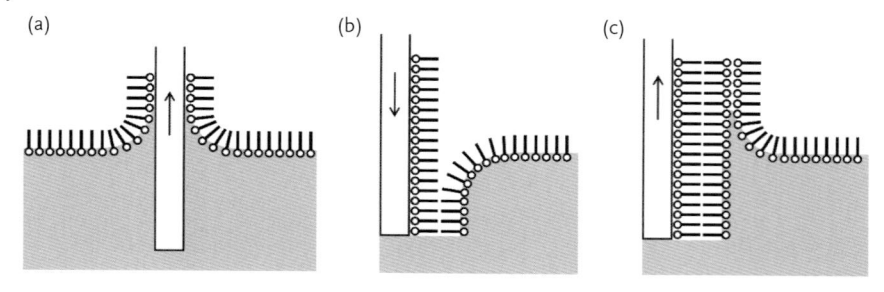

Figure 3.20 Y-type deposition process. A hydrophilic solid substrate is held vertically with respect to the subphase surface and a desired part of the substrate is immersed in the subphase in advance. Then a Langmuir monolayer is spread and compressed to a definite surface pressure. With keeping this surface pressure constant, the substrate is lifted (a) up, (b) down, and (c) up to deposit the first, second, and third layer of LB film, respectively. Processes (b) and (c) are repeated to prepare a multilayer LB film. Note that the left half sides of the substrates in (b) and (c) are not drawn.

X type   Y type   Z type

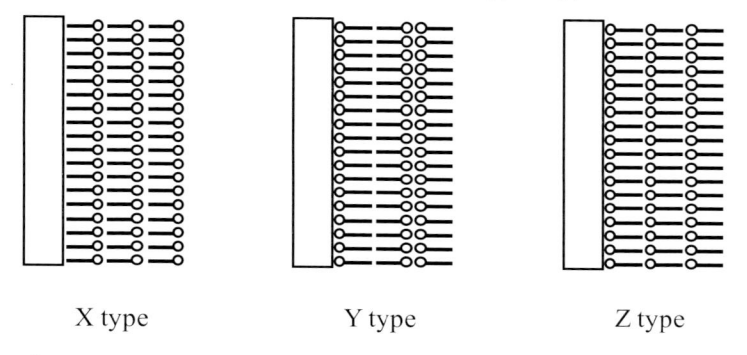

Figure 3.21 X-type, Y-type, and Z-type structures of LB films.

in Z-type structure, all amphiphile molecules direct their hydrophilic groups to the solid substrate. In Y-type structure, amphiphile molecules change their direction, layer by layer alternatively as shown schematically in Figure 3.21. Y-type deposition always gives Y-type structure LB films. However, X-type or Z-type depositions does not always give X- or Z-type LB films, rather they give LB films of Y-type structure, depending on the deposition conditions, such as temperature, controlled surface pressure for transfer, pH and concentration of multivalent metal ion in the subphase, etc., and especially on structures and hydrophilicity of the hydrophilic groups of the amphiphiles. For example, long-chain acids deposit with Y-type deposition and give Y-type structure in general, however, at higher pH with divalent metal ions in the subphase, they deposit with X-type deposition but structures are those of Y-type. Ethyl stearate exhibits X-type deposition and the resultant LB film also has the X-type structure. On the other hand, methyl stearate deposits by Y-type deposition at 10 °C and at 30 mN/m, but by X-type deposition at 20 °C and at 20 mN/m. In both cases, the resultant LB films are of Y-type structure. In conclusion, amphiphiles whose Y-type structures are thermodynamically stable form Y-type structures irrespective of the deposition types, and amphiphiles whose X-type or Z-type structures are stable are very rare. These results described above suggest that amphiphile molecules overturn somewhere during the deposition processes of X-type or Z-type deposition. This problem will be discussed later.

3.2.4.1 Instruments for LB Film Deposition

Barraud and Vandevyer have developed an instrument for fast continuous fabrication of LB films [27]. The rectangular trough is divided into four compartments by three rollers situated at the water surface. The largest compartment 1 is devoted to spreading the amphiphile solution by controlled continuous spreading and to evaporating the spreading solvent. Roller 1 sends spread amphiphile molecules to compartment 2 by a "hydrophilic" mode of operation to precompress the monolayer. Roller 2 sends the precompressed monolayer to compartment 3 by a "hydrophobic" mode of operation to compress the monolayer to the desired surface pressure for deposition in compartment 3. Compartments 2 and 3 are equipped with Wilhelmy-type surface balances and each detected surface pressure is used for feedback control of the surface pressure in compartments 2 and 3. The auxiliary narrow roller 3 works when the surface pressure in the compartment 3 exceeds the set value to exclude the excess amphiphile molecules to the compartment 4 and then sucked by a suction pump at the surface in compartment 4. All the feedback controls including the spreading rate work well and this instrument can deposit LB films ten times faster than with a classical trough. Albrecht *et al.* has also reported development of a fully automatic deposition instrument by controlled spreading of an amphiphile solution, continuous compression by using laminar flow of subphase water, and can deposit continuously over thousands of layers of LB films [28].

Materials with a macroscopic electric polarization display a variety of useful properties, such as piezoelectricity, pyroelectricity, second-order nonlinear optical activity, etc., and LB films of X-type or Z-type structures are expected to be such materials with macroscopic polarization. However, some amphiphiles exhibit X-type or Z-type deposition, but almost all of them finally form Y-type structure LB films because of the instability of X- or Z-type noncentrosymmetric structures. To overcome this difficulty, fabrication of noncentrosymmetric superlattices by combination of two amphiphiles having different dipole moments becomes a good candidate. Fabrication of alternate LB films of two amphiphiles with a single Langmuir trough needs a very troublesome and time-consuming operation. Barraud and his colleagues have developed an automatic instrument for fabricating alternate LB films of two amphiphiles [29]. Figure 3.22 shows a general diagram of the instrument. The apparatus consists of a two-compartment

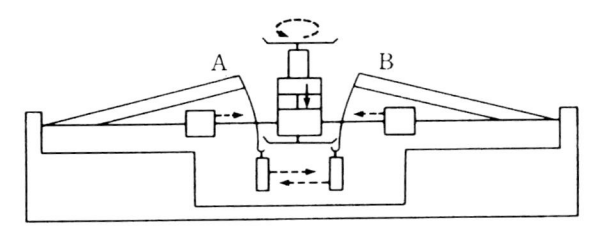

Figure 3.22 A general diagram of the instrument for making alternate LB films of two amphiphiles automatically. Reprinted with permission from Ref. [29]. Copyright 1985 Elsevier.

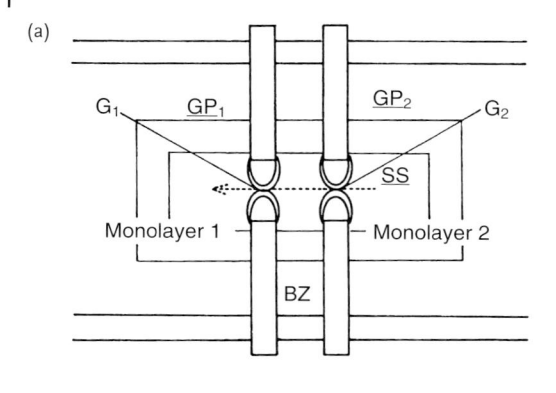

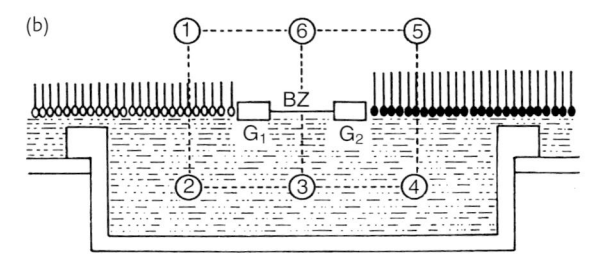

Figure 3.23 (a) Double gates that permit a solid substrate to pass through, but prohibit monolayers. (b) Definition of a deposition process by assigning position numbers to close a cycle (see text). Reprinted with permission from Ref. [30a]. Copyright 1987 The Japan Society of Applied Physics.

Langmuir trough and a system to transfer the substrate from one compartment to the other. The trough contains a single subphase. Its surface is divided into two compartments by a fixed Teflon barrier located above the central well. Each compartment is equipped with a movable barrier to control the surface pressure of a monolayer independently. Each compartment is also equipped with an independently programmable dipping arm. When two dipping arms are at rest, an auxiliary rotating arm picks up the substrate from one of the arms, transfers it to the other compartment by rotation underneath the water or in the air above the central barrier, and hangs it on the other dipping arm. By programming these processes, two amphiphiles can be deposited by desired order on the substrate.

Kato has also developed a microcomputer-controlled instrument for preparing complex (hetero-) LB films fully automatically [30]. Figure 3.23 shows the schematic structure of the central part of the trough, and Figure 3.23b shows definition of a deposition process by assigning a position number to close a deposition cycle. The subphase surface is divided into three compartments by two fixed Teflon barriers. Compartments 1 and 2 are equipped with a movable barrier and a Wilhelmy-type surface balance to control monolayers 1 and 2, independently. At the centers of two fixed barriers, there are two gates that are composed of a looped Teflon thin sheet confronted at the centers of the fixed barriers. The vertically held substrate

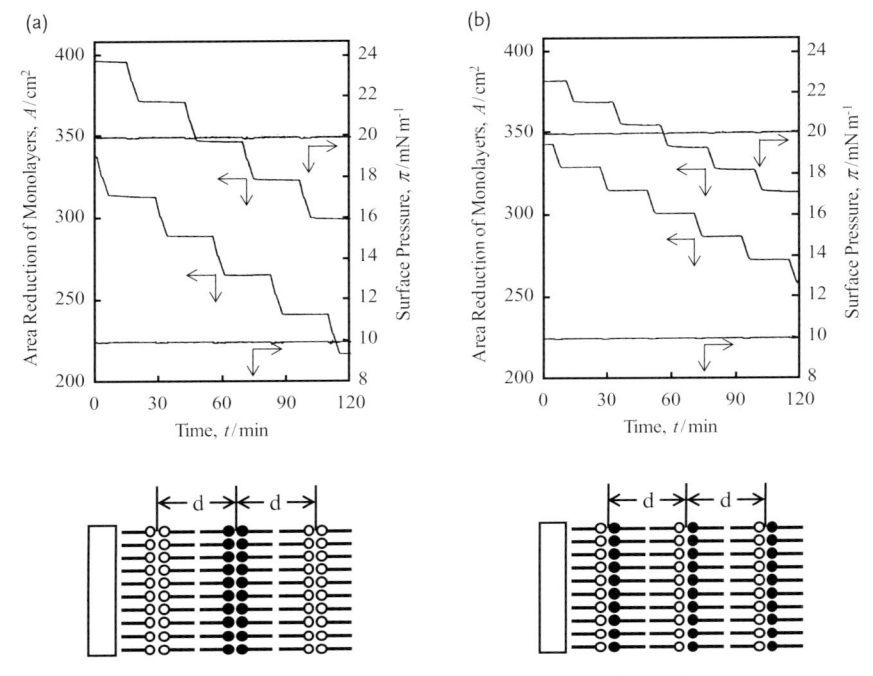

Figure 3.24 Deposition records of (a) an alternate Y-structure LB film and (b) a laminated Y-structure LB film. Schematic illustrations of film structures are also displayed. Reprinted with permission from Ref. [30a]. Copyright 1987 The Japan Society of Applied Physics.

can laterally pass through these gates without leakage of monolayers. Compartment 3 at the center is a buffer zone where substrate can pass through a clean water surface by upstroke or downstroke. Speeds of lateral transfers, downstrokes and upstrokes can also be freely programmed. By programming the processes, we can principally get all desired complex LB films. A regular alternate LB film of a combination of cadmium arachidate monolayer (15 mN/m) and a mixed monolayer of polyoctadecyl acrylate (20 mol%) + cadmium arachidate (80 mol%) was obtained by repeating the deposition cycle of (1)→(2)→(4)→(5)→(1). Figure 3.24a shows deposition records of area reduction of cadmium stearate (10 mN/m) and cadmium behenate (20 mN/m) by repeating the deposition cycle of (1)→(2)→(1) →(5)→(4)→(5) →(1), and Figure 3.24b shows deposition records of another hetero LB film by repeating the deposition cycle of (1)→(2)→(4)→(5)→(1). Very regular complex LB films of the structures shown in Figure 3.24 (bottom) were obtained.

3.2.4.2 Turnover of Amphiphile Molecules during Deposition

With the instrument cited in the preceding paragraph, deposition of a monolayer with forced Z-type deposition can be done by the cycle (6)→(3)→(2)→(1)→(6). When cadmium stearate monolayer was deposited by repeating this compulsory Z-type deposition cycle, deposition record of monolayer area reduction suggested

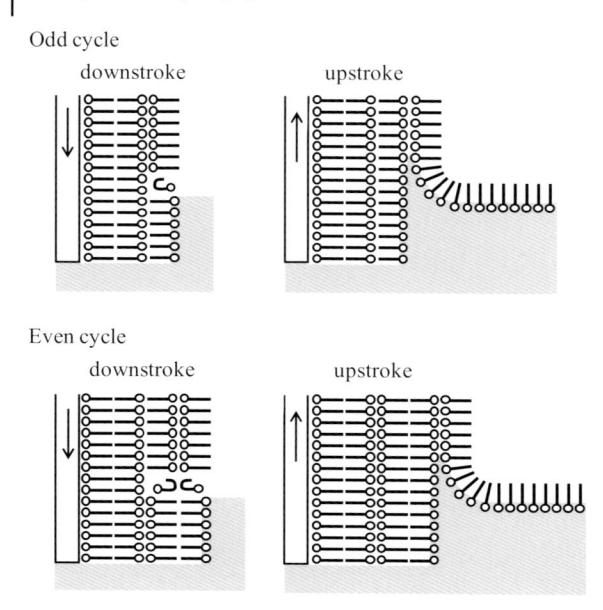

Figure 3.25 Turnover mechanism during Z-type deposition of an amphiphile that finally forms a Y-type LB film. Reprinted with permission from Ref. [30b]. Copyright 1988 The Chemical Society of Japan.

complete Z-type deposition, but the resultant LB film was a very regular Y-type structure, confirmed by X-ray diffraction. This means that cadmium stearate molecules overturn somewhere during deposition or after deposition.

By using a photopolymerizable amphiphile, 10,12-pentacosadiynoic acid monolayer on the subphase containing 10^{-3} M cadmium ion, and by photopolymerization *in situ* by UV irradiation in air and in subphase at every elementary process of the compulsory Z-type deposition cycle, the amphiphile molecules are fixed to suppress turnover. Then, we can guess at where in the Z-type deposition, the amphiphile molecules overturn by measuring the hydrophilicity or hydrophobicity of the phtopolymerized surfaces. Figure 3.25 shows the estimated turnover mechanism during Z-type deposition of the amphiphile that finally forms Y-type LB films. When the substrate is stopped at the middle point of the process of downstroke through a clean water surface, and molecules are fixed by UV irradiation, there should be a clear boundary between hydrophilic areas and hydrophobic area if the proposed overturn mechanism is true. X-ray photoelectron spectroscopy or Auger electron spectroscopy could not differentiate the hydrophilic groups or hydrophobic groups of the outermost layers. However, by cooling the sample to −5 °C in a refrigerator and by exposing to humid air, the boundary was clearly visualized as a horizontal line by the difference of condensation of humidity. This boundary line can be seen even after one year. Thus, we can conclude that the proposed turnover mechanism should be true. Amphiphile molecules deposited by X-type or Z-type deposition should overturn by a flip-flop mechanism at the

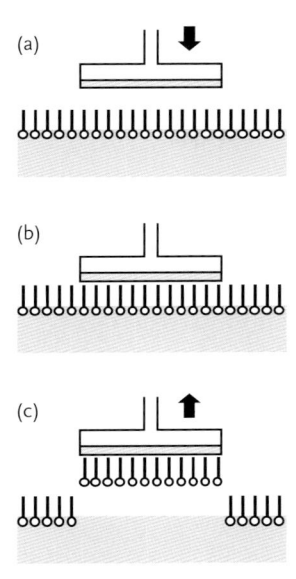

Figure 3.26 The Langmuir–Schaeffer technique for depositing a monolayer on a hydrophobic solid substrate. The solid substrate is (a) held parallel to the subphase surface and (b) brought into contact with the monolayer, and (c) uplifted.

following upstroke or downstroke with the assistance of passing three-phase line of air/water/outermost LB layer, if Y-type structure of the amphiphile is thermodynamically more stable than X- or Z-type structures.

3.2.4.3 Horizontal Lifting-Up Deposition

With very rigid monolayers that are hard to flow, such as polymeric monolayers, protein monolayers, etc., a deposition technique developed by Langmuir and Schaefer (L–S) is useful to lift up one layer of floating monolayer to a horizontal hydrophobic substrate by touching and pulling up, as shown in Figure 3.26. Fukuda and his colleagues have proposed expansion of the L–S technique to deposit multilayers, named horizontal lifting-up (H-L) technique [31]. Figure 3.27 shows schematically the repeating process to form multilayers by this technique. After the horizontally held substrate touches the surface-pressure-controlled monolayer, another barrier is set just beside the substrate to avoid pulling up the monolayer of the surrounding area of the substrate by Y-type deposition, and the cycle is repeated.

Kato has developed a fully automatic horizontal lifting method by applying the instrument in Figure 3.23. Figure 3.28 shows the holder arrangement and the deposition process, schematically. After a horizontally held hydrophobic substrate touches the surface-pressure-controlled monolayer 1, the substrate is sunk into subphase (position [2]) and then the vertical part of the holder passes through the gate 1 to position (3). Then, the substrate is pulled up into the air through the clean water surface to position (6), and transferred to position (1) in air. By repeating this cycle, we can easily get complete X-type structure LB films fully automatically if the X-type structure is thermodynamically stable. Actually, ethyl stearate

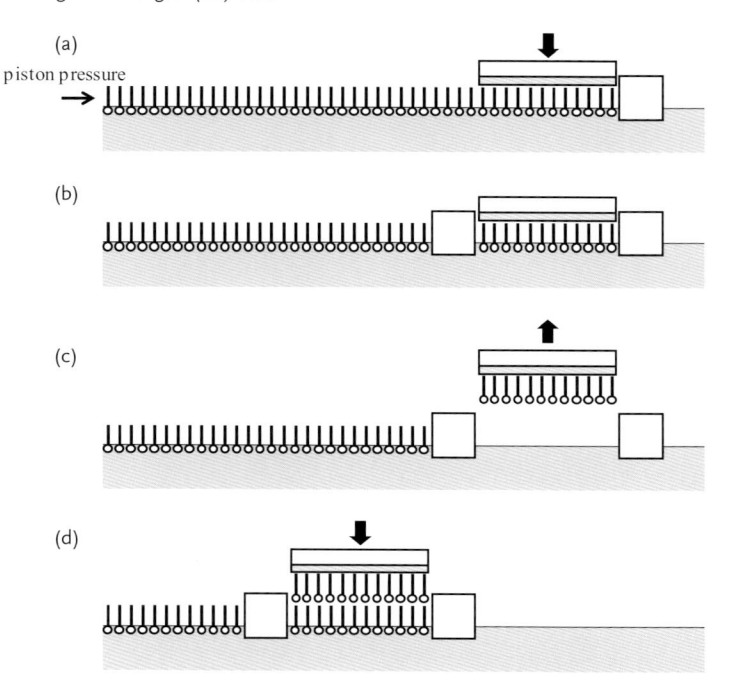

Figure 3.27 Schematic figure of the horizontal lifting-up (H-L) method to form a multilayer LB film. (a) A horizontally held substrate touches to a surface-pressure-controlled monolayer; (b) another barrier is set just beside the substrate to prevent incoming surrounding monolayers; and then (c) the substrate is uplifted to deposit the first layer. (d) Processes (a)–(c) are repeated to deposit multilayers.

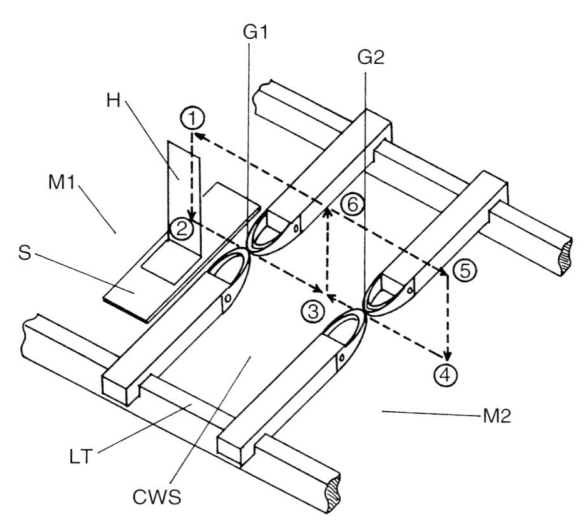

Figure 3.28 Assembly of double gates (G), a holder (H) and a solid substrate (S) on a Langmuir trough (LT). Figures and arrows show automatic horizontal lifting deposition cycles (see text). Reprinted with permission from Ref. [32a]. Copyright 1988 The Japan Society of Applied Physics.

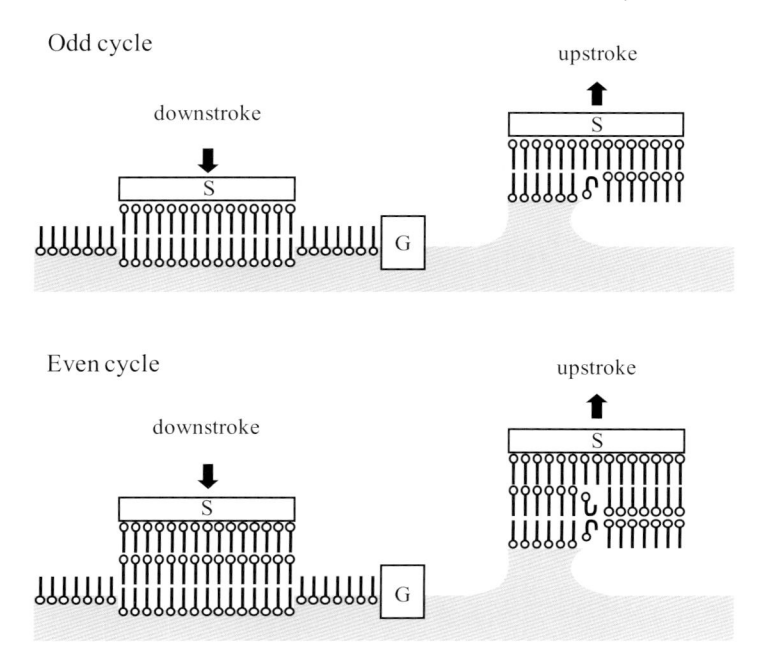

Figure 3.29 Turnover mechanism of amphiphile molecules during deposition by the horizontal lifting-up method. Reprinted with permission from Ref. [32b]. Copyright 1988 The Japan Society of Applied Physics.

gives very regular LB films of the X-type structure. If both of the amphiphiles are stable in X-type structures, we can get complex X-type LB films by the process of (1)→(2)→(3)→(6)→(5)→(4)→(3)→(6)→(1), and by repetition of this cycle. When Y-type structure forming amphiphiles, such as cadmium arachidate are used for the automatic horizontal lifting deposition by the process, (1)→(2)→(3)→(6)→(1), a very regular LB film of Y-type structure was obtained, confirmed by X-ray diffraction. This result suggests that cadmium arachidate molecules overturn just at the point of detouching clear water surface with receding three phase line of water/air/the outermost LB layer, like domino toppling, as shown in Figure 3.29 [32].

Iwahashi and his colleagues have reported a modified horizontal lifting-up method to form Y-structured LB films [33]. Figure 3.30 schematically shows the deposition process. After a hydrophobic substrate touches a surface-pressure-controlled monolayer, the substrate is pulled up slowly to the rotating direction of the holder to deposit the second layer. These processes are repeated to deposit multilayers. This method is suitable for highly valuable amphiphiles such as tritium-substituted stearic acid and physiologically active substances.

3.2.4.4 Horizontal Scooping-Up

To observe structures of Langmuir monolayers by atomic force microscopy, it is necessary to transfer them to smooth substrates without changing structures. LB deposition is not suitable for this object because LB deposition accompanies flow

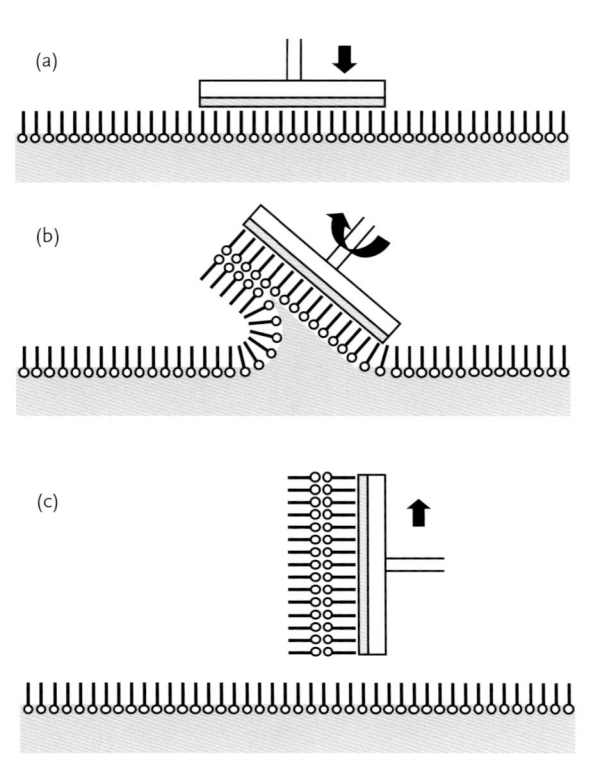

Figure 3.30 The modified horizontal lifting-up method for Y-structured LB films. (a) A hydrophobic substrate touches a surface-pressure-controlled monolayer, and (b, c) the substrate is pulled up to the rotating direction of the holder to deposit the second layer. Processes (a)–(c) are repeated to deposit multilayers. Reprinted with permission from Ref. [33]. Copyright 1985 The Chemical Society of Japan.

of the Langmuir monolayer during deposition and flow may change the microstructures of monolayers. The horizontal scooping-up method was devised to transfer one layer of Langmuir monolayers to solid substrate without changing structures at the subphase surface by flow of the monolayer [34]. The transfer processes are as follows; (i) use of hydrophilic and atomically smooth solid substrate, (ii) set the substrate almost horizontally just beneath the water surface before monolayer spreading by using a thin platinum wire support. (iii) after compression of a monolayer to a desired surface pressure, raise the substrate as slowly as possible, keeping it almost horizontal. Some tips for this technique to get one layer LB films without changing structures of Langmuir films by this transfer process are, (i) use of completely hydrophilic substrates, such as hydrophilized silicon wafer or cover glass made by the floating process. (ii) Slightly incline the substrate (around 5°) from a complete horizontal arrangement so that one corner of the substrate emerges first from subphase surface. You can see that the monolayer adheres to the solid substrate from this corner and a thin water layer between the substrate and the transferred monolayer is completely excluded.

3.3 Functions and Applications

If the substrate is raised keeping a completely horizontal arrangement, a large amount of subphase water is inserted between the substrate and the transferred monolayer and the structure of the transferred monolayer will be changed by the drying process of the inserted water. (iii) The raising speed of the substrate should be less than 1 mm/min. This technique is also suitable for preparing TEM samples of monolayers by setting collodion thin membrane covered copper meshes on the hydrophilic substrate.

3.3
Functions and Applications

3.3.1
Molecular Recognition

Molecular recognition is achieved between two or more molecules through specific noncovalent interactions such as hydrogen bond, coordination bond, van der Waals interaction, hydrophobic force, electrostatic interaction, π–π stacking interaction, and so on, under steric restriction (Figure 3.31) [35]. In general, some of these interactions need to act cooperatively in order to attain efficient molecular recognition. This phenomenon is crucial in biological systems where numerous molecular recognitions continually occur with extremely high selectivity and accuracy. Studies on molecular recognition, therefore, provide deeper insight and understanding on biological functions in the living organisms. On the other hand, the molecular recognition has contributed to development of new fields in chemistry, namely host–guest chemistry and supramolecular chemistry, where the selective noncovalent bindings are utilized to elaborate well-designed molecular self-assemblies or potential molecular devices for practical applications.

3.3.1.1 Molecular Recognition by Hydrogen Bonding and Electrostatic Interaction at the Air/Water Interface

Hydrogen bonding and electrostatic interactions are major driving forces for efficient molecular recognition [36, 37]. Such interactions are not favored in highly polar aqueous medium, but enhanced at the air/water interface with a lipid

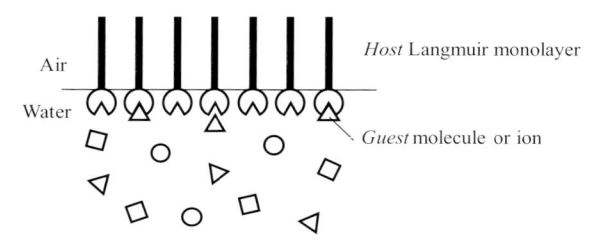

Figure 3.31 Molecular recognition at the air/water interface with a Langmuir monolayer as host. The film molecules as host recognize the guest molecules in the aqueous subphase through the chemical interactions and the steric coincidence (shape and dimension).

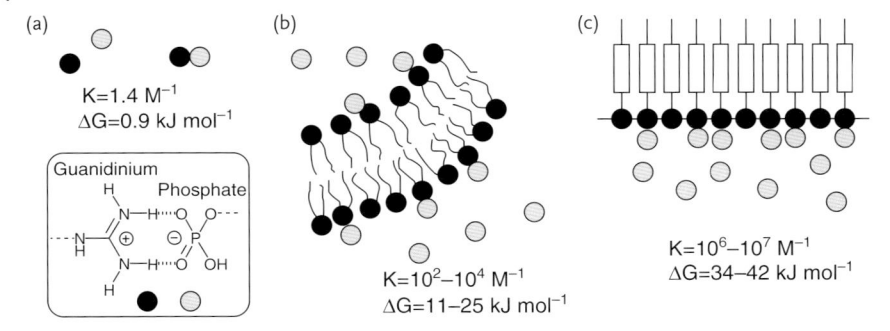

Figure 3.32 Typical binding constant (K) and binding energy (ΔG) between guanidinium and phosphate at various interfaces: (a) molecular interface (molecular dispersion); (b) microscopic interface (surfaces of micelles or lipid bilayer); (c) macroscopic interface (air/water interface). An inset displays hydrogen-bonded ion pairs of guanidinium and phosphate. Reprinted with permissions from Ref. [37a]. Copyright 2006 The Royal Society of Chemistry.

monolayer. An example is shown in Figure 3.32, where binding constants and energies for guanidium (host)–phosphate (guest) interaction are compared at (a) a molecular interface (in bulk water), (b) a microscopic interface (surface of micelles and bilayers), and (c) a macroscopic interface (air/water interface) [38]. The binding constants observed for the surface of a bilayer and micelle of amphiphilic guanidium are in the range from 10^2–10^4 M^{-1}, much larger than that of 1.4 M^{-1} reported for monomerically dispersed guanidium chloride and simple phosphate in bulk water. However, further enhancement of binding efficiency is attained at the air/water interface, where phosphate species dissolved in an aqueous subphase binds to guanidium groups of guanidium-functionalized lipid monolayer with the binding constants of 10^6–10^7 M^{-1}. This binding efficiency enhancement is interpreted as being related to a dielectric property of the interface on the basis of theoretical aspects [39]. In the calculations, a guanidium host and a phosphate guest were placed in various positions at a model interface consisting of a lipid layer (dielectric constant $\varepsilon = 2$) and water ($\varepsilon = 80$). The calculated binding energies significantly depend on the position of the hydrogen-bonding site relative to the two-phase boundary; a large binding constant is obtained when the site is located in the lipid phase, but even when the site is positioned in the aqueous phase close to the interface the intermolecular hydrogen bonding and electrostatic interactions are strengthened due to the influence of the low-dielectric lipid layer. These results indicate that the field beneath a sufficiently extended low dielectric and hydrophobic surface is favorable for efficient molecular recognition.

A heterogeneous dielectric environment at a lipid/water boundary resembles that at cell membrane surfaces. Since Kitano and Ringsdorf first reported the nucleobase recognition at the air/water interface [40], various kinds of biomolecular recognition have been investigated. Figure 3.33 displays selected examples of molecular recognition systems through specific hydrogen-bonding interactions at the air/water interface [41]. As one can recognize, multiple hydrogen bondings are

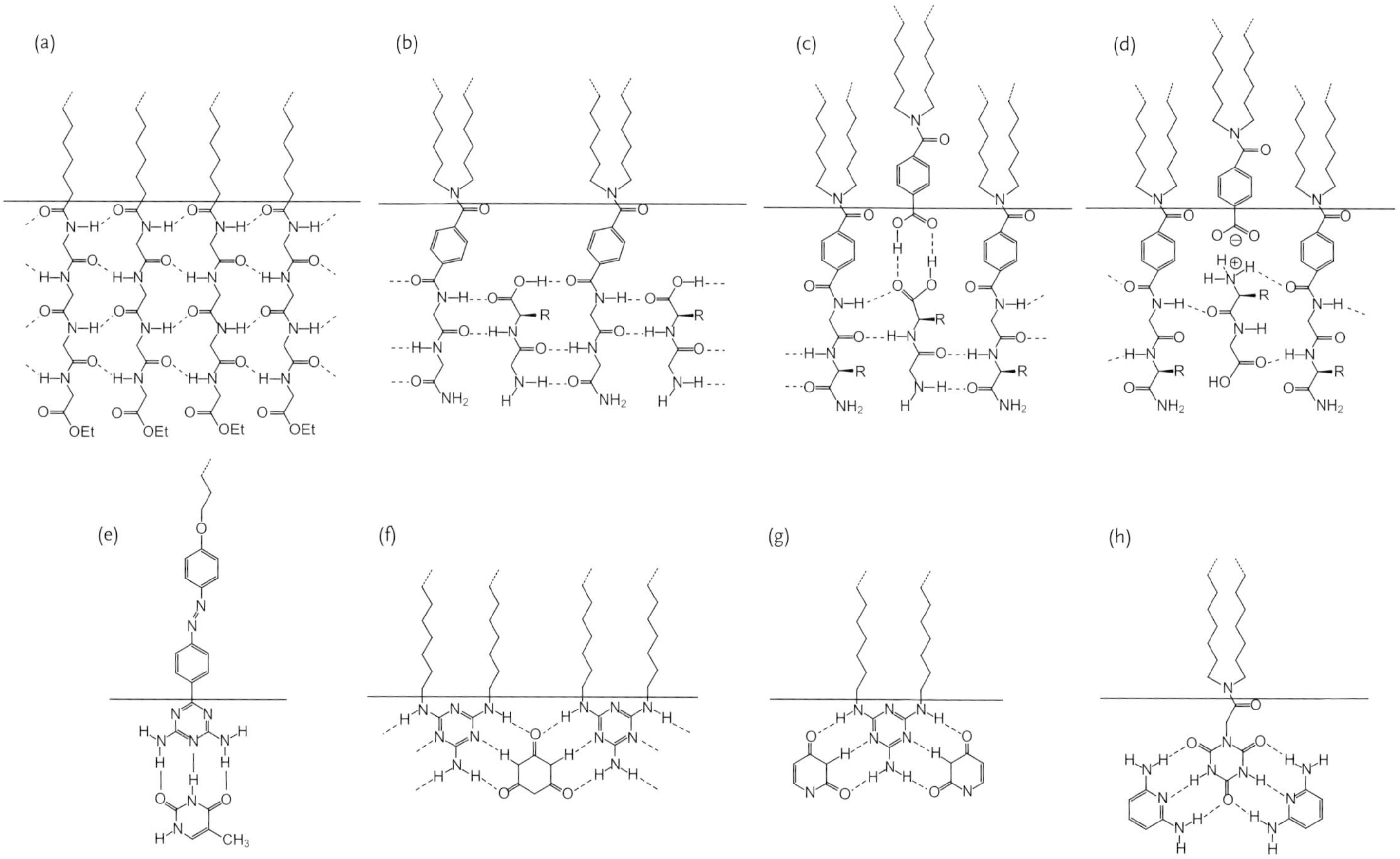

Figure 3.33 Some molecular recognition systems at the air/water interface. (a)–(d) peptide-functionalized host monolayers and aqueous guest dipeptides, (a) a monolayer of a single-chain oligoglycine amphiphile, (b) a double-chain oligoglycine amphiphile and a dipeptide by C-terminal insertion, (c) 1:1 mixed monolayer of dialkyl peptide/benzoic acid and dipeptides by (c) C-terminal insertion and (d) N-terminal insertion [41a], (e) a diaminotriazine amphiphile and thymine [41b], (f) and (g) amphiphilic melamine and (f) barbituric acid [41c, d] and (g) uracil [41d], (h) a cyanurate amphiphile and 2,6-diaminopyridine [41e]. The dotted lines represent hydrogen bonds.

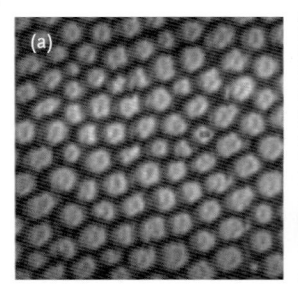

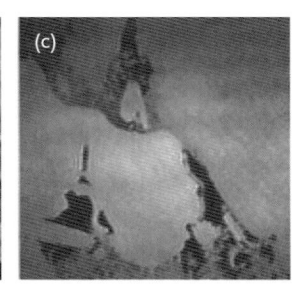

Figure 3.34 BAM images observed for 2C$_{11}$H$_{23}$-elamine monolayers on (a) pure water at 20 °C, 0.41 nm^2/molec., (b) 0.2 mM uracil subphase at 20 °C, 0.44 nm^2/molec., and (c) 0.01 mM barbituric acid subphase at 20 °C, 0.634 nm^2/molec. Image sizes are 400 × 400 µm^2. Reprinted with permission from Ref. [41d]. Copyright 2005 The American Chemical Society.

typically necessary to keep a stable host–guest binding state in the monolayers. The binding of aqueous guest molecules to host monolayers is found most easily by the changes in π–A isotherms: various changes induced by the interfacial molecular recognition have been reported, such as expansion or condensation of monolayers, disappearance of phase transition, decrease of collapse pressures. Direct evidence on intermolecular interactions including hydrogen-bond formation can be gained using FT-IR spectroscopy. The transmission and RAS technique are applicable to LB films on solid substrates [41a, 42], and the RAS or modified RAS methods to the monolayers on an aqueous subphase [43]. Spectral changes between a monolayer on the pure water surface and that on the guest-dissolved subphase, for example, shift and emergence/disappearance of infrared absorption peaks provide information on bonded species, chemical state and orientation of functional groups, the secondary structure of large biological molecules such as peptides, and so on, helps in the understanding of molecular interactions at the interfaces. In terms of measurement of the secondary structure of peptides and proteins, circular dichroism spectroscopy has been frequently applied [42a]. Brewster-angle microscopy is also useful to visually recognize morphological changes induced by the interfacial molecular recognition. Figure 3.34 shows, as examples, BAM images of 2,4-di(n-undecylamino)-6-amino-1,3,5-triazine (2C$_{11}$H$_{23}$-melamine) monolayers spread on (a) pure water, and aqueous solutions of (b) uracil (c) barbituric acid [41d]. Different morphological features of condensed-phase domains depending on subphase conditions reflect different molecular interactions of the guest with the melamine monolayer. The pure monolayers show only small compact, nontextured condensed-phase domains, whereas on the barbituric acid solution (corresponding to Figure 3.33f), large homogeneous islands of condensed phase are formed. In contrast, the 2C$_{11}$H$_{23}$-melamine-uracil assemblies (Figure 3.33g) develop well-shaped circular condensed-phase domains with an inner texture due to parallel orientation of the alkyl chains to the domain periphery. For LB films on solid supports, atomic force microscopy (AFM) allows direct characterization of the surface structures of molecular assemblies in

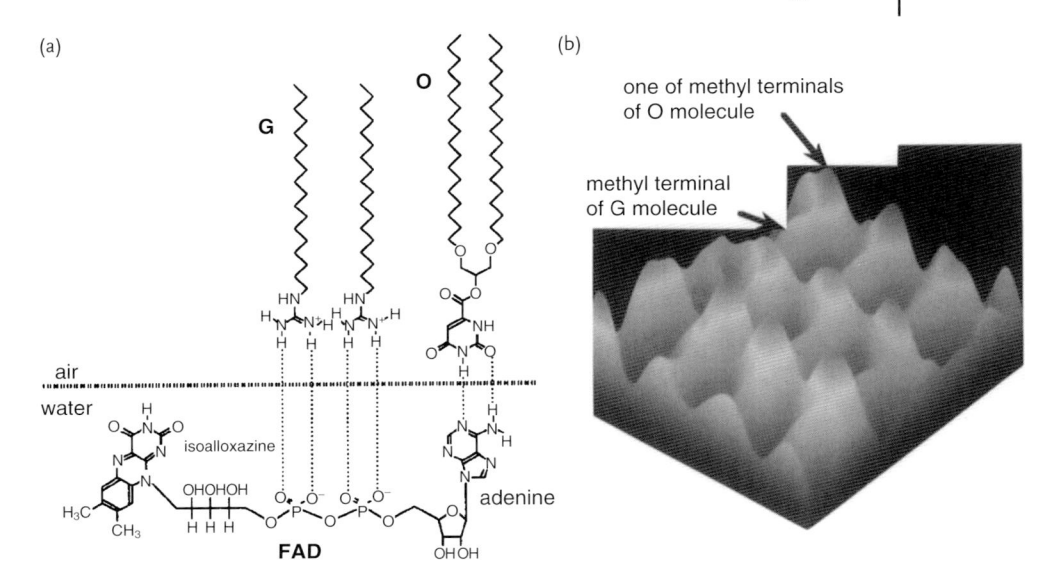

Figure 3.35 (a) Multisite molecular recognition between flavin adenine dinucleotide (FAD) and a mixed monolayer of 1-octadecylguanidinium *p*-toluenesulfonate (G)/1′,3′-bis(octadecyloxy)isopropyl orotate (O) at the air/water interface, and (b) a three-dimensional AFM image of the G/O-FAD film transferred on mica. Reprinted with permission from Ref. [44]. Copyright 1997 The American Chemical Society.

the range from approximately a hundred micrometers to a few nanometers. An example is shown in Figure 3.35, where the periodic oblique pattern composed of two kinds of peaks, corresponding to the methylene terminals of guanidium (G) and orotate (O) amphiphiles in their mixed monolayers transferred on mica from aqueous flavin adenine dinucleotide (FAD), are clearly observed with high resolution [44].

3.3.1.2 Chiral Discrimination at the Air/Water Interface

The chiral recognition/discrimination plays an important role in biological systems. In lipid monolayers at the air/water interface as simple model systems of biological membranes, chiral effects appear in monolayer properties and morphologies, originating from molecular arrangements. In this viewpoint, combined approaches with fluorescence microscopy or BAM and FT-IR or GIXD, in addition to film-balance experiments, have been frequently applied for the experimental characterization of chiral discrimination in amphiphilic monolayers [45]. Many film substances, such as phospholipids [46], amphiphilic amino acids [47], and acid amides [48], have been used for the monolayer experiments and shown chiral recognition/discrimination behaviors. It is worth mentioning here that, even in the chiral recognition, the hydrogen bonds at polar head groups play essentially an important role. For instance, monolayers of pure enantiomers and a racemic mixture of amphiphilic *N*-dodecylgluconamide exhibited striking chiral

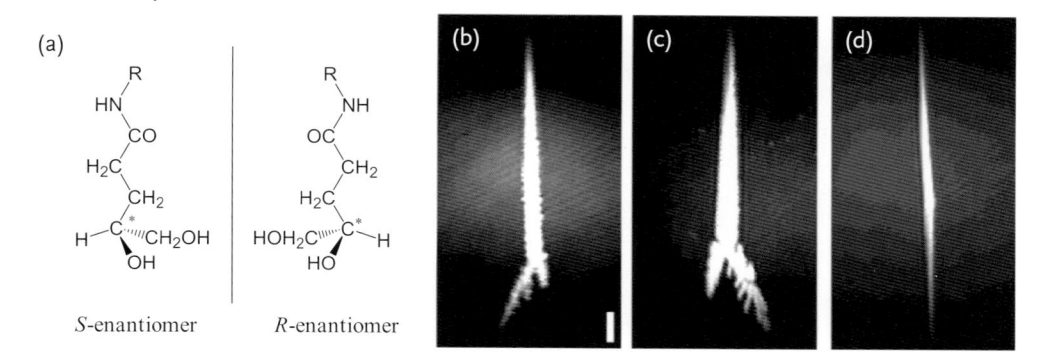

Figure 3.36 (a) Molecular structures of enantiomers of N-tetradecyl-γ,δ-dihydroxypentanoic acid amide (R = $C_{14}H_{29}$), and chiral discrimination of the condensed-phase textures at the air/water interface observed by BAM: (b) S-enantiomer, (c) R-enantiomer, (d) 1:1-R,S racemate. (a) and (b)–(c) are reprinted with permission from Ref [48]. (Copyright 1997 The American Chemical Society) and [45a] (Copyright 2003 The American Chemical Society), respectively.

discrimination in surface-pressure isotherms, molecular area relaxation kinetics at constant surface pressures, and monolayer morphology [49]. BAM observation demonstrated that D- and L-N-dodecylgluconamides formed identical dendritic crystals growing anisotropically with straight main axes and numerous straight side branches, while the corresponding racemate, N-dodecyl-D,L-gluconamide, formed an isotropic monolayer of condensed phase. On the basis of consideration on the crystal structures of D-gluconamides, a highly developed hydrogen-bonding network between neighboring hydroxyl groups of the sugar head groups and amide groups of enantiomeric N-dodecylgluconamide give rise to molecular orientation order, and consequently are responsible for the dendritic growth of condensed-phase domains. In the racemic mixture, the hydrogen-bonding networks are disturbed by equivalent packing of the enantiomers.

The pronounced difference between enantiomers hardly appear in surface-pressure experiments but have been seen in the monolayer morphology. BAM images showing the chiral discrimination in Langmuir monolayers of N-tetradecyl-γ,δ-dihydroxypentanoic acid amide (TDHPA) are depicted in Figure 3.36 [48]. π–A isotherms for the racemate and the enantiomers of TDHPA were identical within the experimental accuracy. However, the BAM images for the enantiomers indicate that condensed-phase domains grow with dendritic shapes, and the growth axes of main arms bear a mirror image relationship. On the other hand, the domains of the racemic mixture reveal no mirror symmetry, with two main growth directions at an angle of nearly 180°. Here again, the dendritic growth with preferred directions observed in the enantiomer monolayers was understood by directed interactions of the amide part of the amphiphile through hydrogen bonds, as supported by GIXD results. Similar chiral discriminable differences in monolayer morphology have also been reported, and among them some selected examples are displayed in Figure 3.37 [45].

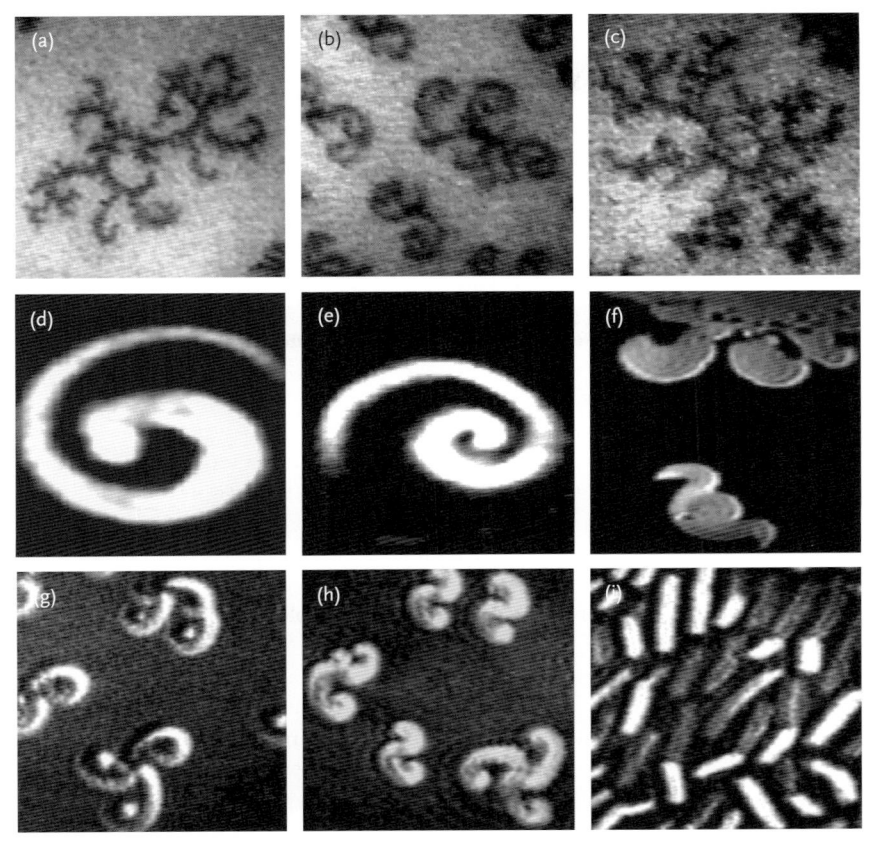

Figure 3.37 Various two-dimensional assemblies formed in chiral amphiphilic monolayers: fluorescence microscopic images of 1-stearylamine glycerol monolayers; (a) S-enantiomer, (b) R-enantiomer, (c) 1:1 racemate, BAM images of N-stearoylserine methyl ester monolayers; (d) D-enantiomer, (e) L-enantiomer, (f) 1:1 racemate, and of dipalmitoylphosphatidylcholine (DPPC); (g) D-enantiomer, (h) L-enantiomer, (i) 1:1 racemate. Reprinted with permission from Ref. [45a]. Copyright 2003 The American Chemical Society.

3.3.1.3 Macrocyclic Hosts

The molecular recognition with a macrocyclic host cavity is also important in host–guest chemistry. So far, many macrocyclic compounds, including cyclodextrins (CDs), calixarenes (CAs), crown ethers, macrocyclic polyamines, cyclophanes, etc., have been used as molecular recognition scaffolds. In the molecular-recognition system with such macrocyclic hosts, guest molecules or ions are encapsulated into the host cavities through hydrogen bonds, electrostatic interactions, coordination bonds, and/or hydrophobic interactions to form noncovalent inclusion complexes. As expected, size, shape, and/or charge distribution of the host cavity should be complementary to those of the guest for the efficient molecular recognition.

The CDs are cyclic oligosaccharides consisting of six (α-CD), seven (β-CD), or eight (γ-CD) glucose units linked by α-1,4-glycosidic bonds (Figure 3.38) [50]. Because of the characteristic cylindrical structure with a hydrophilic outer surface and a hydrophilic central cavity, the CDs possess an ability of inclusion for a wide variety of organic and inorganic guest compounds. The native CDs are water soluble and have parallel rims with different diameters at the opposite ends. Primary and secondary hydroxyl groups, placed at the narrow and wide ends of the CD cavities, respectively, are substituted to lipophilic groups to provide the amphiphilic nature necessary for preparation of stable Langmuir and LB films with and/or without hosts [52], or to polar functional groups for introducing further functions such as additional electrostatic interaction sites [53]. The calyx[n] arenes, CAs, formed by oligomerization of phenol and formaldehyde, are also one of the most useful types of macrocyclic host in host–guest chemistry. These hosts have a cup-like shape with defined upper and lower rims, and the vacancy can be varied at least from 1 to 8 [51]. The stereochemical orientation of ligating arms can be turned: for example, the calix[4] arene has four possible conformations with different orientational combinations of the arms (cone, partial cone, 1,2 and 1,3 alternates), and each conformer encapsulates different ion species. The ease of chemical functionalization at either the upper and/or lower rims as well as the adjustable vacancy size gives various derivatives with different selectivities for many kinds of guest ions and small molecules including fullerenes [54]. Furthermore, Langmuir monolayers of amphiphilic calixarenes with additional molecular-recognition moieties successfully captured sugars [55a], nucleotides [55b], dopamine [55c], or showed enantioselective recognition for phenylalanines [55d, e]. The crown ethers and their analogs in amphiphilic forms work as host for metal cations in aqueous medium [56]. The crown ethers themselves catch cations of alkali metals and alkaline-earth metals with fairly high selectivity through formation of the complexes. When nitrogen and/or sulfur atoms are inserted into their structure they become sensitive to other ions such as Ag^+, Hg^+, and Cd^+ ions [57]. In addition, the protonated nitrogen in the macrocyclic polyamines provides the complexing ability for organic anions such as nucleotides [58].

3.3.1.4 Dynamic Host Cavity

The reversible control of binding-release processes by an external stimulus has been an exciting challenge in the field of molecular recognition. Ariga *et al.* realized mechanically controlled molecular recognition at the air/water interface through dynamic formation of a cavity structure using steroid cyclophanes as host, which contain a cyclic core connected to four steroid moieties through a flexible L-lysine spacer with cholic acid (Figure 3.39) [37a, 59]. The host molecule forms an open conformation at low surface pressure in order to contact the hydrophilic face efficiently to the water phase, whereas compression of the monolayer induces a transition to a more compact conformation of the steroid cyclophane molecule resulting in a cavity formation. The reversible capture and release of an aqueous

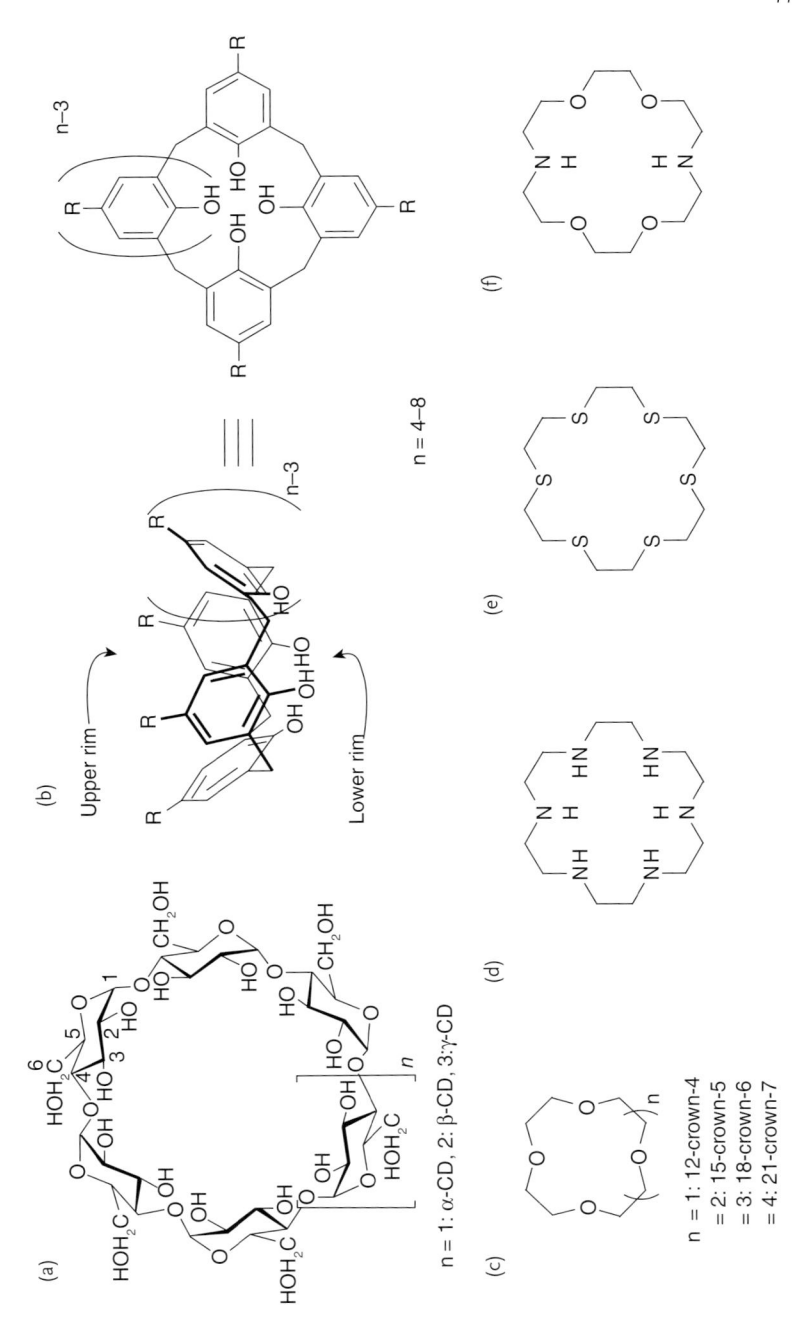

Figure 3.38 Structures of some representative macrocyclic molecules: (a) cyclodextrin, (b) calix[n]arene, (c) crown ether, and (d) azacrown ether [18],ane-N6, (e) thiacrown ether [18],ane-S6, and (f) 4, 13, diaza-18-crown-6-ehter. (a) and (b) are reprinted with permission from Ref [50b]. (Copyright 1997 The American Chemical Society) and [51a] (Copyright 2006 The Royal Society of Chemistry), respectively.

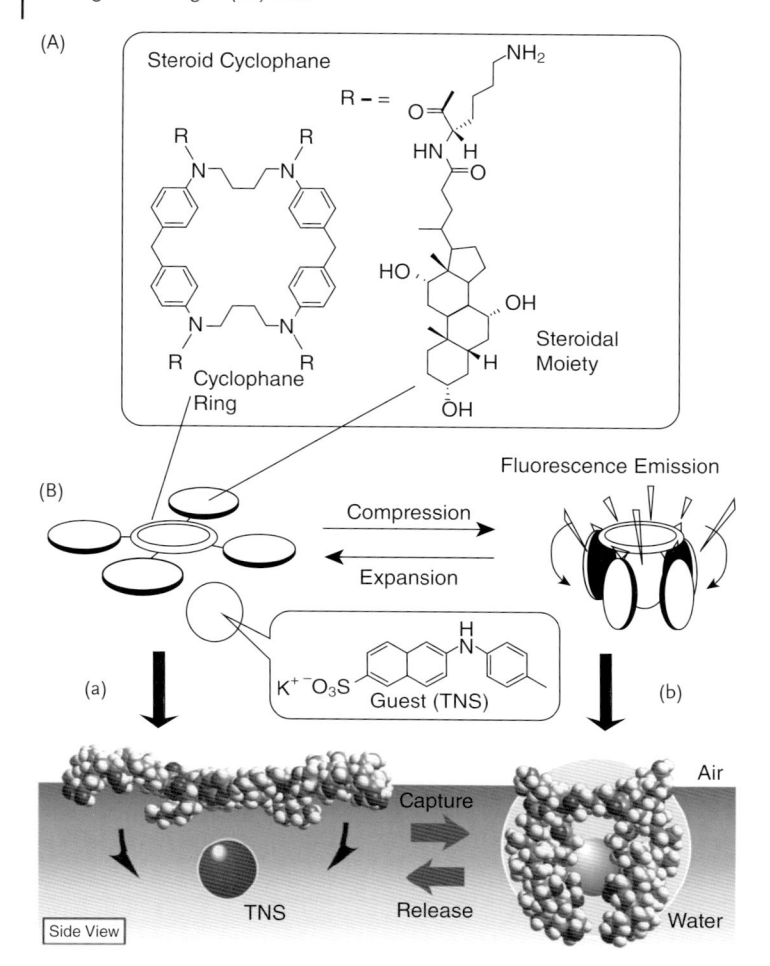

Figure 3.39 (A) Chemical structure of the steroid cyclophane as host. (B) Plausible models for the conformational change of the steroid cyclophane upon monolayer compression and reversible capture of the guest TNS (6-(p-toluidino)naphthalene-2-sulfonate): (a) open conformation at low pressures; (b) cavity conformation at high pressures. Reprinted with permission from Ref. [37a]. Copyright 2006 The Royal Society of Chemistry.

florescent naphthalene guest were monitored by a periodic change in the fluorescence intensity on repeated compression and expansion cycles.

A dynamic formation of nanobox structure at the air/water interface was also reported [60]. An amphiphilic terpyridine ligand and a biphenyl-type guest were spread on an aqueous Pd(II) complex solution. The box-shaped host–guest complex was formed through self-assembly of the amphiphilic ligands with end-capped Pd(II) components in the presence of the rod-like guest. Other examples of the dynamic conformational changes of the host molecules associated with the molecular recognition at the air/water interface involve resorcin [4]arene cavitated-based molecular switches [61] and enatioselective amino acid recognition by dynamic motion of a polycholesterol-substituted cyclen complex [62].

3.3.2
Multilayer Films for Photoelectronic Functions

The LB technique enables the controlled construction of highly ordered monomolecular assemblies with the intended number of layers on various solid substrates including metal and semiconductor electrodes. This feature is beneficial for molecular engineering to fabricate artificial self-assemblies or devices with desired functions. Since the pioneering works with fruitful concepts and key theories on the photoconductivity within LB multilayers by Kuhn and coworkers [63], much interest has been attracted to the construction of photoelectrically functionalized molecular architectures by the LB technique. The understanding of photophysical processes taking place in the organized LB films is crucial for designing molecular photoelectronic devices with specific characteristics. One of the potential application of such films is the use of photoinduced vectorial electron transfer, the goal of which is to develop photoelectronic components such as photodiodes, phototransistors, photoswitches, organic solar cells, and so on.

3.3.2.1 Molecular Photodiode
In biological photosynthetic process, the light energy harvested by the antenna pigments in light-harvesting proteins is funneled to the reaction center where the photoexcited reaction center chlorophylls transfer an electron to an electron acceptor and are simultaneously reduced (electron donated) by a donor [64]. Mimicking the biological charge-separation process, artificial molecular assemblies with layered electron donor–sensitizer–acceptor systems or donor–acceptor systems where either the donor or acceptor works as a sensitizer as well, have been extensively studied. The essential mechanism of light-induced vectorial charge separation is the "light-driven electron pump" proposed by Kuhn and coworkers [63], where electrons of an excited dye are transferred to an acceptor and the dye is recovered by electron tunneling through a high and narrow potential barrier (Figure 3.40). Based on this mechanism, LB film photodiodes were first fabricated by Fujihira *et al.* [65]. Figure 3.41 displays the film structure of the A/S/D molecular photodiode and the energy diagram as a function of distance across the film, where A, S, and D are electron acceptor, sensitizer, and electron donor moiety, respectively. If the forward electron-transfer processes (solid line arrows, i–iii) are accelerated and the back-electron-transfer processes (dashed lines, iv–vi) are retarded by setting the distances and the energy levels appropriately, the photoinduced vertical electron flow can be achieved. Once an electron–hole pair is separated successfully, the recombination of the pair across the large separation by the LB film (process vi) is hindered. The A/S/D LB film on a gold optically transparent electrode (Figure 3.41a) was used as a working electrode in a photoelectrochemical cell, and the photoinduced vectorial flow of

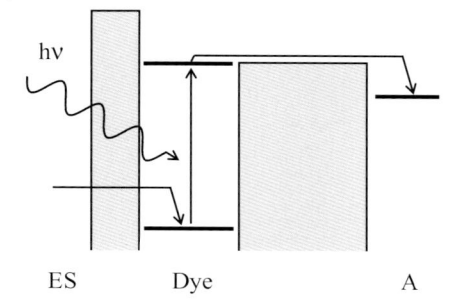

Figure 3.40 Schematic explanation of light-driven electron pump. Dye (dye) is exited by light, and the low barrier conducts the electron which is then captured by A (electron acceptor). The barrier is sufficiently broad to prevent the electron from back tunneling. The high barrier is sufficiently narrow to allow an electron from ES (electron source) to tunnel to oxidized dye Dye$^+$. Reproduced with permission from Ref. [63c]. Copyright 1979 Elsevier.

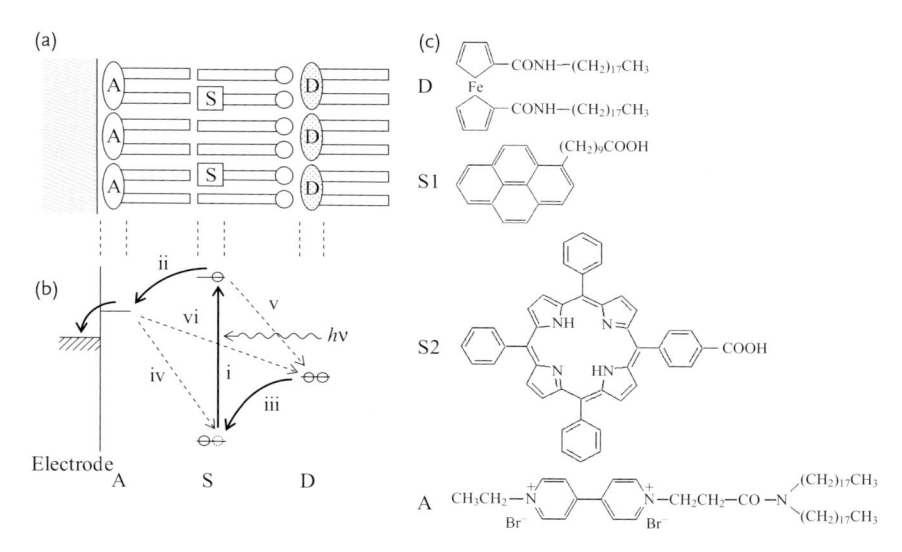

Figure 3.41 Molecular photodiode with heterogeneous A/S/D LB films on gold optically transparent electrode: (a) structure of LB film, (b) energy diagram, and (c) structural formula of film molecules. The second layer is a mixed monolayer of S and arachidic acid (1:2). Reproduced with permission from Ref. [65c]. Copyright 1995 Elsevier (Academic Press).

electrons was detected as anodic photocurrent. On the other hand, when the D, S, and A monolayers were deposited in this order onto the electrode surface (i.e., D/S/A molecular photodiode, Figure 3.42a), cathodic photocurrents were observed. Besides, mixed monolayers of triad compound (Figure 3.42b, c) and behenic acid (1:10) also showed anodic photocurrent generation. The light-driven electron pump is the core mechanism of photodiodes with layered organic assemblies although many other photophysical processes should also be taken into

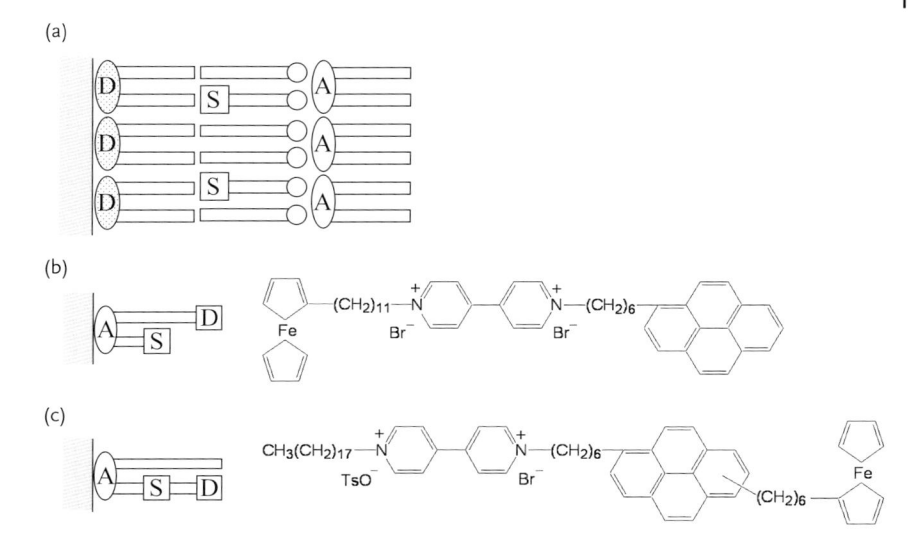

Figure 3.42 Schematic presentation of (a) D/S/A LB film, (b) folded-type S–A–D triad molecule, and (c) linear-type A–S–D triad molecule on an electrode. The molecular structures are also displayed for the triad molecules. (b) and (c) are reprinted with permission from Ref. [65b]. Copyright 1989 Elsevier.

account in order to understand the whole process of photoinduced electron-transfer systems.

3.3.2.2 Fullerene C_{60} Containing LB Film

Fullerenes C_{60} has particularly excellent electron-acceptor properties and is thus frequently used in organic donor-acceptor systems [66]. Incorporation of C_{60} units into LB film systems is a promising way to achieve an enhanced increase in photoinduced electron transfer. However, it was not so easy to prepare homogeneous and stable Langmuir monolayers with C_{60} because pristine C_{60} has low solubility in common organic solvents useable for spreading solution, and even if spread, C_{60} units tend to readily self-aggregate at the air/water surface due to strong intermolecular π–π interaction [67]. Figure 3.43 represents examples of C_{60} derivatives that are reported to form electron donor–acceptor LB films with high quality [67, 68]. A bilayer LB film was constructed on a gold electrode surface by successive deposition of spread monolayers of C_{60}-cyclic peptide-poly(ethylene glycol) conjugate (Figure 3.43a) and a pyrene derivative from the air/water interface [67]. Anodic photocurrents were generated in aqueous solution of triethanolamine through stepwise electron transfer from the excited pyrene to the electrode via the C_{60} unit. In this molecule, the cyclic peptide moiety acted as a scaffold to prevent the C_{60} units from self-aggregation, and accordingly allowed ordered molecular orientation leading to homogeneous and stable monolayer formation. Molecules in Figures 3.43b and c are amphiphilic donor–acceptor dyads [68]. In these molecules, the porphyrin donor and the C_{60} acceptor are covalently double-bridged to

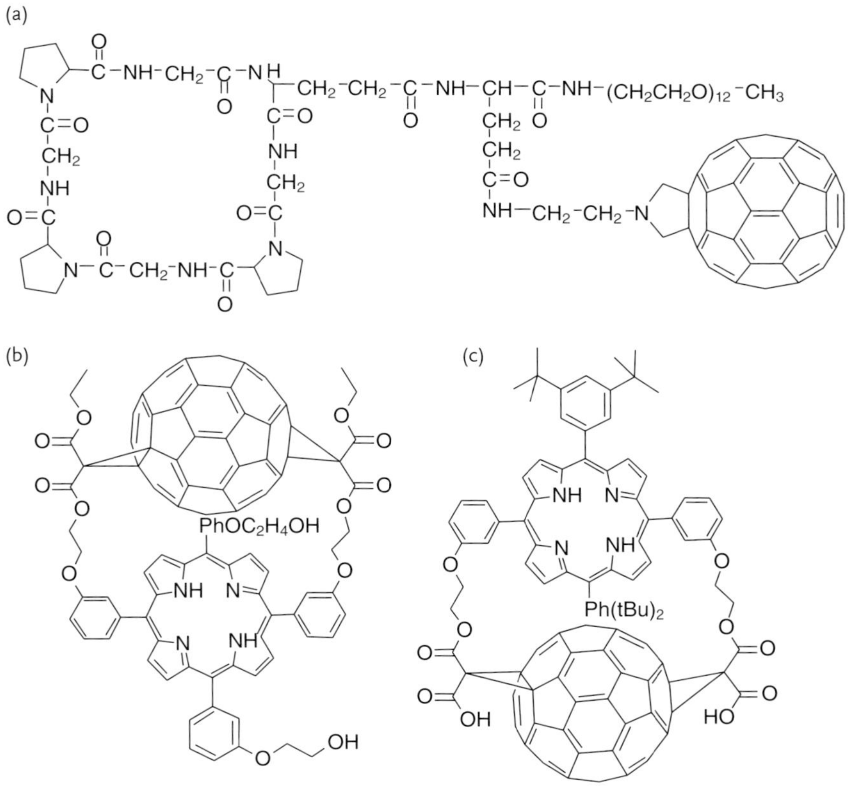

Figure 3.43 Structures of (a) C_{60}-cyclic peptide-poly(ethylene glycol) conjugate [67b], and (b) (c) porhyrin-fullerene dyads [68b]. (a) and (b)–(c) are reprinted with permission from Ref [67b]. (Copyright 2008 The American Chemical Society) and [68b] (Copyright 2005 The American Chemical Society), respectively.

fix their spatial arrangement. Difference between (B) and (C) is the position of the hydrophilic end, located on the porphyrin or fullerene moiety, respectively. These molecules formed Langmuir monolayers with opposite orientation of donor–acceptor pairs, as expected from the hydrophilic group position in the molecules, and the monolayers could be transferred onto solid substrates without change in the orientation. As a result, the LB films showed the vectorial photoelectron transfer to opposite directions. Furthermore, the orientation could also be altered by depositing the dyad monolayers in either the upward or downward direction. Therefore, the direction of the photoinduced vectorial electron transfer could be controlled by the selection of donor–acceptor dyad molecules and by the deposition direction of the LB monolayer. The effects of introduction of secondary electron acceptors or donors in the oriented porphyrin–fullerene dyad systems to photoelectric properties have also been discussed [68c–e]. Another example is electron donor–acceptor dyad ensembles of a water-soluble cationic zinc porphyrin and a

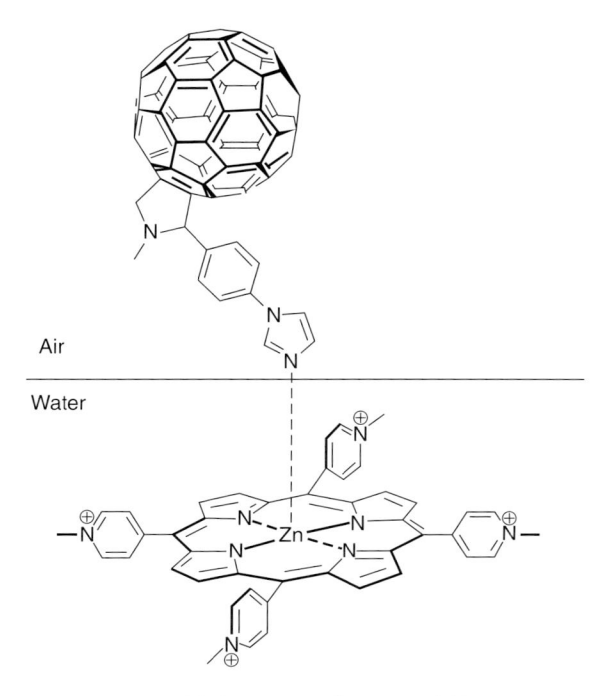

Figure 3.44 Simplified structure of a water-soluble zinc porphyrin axially coordinated to the fullerene adduct forming donor–acceptor dyad ensembles at the air/water interface. Reprinted with permission from Ref. [69]. Copyright 2007 The American Chemical Society.

C_{60} derivative with an imidazole ligand at the air/water interface (Figure 3.44) [69]. When the C_{60} derivative was spread from a chloroform solution on the porphyrin solution surface, the porphyrin axially coordinated to the C_{60} derivative through the imidazole ligand. Here again, the monolayer deposition direction (i.e., upstroke and downstroke deposition) on a solid substrate changes the donor–acceptor orientation, and thus the photocurrent flow direction.

3.3.2.3 Optical Logic Gate/Photoswitch

As mentioned above, the special arrangement of photofunctional layers or molecules enables the control of photocurrent flow direction in the assemblies. This was applied to construction of optical logic gates by combining LB film photodiodes of photofunctional polymers (Figure 3.45) [70]. A phenanthrene-based (Phen) and an anthracene-based (An) polymer were employed as sensitizers for each photodiode. Chromophores of Phen and An can be selectively excited since there is no overlap in absorption spectra, and so that combination of excitation wavelength of chromophore ($\lambda_1 = 300$ nm for Phen and $\lambda_2 = 380$ nm for An) could be used as input signals. In a structure shown in Figure 3.45a where two LB-film photodiodes are connected in series, when the assembly was irradiated by light with each wavelength individually, corresponding to input "10" or "01", only either photodiode

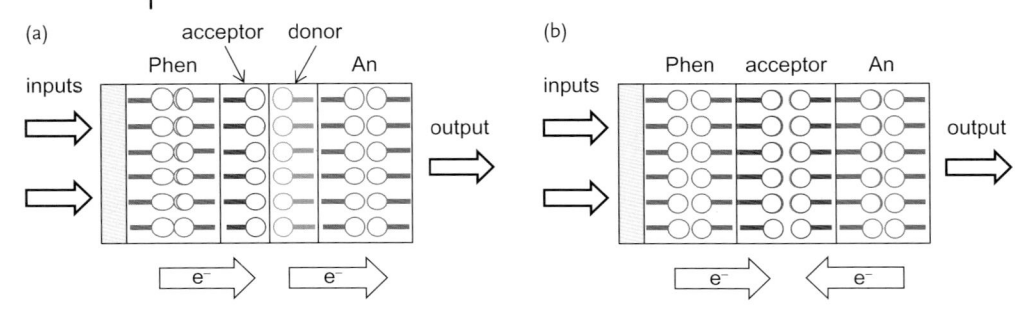

Figure 3.45 Photofunctional polymer LB films for optical (a) AND and (b) EXOR logic gates. Although film molecules used were amphiphilic polymers, they are drawn like normal amphiphiles for convenience here. Phen: phenanthrene-based polymer, An: anthracene-based polymer, acceptor: dinitrobenzene-based polymer, donor: dimethylaminobenzene-based polymer. See Refs [70a]. and [70b] for exact molecular structures.

Table 3.1 Truth table for the AND logic gates (see Figure 3.45) [70c].

Input 1 (λ_1)	Input 2 (λ_2)	Output	Total photo-current / pA
0	0	0	0
1	0	0	70
0	1	0	70
1	1	1	190

operated to generate photocurrent (Table 3.1). The magnitude of the total photocurrent was low under these excitation conditions because the unexcited layer acts as an insulating layer. On the other hand, when the light of both wavelengths irradiated simultaneously (input "11"), both photodiodes not only generate photocurrent but also transfer the charges to the electrode, accordingly enhancing the total electron flow. When there was no irradiation (input "00"), no photocurrent was observed. In this "optical AND logic gate", the high and low values were separated by a factor of 2.6 because of the photocurrent enhancement. In the Figure 3.45b, an acceptor layer was sandwiched by Phen and An films, so the photocurrent flows to opposite directions. In this assembly, the change of irradiation wavelength (i.e., "10" and "01") controls the photocurrent direction, and the simultaneous excitation of both chromophores ("11") reduced the total photocurrent. Therefore, the absolute total photocurrent values can be the output signals (Table 3.2).

3.3.3
Biomimetic Functions

The basic structure of biological membranes is the bilayer, namely two phospholipid monolayers attach back-to-back with their hydrophobic tails and the hydrophilic

Table 3.2 Truth table for the EXOR logic gates (see Figure 3.45) [70c].

Input 1 (λ_1)	Input 2 (λ_2)	Output	Total photo-current / pA
0	0	0	0
1	0	0	−150
0	1	0	190
1	1	1	20

heads direct outward interacting with the aqueous environment. Biological membranes contain a variety of lipids and proteins, and play a critical role in many biological processes. However, it is not so easy to understand the detailed roles of each membrane component because of the complex composition in the actual biomembranes. Simplified biomembrane model systems help us to obtain this insight. Langmuir monolayers of lipids on aqueous subphases correspond to leaflets of the biomembranes, and thus have been recognized as good model systems to study lipid-associated physical phenomena such as phase transitions, domain formation, mixing interaction with other lipids, and so on. The molecular recognition described in Section 3.3.1 is also an example of the biomimetic function of the Langmuir monolayer systems. Besides, Langmuir monolayers and LB films have been used for adsorption of biomaterials and biomineralization as convenient model systems mimicking biological surfaces.

3.3.3.1 Biomembrane Models – Langmuir Monolayers of Lipids

Many types of lipids make up cell membranes, but they are categorized into three main groups: glycerophospholipids, sphingolipids, and cholesterol. Molecular structures of these lipids are shown in Figure 3.46 [71]. The glycerophospholipids consist of a glycerol backbone, two ester-linked acyl chains, and a polar head group. There are a variety of combinations of the acyl chains and the head group. Typically, one long hydrophobic tail containing R_1 is a C_{16} or C_{18} saturated fatty acid chain and the other involving R_2 is a C_{16}–C_{20} unsaturated fatty acid chain [71]. The major head group classes are phosphoethanolamine (PE), phosphatidylcholine (PC), phosphoserine (PS), phosphoinositol (PI), and phosphatidylglycerol (PG). The sphingolipids have a sphingosine backbone connected to a single amide-linked saturated acyl chain, and either to a phosphorylated alcohol (usually phosphorylcholine, i.e., sphingomyelin (SM)), or sugar molecules including sialic acid (i.e., ganglioside GM1, GM2, GM3). The sphingosine chain involves a *trans*-double bond, and the chain length is usually fixed. For SM, the most common sphingosine base is the $18:1^{trans\Delta 4}$ in nature [72a]. In contrast, the acyl chain of SM includes different length and saturation, but is most typically a saturated long chain of C16, 18, 22, or 24, or an unsaturated C24 chain with a single *cis*-double bond [72b]. Additionally, polar hydroxyl and amino moieties of SM can act as a hydrogen-bond donor or acceptor. These structural features are closely associated with the agglomerating property of SM in the membranes. Cholesterol has a unique molecular

(a)

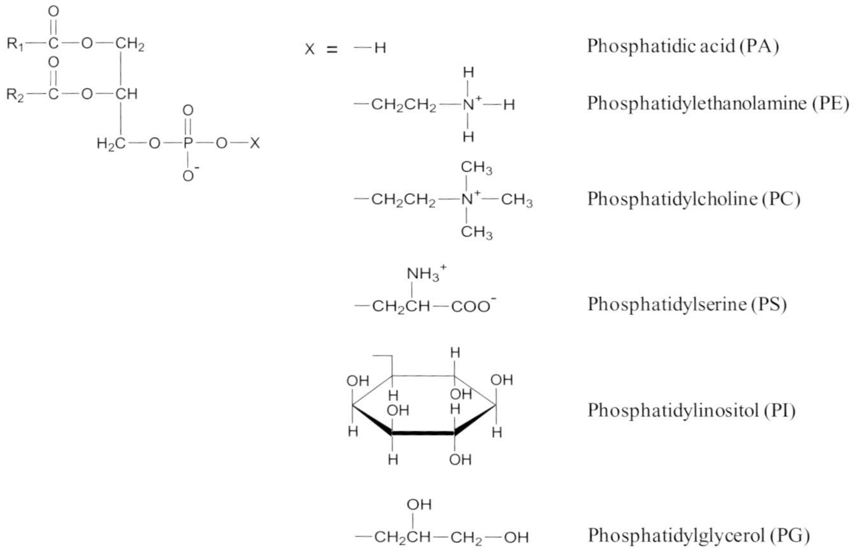

Phosphatidic acid (PA) X = —H

Phosphatidylethanolamine (PE) —CH$_2$CH$_2$—N$^+$H$_3$

Phosphatidylcholine (PC) —CH$_2$CH$_2$—N$^+$(CH$_3$)$_3$

Phosphatidylserine (PS) —CH$_2$CH(NH$_3^+$)—COO$^-$

Phosphatidylinositol (PI)

Phosphatidylglycerol (PG) —CH$_2$CH(OH)—CH$_2$—OH

(b)

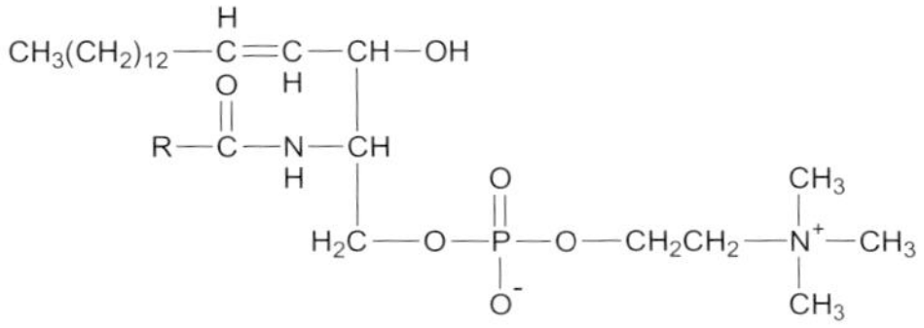

(c) G$_{M3}$, G$_{M2}$, G$_{M1}$

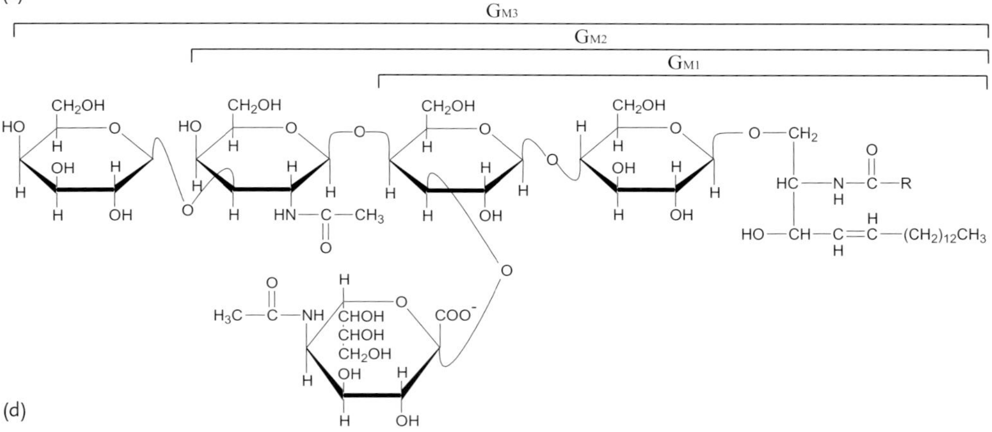

(d)

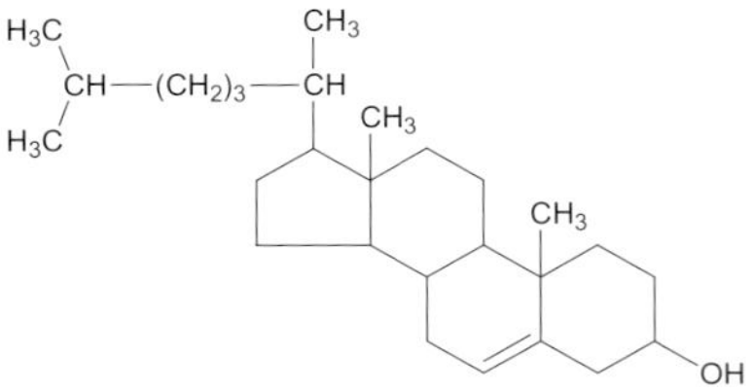

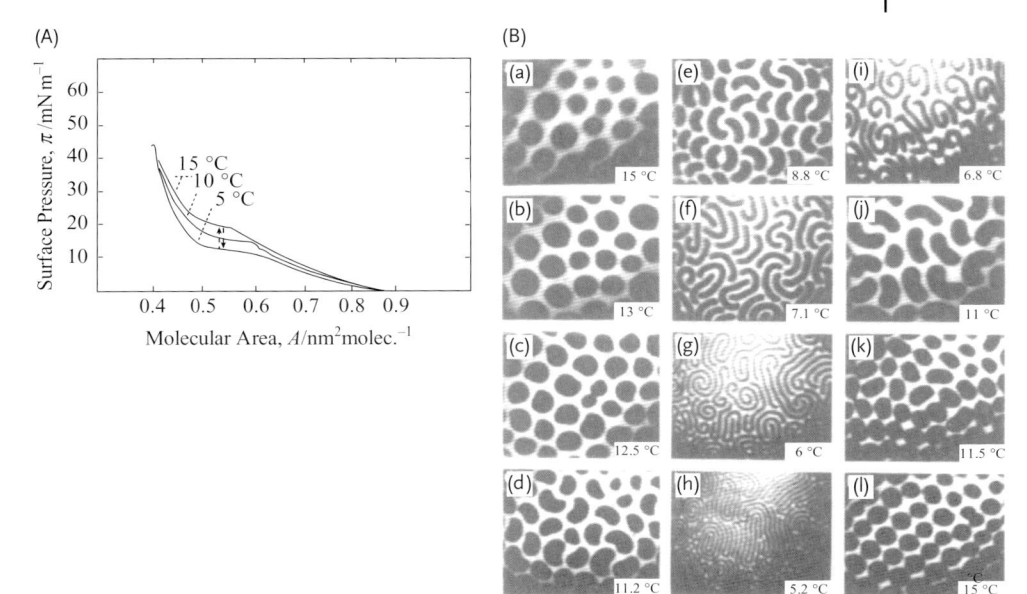

Figure 3.47 (A) π–A isotherms of a phospholipid L-α-dimyristoylphosphatidic acid (DMPA) monolayer at different temperatures, and (B) fluorescence micrographs of upon decreasing the temperature ((a) to (h)) and increasing the temperature ((i) to (l)) at a fixed molecular are of 0.54 nm^2/molec in the LE/LC transition region (indicated by arrows in (A)). The monolayer contains 1 mol% dye, dipalmitoyl-nitro-benzooxadiazolphosphatidylethanolamine, and 2 mol% cholesterol. Subphase conditions: pH 11.4, 1×10^{-5} M ethylenediamine tetraacetic acid, sodium salt. Reprinted with permission from Ref. [75a]. Copyright 1986 The Deutsche Bunsen-Gesellschaft.

structure that is completely different from other major lipids. It consists of an isooctyl side chain and a rigid cyclic four-ring with a single hydroxyl group.

One of the most interesting phenomena gained through lipid monolayer studies is the phase transition due to compression or temperature change [73]. Upon compression, the slope in the π–A isotherms becomes nearly horizontal, suggesting a first-order phase transition where condensed-phase domains are formed in a liquid-expanded phase, as directly visualized by fluorescence microscopy or BAM. The decrease of temperature also induces formation of the condensed-phase domains from the liquid-expanded phase when the molecular area is in the LE/LC coexistent range at the decreased temperature (Figure 3.47). Shape, size, and inplane distribution of such domains are very diverse, but their morphological features have been basically interpreted in terms of a competition between line

---

Figure 3.46 Structures of lipid molecules: (a) glycerophospholipid, (b) sphingomyelin (SM), and (c) ganglioside GM1, GM2, and GM3 as sphingoglycolipids, and (d) cholesterol. Molecular structures are drawn in reference to Ref. [71].

tension at the domain boundary and dipole–dipole electrostatic repulsion between film molecules within and also between domains [74–76]. The line tension corresponds to the energy per length of domain boundary and favors a compact domain shape to minimize the boundary length. In contrast, the intermolecular electrostatic repulsion, due to an arrangement of molecular dipoles parallel to the surface normal, tends to maximize the distance between film molecules and favors noncircular domain shapes. Additionally, the interdomain repulsion enforces regular arrangement of rather uniformly sized domains in hexagonal or lamellar stripes.

Phase diagrams for lipid monolayers give fruitful information on compositions and intermolecular interactions of lipid molecules at the air/water interface. Many phase diagrams have been reported for the number of monolayer systems through film-balance experiments, fluorescence or BAM observations, and X-ray diffraction experiments, etc. Among them, mixed monolayers containing cholesterol as one component have attracted long-standing interest because of their indispensible roles in biomembrane functions [77]. Systematic investigations have been performed by McConnell and collaborators for mixed monolayers of cholesterol and phospholipids [78] with saturated acyl chains in different lengths on the basis of surface balance experiments and epifluorescence microscopic observation. Binary mixed monolayers of cholesterol with phosphatidylcholines (PCs) having saturated shorter acylchains tended to form two coexisting liquid phases at low surface pressures (e.g., di(14:0)PC in Figure 3.48a). At higher surface pressures, the two liquid phases merged into one phase, and the diagram displays an upper miscibility critical point. In contrast, when cholesterol was mixed with a phosphatidylcholine having saturated longer acylchains, two upper critical points appeared (e.g., di(15:0)PC in Figure 3.48b). This unique phase diagram has been interpreted in terms of complex formation between cholesterol and phospholip-

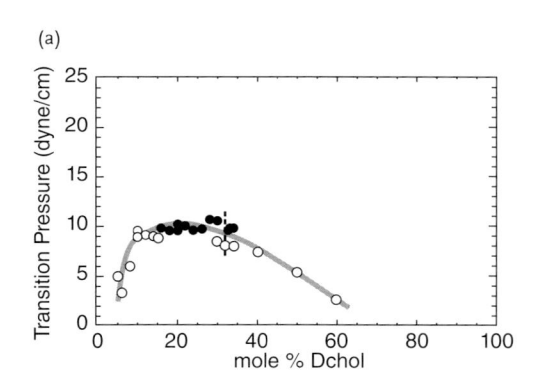

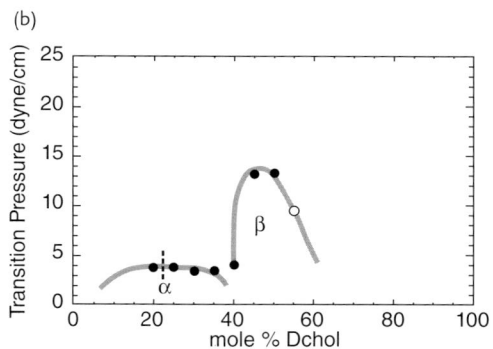

Figure 3.48 Phase diagrams for mixed monolayers of cholesterol and (a) di(14:0)PC at 23 °C and (b) di(15:0)PC at 33 °C. Here, dihydrocholesterol (Dchol) was used instead of cholesterol since Dchol is more resistant to air and photooxidation [78b, 79a]. It has been found that using Dchol instead of cholesterol yields slightly lower values of the phase-transition pressures, but the phase behavior of the two steroids is virtually identical [74f, 78e]. Reprinted with permission from Ref. [78b]. Copyright 2000 The American Chemical Society.

ids: the two two-phase regions, named α and β in order of increasing cholesterol concentration, were thought to correspond to the coexistent state of a cholesterol–PC complex and pure PC and that of the complex and pure cholesterol, respectively [79]. It should be noted here that the phase diagrams with a single, well-defined upper miscibility critical point have also been reported for other mixtures of cholesterol with PCs having two unsaturated fatty acid chains [79c], while the phase diagrams with two upper miscibility critical points have been observed for mixed monolayers of cholesterol with some SMs [79d, e]. The combinations of cholesterol and SMs are especially interesting in relation to the possible existence of membrane microdomains, called lipid rafts. It is believed that the lipid rafts are dynamic nanoscale assemblies enriched in cholesterol, sphingolipids, and proteins, and have important roles in various cell functions such as signaling and trafficking through membranes [80]. In order to understand molecular interactions and biological roles of lipids in the rafts as well as functions of lipid raft microdomains themselves from the viewpoint of physical chemistry, strenuous studies have continually taken place with the monolayer systems including ternary mixtures [81].

Lipid monolayers on solution surfaces, LB films immersed into solutions as well, have also been employed as platforms for adsorption of biomaterials such as peptides (e.g antimicrobial peptides [82], amyloid β peptides [83]), proteins [84], enzymes (e.g., lipase [85], alkaline phosphatase [86], and so on [87]). Biophysical investigations and mechanistic studies with such simple model systems contribute to fundamental understanding of interactions and/or reactions of biomaterials with lipid membranes and of biological processes related to biomembrane surfaces.

3.3.3.2 Lung Surfactants

There is a thin film of liquid inside the alveolus of the lung. Mammalian lung surfactant, a complex mixture of lipids and proteins (a lipo-protein complex), is a highly surface active material that forms a monolayer at the air/alveolus liquid interface. The lung surfactant prevents alveolar collapse during the breathing cycle by lowering the air/liquid surface tension during expiration (corresponding to monolayer compression) and respreading easily on inspiration [88]. A lack of the surfactant leads to neonatal respiratory distress syndrome (NRDS). The main component of the natural lung surfactant is DPPC, and the remainder is other monosaturated and unsaturated phospholipids and cholesterol [88a]. Additionally, there are four surfactant proteins (SPs), SP-A, SP–B, SP–C, and SP-D, playing important roles in lung functions [89]. SP-A and SP-D are large hydrophilic proteins and are believed to defend the lung against infections. In addition, SP-A has a crucial function of facilitating the rate of surfactant adsorption onto the air/liquid interface by participating in the formation of tubular myelin structures. SP–B and SP–C are hydrophobic small peptides, and thereby are highly surface active. They contribute to lowering the surface tension and are also important for facilitated adsorption of phospholipids. Two-dimensional structures and properties of lung surfactants at the air/water interface including monolayer stability and two-dimensional viscosity [90], and mechanical stimuli-induced dynamic behaviors such as monolayer-to-multilayer transition or squeeze-out during compression and contrary respreading during expansion are matters of much account [91]. Knowledge obtained from such researches will be fruitful for future development of surfactant therapies for NRDS.

3.3.3.3 Biomimetic Mineralization

Living organisms produce inorganic crystalline minerals with defined shape, size, and crystallographic orientation. The formation of biominerals is fairly complicated in nature, but it is virtually assured that organic matrix surfaces have crucial roles in oriented nucleation and growth, and thus final morphologies and functions of the crystals. A Langmuir monolayer at the air/water interface has served as an ideal platform for biomimetic mineralization because its simplified flat form and easy control of two-dimensional composition and density, are advantageous for mimicking elementary biomineralization processes in an artificial system (Figure 3.49). Studies are motivated not only by scientific issues aimed at elucidation of biomineralization mechanisms on membrane-like monolayers but also by their relevance in controlled development of biomimetic materials for applications in material science and engineering. A major component of mollusc shell, calcium carbonate ($CaCO_3$), is among the most intensively studied biominerals with the monolayer system. There exist three $CaCO_3$ polymorphs; calcite, aragonite, and vaterite. Highly oriented $CaCO_3$ crystal layers, which are interspread with thin sheets of an organic matrix to form a microlaminate structure in the mollusc shell, are usually composed of either aragonite or calcite [92]. Vaterite is an unstable polymorph and unusually related to biomineralization events such as shell regen-

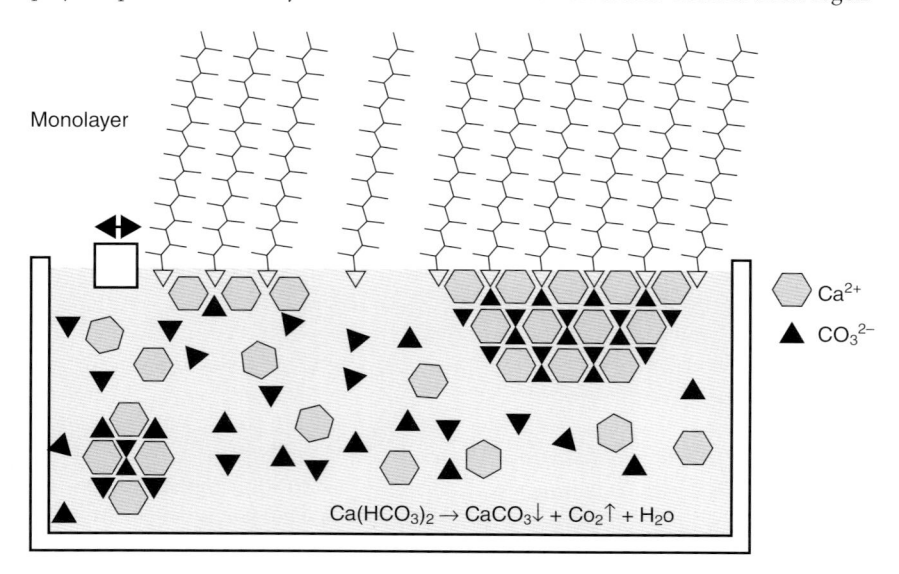

Figure 3.49 A schematic illustration of biomimetic crystallization of $CaCO_3$ under a Langmuir monolayer. Reprinted with permission from Ref. [92]. Copyright 2006 Springer.

eration, pearls and initial stages of shell formation [93]. So far, a variety of film substances, involving amphiphilic fatty acids [94], sulfate and phosphase [95], amines [94d, 96], phospholipids [94d, 97], and macrocyclic polyacids (calixarenes and resorcarenes) [92, 98], have been reported to successfully form $CaCO_3$ crystals beneath their monolayers. In a typical protocol for the $CaCO_3$ crystal growth experiment, the monolayer is spread onto a $Ca(HCO_3)_2$ solution prepared by bubbling CO_2 gas through a stirred aqueous solution of $CaCl_2/NaHCO_3$. The $CaCO_3$ crystals are then grown with a preferential orientation relative to the surface plane underneath the monolayer. As expected, the crystal polymorph, orientation, and morphology are largely influenced by experimental parameters such as the time elapsed after monolayer spreading (reaction time), concentration and ratio of ionic species including coexistent additives (e.g., Mg^{2+}) in the reaction solution, film materials forming the monolayers, the degree of monolayer compression, and so on. Alternatively the strict adjustment of these parameters leads to controlled formation of the crystals with well-defined crystallographic features and morphologies. In spite of a large number of earnest works, however, a universal mechanism for the monolayer-mediated oriented growth of $CaCO_3$ crystals is still under enthusiastic argument. An important aspect commonly derived from researches to date is that Langmuir monolayers work as soft dynamic templates for the crystallizations: the organic monolayer surfaces are more flexible than the inorganic ones, and thus film molecules in the monolayers would be rearranged at the mineralization interface to comply with the geometrical and electrostatic requirements of calcium ions for the crystallization in the reaction solution. Moreover, recent studies with a series of macrocyclic polyacid monolayers that form liquid-like phases on aqueous subphases have pointed out that nondirectional electric parameters, such as the average charge density or the mean dipole moment of the monolayer, are crucial to determine the orientation and the crystalline structures of the crystals (Figure 3.50), rather than the classical interpretation in terms of a geometrical and stereochemical complementarity between the head groups in the organic monolayer matrix and the calcium ions in the nucleated crystal face [92]. It has been also proposed, through a $CaCO_3$ crystallization study under monolayers of octadecanoic acid and its mixtures with octadecanol, that a cation-mediated hydrogen-bonded network including the surfactants, counter- and co-ions, is responsible for a face-selective nucleation of the crystals (Figure 3.51) [99].

3.3.4
Advanced Applications

3.3.4.1 Sensors
Voltammetric sensors responsive to anionic guests were constructed with LB films containing macrocyclic polyamine or cyclodextrin polyamine as anion receptor on glassy carbon electrode [100]. Binding of organic anion guests such as nucleotides, positional and geometric isomers of phthalate, and inorganic anions to the receptor molecules induced the change of marker-ion permeability through the film, which was detected by cyclic voltammetry. Although both LB films showed a

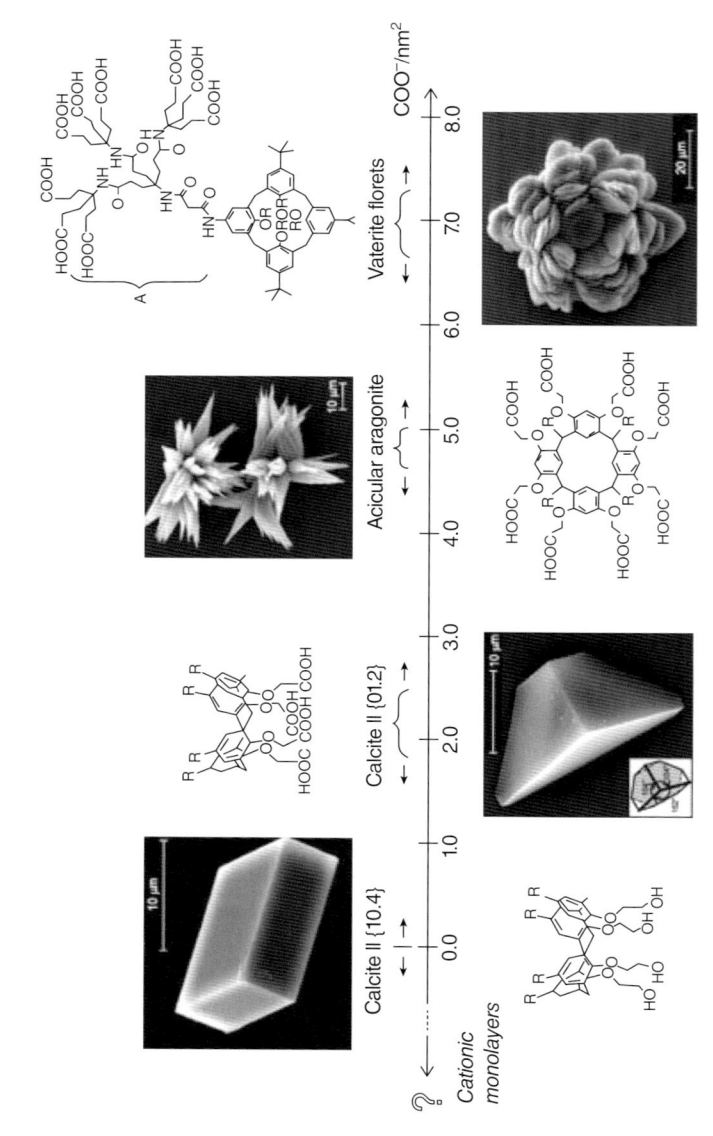

Figure 3.50 Overview of macrocyclic polyacid amphiphiles employed in studies on the $CaCO_3$ crystal growth beneath monolayers, and $CaCO_3$ crystals arranged according to the increasing negative charge density gained from Langmuir isotherms. Reprinted with permission from Ref. [92]. Copyright 2006 Springer.

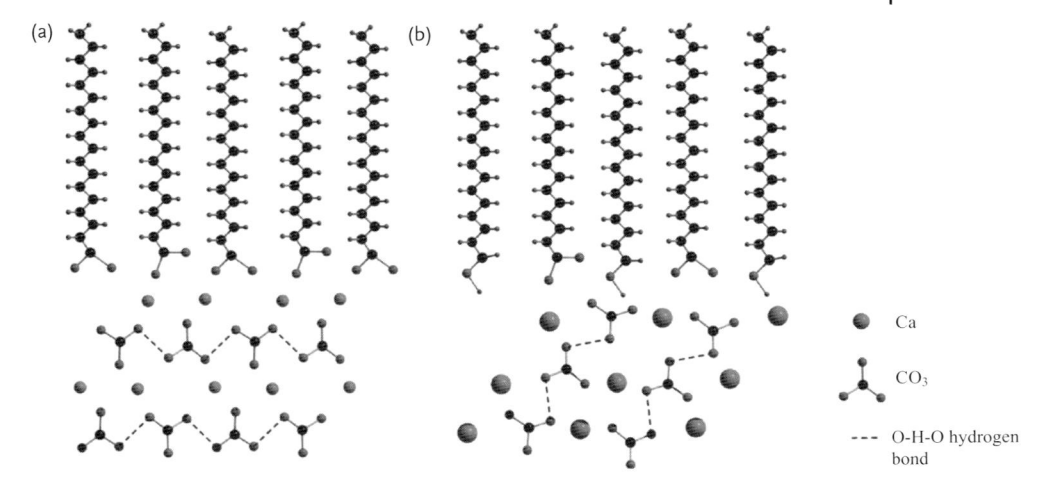

Figure 3.51 Schematic presentations of cation-mediated hydrogen-bonded network ideally formed under monolayers of (a) octadecanoic acid, and (b) a mixture of octadecanoic acid/octadecanol. Reprinted with permission from Ref. [99]. Copyright 2009 The American Chemical Society.

- Ca
- CO_3
- --- O-H-O hydrogen bond

similar trend of selectivity for most of the anionic guests examined, the selectivity for the positional isomers of phthalate was different, possibly due to the host–guest interaction involving the cyclodextrin cavity. Enzyme-incorporated LB films have also long been applied for development of thin-film sensors [101]. A glucose sensor with a lipid-coated enzyme (glucose oxidase) was prepared by the LB technique [102]. A benzene solution of the lipid–enzyme complex was spread on a water subphase, then the 2-layer LB film was deposited on a Pt electrode. The complex showed a good amperometric response and reproducibility, without causing denaturation, even after three months. Other examples involve LB film sensors fabricated with peptide lipids for selective detection of copper ions [103a], an isothiouronium-functionalozed amphiphile for dihydrogen phosphate ion [103b], mixtures of prophyrin and calyx [8]arene derivatives for nitrogen dioxide gas [103c], and mixed phytase-lipid films for phystic acid [103d].

As an interesting approach for the development of sensing elements with the LB technique, a molecular recognition-based electrochemical system for selective detection of a redox-active sugar derivative was fabricated with a mixed LB monolayer containing a boronic acid-functionalized electroconductive amphiphile on an electrode (Figure 3.52) [104]. Successive doping of the monolayer with iodine made the corresponding amphiphilic component conductive. As the redox-active sugar derivative, a mannose with a nitrobenzene moiety as the redox-active site, bound to the terminal boronic acid group through the molecular recognition, the redox current was selectively observed. This means that the amphiphile works as the molecular wire with the connecting terminal, and the system generates the electrical output signal through the wire when the terminal sensed the specific chemical species. This strategy would open up

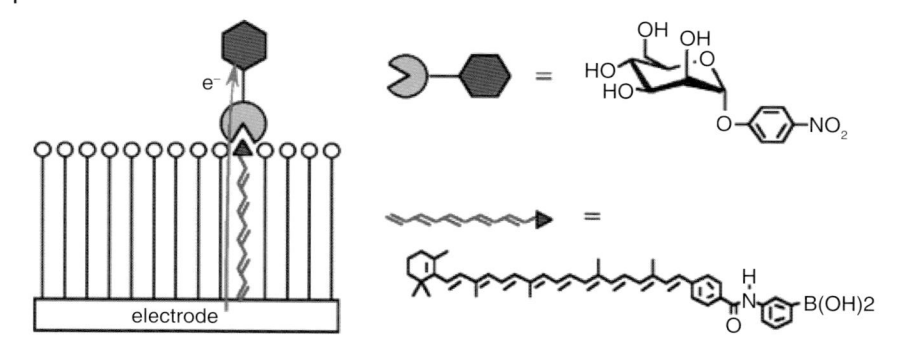

Figure 3.52 Sugar guest recognition by boronic acid-functionalized LB film on an electrode surface. Reprinted with permission from Ref. [104]. Copyright 2004 The American Chemical Society.

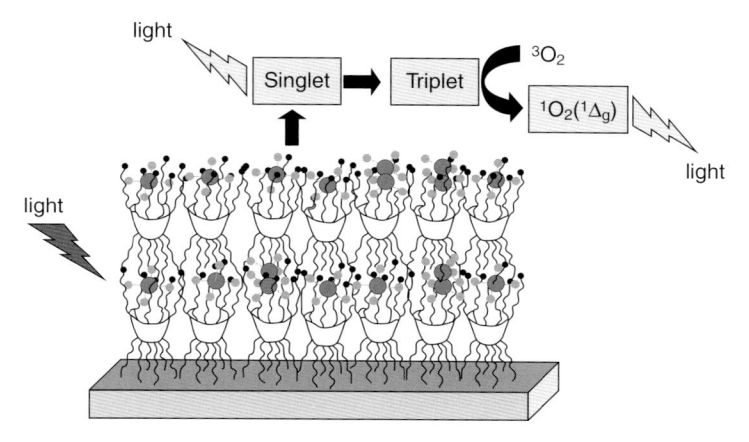

Figure 3.53 Schematic illustration of photoresponsive multilayer films fabricated by self-assembly of cationic amphiphilic CD and anionic porphyrins. Reprinted with permission from Ref. [107]. Copyright 2007 The Royal Society of Chemistry.

possibilities for designing molecular electronic sensors or devices based on molecular assembly at the interfaces.

3.3.4.2 Photoresponsive Films

Azobenzene derivatives are known to show *trans–cis* photoisomerization. Seki *et al.* evidenced that the conformational change of the *cis-trans* photochromic reaction in the LB films of side-chain-type azobenzene amphiphilic polymers induces a homeotropic–parallel alignment transition in the alignment of liquid crystal (so-called "command surface") [105]. Photochemical switching in conductive LB films fabricated amphiphilic molecules bearing with TCNQ and azobenzene unit was also reported, in which the photoconductivity of LB films was reversibly controlled by the photoinduced conformational change in the azobenzene unit [106].

Photoresponsive multilayer films were fabricated by self-assembling cationic amphiphilic CD and anionic porphyrins at the air/water interface [107]. The film deposited on hydrophobized quartz slides exhibited a good response to light excitation, as proven by fluorescence emission, triplet–triplet adsorption, and singlet oxygen photogeneration (Figure 3.53). The CD macrocycles were utilized for trapping the aqueous porphyrin molecules via electrostatic binding, and thus the CD cavity was not involved in the binding with the porphyrin. This means that the cavity is still available for incorporation of additional guest molecules, to make the present architectures intriguing platforms for the fabrication of more complex supramolecular assemblies on two-dimensional surfaces.

References

1 Henon, S., and Meunier, J. (1993) *J. Chem. Phys.*, **98**, 9148–9154.
2 Melzer, V., and Vollhardt, D. (1996) *Phys. Rev. Lett.*, **76**, 3770–3773.
3 (a) Vollhardt, D., and Melzer, V. (1997) *J. Phys. Chem. B*, **101**, 3370–3375; (b) Melzer, V., Vollhardt, D., Brezesinski, G., and Möhwald, H. (1998) *J. Phys. Chem. B*, **102**, 591–597.
4 (a) Hossain, M.M., Yoshida, M., and Kato, T. (2000) *Langmuir*, **16**, 3345–3348; (b) Hossain, M.M., Suzuki, T., and Kato, T. (2000) *Langmuir*, **16**, 9109–9112; (c) Hossain, M.M., and Kato, T. (2000) *Langmuir*, **16**, 10175–10183.
5 (a) Islam, M.N., and Kato, T. (2002) *J. Colloid Interface Sci.*, **252**, 365–372; (b) Islam, M.N., and Kato, T. (2005) *J. Colloid Interface Sci.*, **282**, 142–148; (c) Islam, M.N., and Kato, T. (2004) *Langmuir*, **20**, 6297–6301.
6 Gains, G.L., Jr. (1966) *Insoluble Monolayers at Liquid-Gas Interface*, Chap. 6, Interscience Publishers, New York, p. 283.
7 (a) Albrecht, O., Gruler, H., and Sackmann, E. (1978) *J. Phys. (Paris)*, **39**, 301–313; (b) Albrecht, O. (1983) *Thin Solid Films*, **99**, 227–234.
8 (a) Kato, T., Wakana, T., Ohshima, K., and Suzuki, K. (1989) *Bull. Chem. Soc. Jpn.*, **62**, 2492–2496; (b) Kato, T., Ohshima, K., and Suzuki, K. (1989) *Thin Solid Films*, **178**, 37–45; (c) Kato, T., Akiyama, H., and Yoshida, M. (1992) *Chem. Lett.*, **21**, 565–566.
9 Kato, T., Tatehana, A., Suzuki, N., Iimura, K., Araki, T., and Iriyama, K. (1995) *Jpn. J. Appl. Phys.*, **34**, L911–L914.
10 Shimizu, M., Yoshida, M., Iimura, K., Suzuki, N., and Kato, T. (1995) *Colloid Surface*, **102**, 69–73.
11 Mingotaud, A.-F., Mingotaud, C., and Patterson, L.K. (1993) *Handbook of Monolayers*, vol. 1 and 2, Academic Press.
12 Iimura, K., Yukari, Y., Tsuchiya, Y., Kato, T., and Suzuki, M. (2001) *Langmuir*, **17**, 4602–4609.
13 (a) Kato, T. (1990) *Langmuir*, **6**, 870–872; (b) Kato, T., Hirobe, Y., and Kato, M. (1991) *Langmuir*, **7**, 2208–2212.
14 Kato, T. (1990) *Jpn. J. Appl. Phys.*, **29**, L2102–L2104.
15 (a) Henon, S., and Meunier, J. (1991) *Rev. Sci. Instrum.*, **62**, 936; (b) Hoenig, D., and Moebius, D. (1991) *J. Phys. Chem.*, **95**, 4590.
16 Allara, D.L., and Nuzzo, R.G. (1985) *Langmuir*, **1**, 52–66.
17 Umemura, J., Kamata, T., Kawai, T., and Takenaka, T. (1990) *J. Phys. Chem.*, **94**, 62–67.
18 (a) Hasegawa, T., Takeda, S., Kawaguchi, A., and Umemura, J. (1995) *Langmuir*, **11**, 1236–1243; (b) Hasegawa, T. (2002) *J. Phys. Chem. B*, **106**, 4112–4115; (c) Hasegawa, T. (2007) *Anal. Bioanal. Chem.*, **388**, 7–15; (d) Hasegawa, T., Itoh, Y., and Kasuya, A. (2008) *Anal. Chem.*, **80**, 5630–5634.
19 (a) Dluhy, R.A., and Cornell, D.G. (1985) *J. Phys. Chem.*, **89**, 3195–3197; (b) Dluhy, R.A. (1986) *J. Phys. Chem.*, **90**, 1373–1379; (c) Dluhy, R.A., Wright, N.A., and Griffiths, P.R. (1988) *Appl. Spectrosc.*, **42**, 138–141; (d) Dluhy,

R.A., Mitchell, M.L., Pettenski, T., and Beers, J. (1988) *Appl. Spectrosc.*, **42**, 1289–1293.

20 (a) Blaudez, D., Buffeteau, T., Cornut, J.C., Desbat, B., Escafre, N., and Turlet, J.M. (1993) *Appl. Spectros.*, **47**, 869–874; (b) Blaudez, D., Buffeteau, T., Cornut, J.C., Desbat, B., Escafre, N., and Pezolet, M. (1994) *Thin Solid Films*, **242**, 146–150; (c) Blaudez, D., Turlet, J.-M., Dufourcq, J., Bard, D., Buffeteau, T., and Desbat, B. (1996) *J. Chem. Soc. Faraday Trans.*, **92**, 525–530; (d) Huo, Q., Dziri, L., Desbat, B., Russell, K.C., and Leblanc, R.M. (1999) *J. Phys. Chem. B*, **103**, 2929–2934; (e) Bellet-Amalric, E., Blaudez, D., Desbat, B., Graner, F., Gauthie, F., and Renault, A. (2000) *Biochim. Biophys. Acta*, **1467**, 131–143.

21 (a) Flach, C.R., Xu, Z., Bi, X., Brauner, J.W., and Mendelsohn, R. (2001) *Appl. Spectrosc.*, **55**, 1060–1066; (b) Bi, X., Taneva, S., Keough, K.M.W., Mendelsohn, R., and Flach, C.R. (2001) *Biochemistry*, **40**, 13659–13669; (c) Brauner, J.W., Flach, C.R., Xu, Z., Bi, X., Lewis, R.N.A.H., McElhaney, R.N., Gericke, A., and Mendelsohn, R. (2003) *J. Phys. Chem. B*, **107**, 7202–7211; (d) Flach, C.R., Brauner, J.W., Taylor, J.W., Baldwin, R.C., and Mendelsohn, R. (1994) *Biophys. J.*, **67**, 402–410; (e) Mendelsohn, R., Mao, G., and Flach, C.R. (2010) *Biochim. Biophys. Acta*, **1798**, 788–800.

22 Weers, J.G., and Scheuing, D.R. (1991) in *Fourier Transform Infrared Spectroscopy in Colloid and Interface Science* (ed. D.R. Scheuing), American Chemical Society, Boston, p. 91.

23 Shimomura, M., Song, K., and Rabolt, J.F. (1992) *Langmuir*, **8**, 887–893.

24 Lang, P. (1999) in *Amphiphiles at Interfaces Studied by Surface Sensitive X-Ray Scattering, in Modern Characterization Methods of Surfactant Systems* (ed. B.P. Binks), Marcel Dekker, pp. 377–415.

25 Als-Nielsen, J., Jacquenmain, D., Kjear, K., Leveiller, F., Lahav, M., and Leiserowitz, L. (1994) *Phys. Rep.*, **246**, 251–313.

26 Kenn, R.M., Biihm, C., Bib, A.M., Peterson, I.R., Mohwald, H., Als-Nielsen, J., and Kjaer, K. (1991) *J. Phys. Chem.*, **95**, 2092–2097.

27 Barraud, A., and Vandevyer, M. (1983) *Thin Solid Films*, **99**, 221–225.

28 (a) Albrecht, O., Matsuda, H., Eguchi, K., and Nakagiri, T. (1992) *Thin Solid Films*, **221**, 276–280; (b) Albrecht, O., Matsuda, H., Eguchi, K., and Nakagiri, T. (1996) *Thin Solid Films*, **284–285**, 152–156.

29 Barraud, A., Leloup, J., Gouzerh, A., and Palacin, S. (1985) *Thin Solid Films*, **133**, 117–123.

30 (a) Kato, T. (1987) *Jpn. J. Appl. Phys.*, **26**, L1377–L1380; (b) Kato, T. (1988) *Chem. Lett.*, **17**, 1993–1996.

31 Fukuda, K., Nakahara, H., and Kato, T. (1976) *J. Colloid Interface Sci.*, **54**, 430–438.

32 (a) Kato, T. (1988) *Jpn. J. Appl. Phys.*, **27**, L1358–L1360; (b) Kato, T. (1988) *Jpn. J. Appl. Phys.*, **27**, L2128–L2130.

33 Iwahashi, M., Naito, F., Watanabe, N., Seimiya, T., Morikawa, N., Nogawa, N., Ohshima, T., Kawakami, H., Ukai, K., Sugai, I., Shibata, S., Yasuda, T., Shoji, Y., Suzuki, T., Nagafuchi, T., Taketani, H., Matsuda, T., Fukushima, Y., Fujioka, M., and Hisatake, K. (1985) *Bull. Chem. Soc. Jpn.*, **58**, 2093–2098.

34 Kato, T., Matsumoto, N., Kawano, M., Suzuki, N., Araki, T., and Iriyama, K. (1994) *Thin Solid Films*, **242**, 223–228.

35 (a) Meyer, E.A., Castellano, R.K., and Diederich, F. (2003) *Angew. Chem. Int. Ed.*, **42**, 1210–1250; (b) Leblanc, R.M. (2006) *Curr. Opin. Chem. Biol.*, **10**, 529–536. Ariga, K., and Kunitake, T. (2006) *Supramolecular Chemistry -Fundamentals and Applications, Advanced Textbook*, Springer. Schneider, H.-J. (2009) *Angew. Chem. Int. Ed.*, **48**, 3924–3997.

36 Paleos, C.M., and Tsiourvas, D. (1997) *Adv. Mater.*, **9**, 695–710.

37 (a) Ariga, K., and Nakanishi, T. (2006) *J. P. Hill Soft Matter*, **2**, 465–477; (b) Ariga, K., Hill, J., and Endo, H. (2007) *Int. J. Mol. Sci.*, **8**, 864–883.

38 Onda, M., Yoshihara, K., Koyano, H., Ariga, K., and Kunitake, T. (1996) *J. Am. Chem. Soc.*, **118**, 8524–8530.

39 Sakurai, M., Tamagawa, H., Inoue, Y., Ariga, K., and Kunitake, T. (1997) *J. Phys. Chem.*, **101**, 4810–4816.

40 Kitano, H., and Ringsdorf, H. (1985) *Bull. Chem. Soc. Jpn.*, **58**, 2826–2828.

41 (a) Cha, X., Ariga, K., and Kunitake, T. (1996) *J. Am. Chem. Soc.*, **118**, 9545–9551; (b) Kurihara, K., Ohto, K., Honda, Y., and Kunitake, T. (1991) *J. Am. Chem. Soc.*, **113**, 5077–5079; (c) Koyano, H., Bissel, P., Yoshihara, K., Ariga, K., and Kunitake, T. (1997) *Langmuir*, **13**, 5426–5432; (d) Vollhardt, D., Liu, F., and Rud, R. (2005) *J. Phys. Chem. B*, **109**, 17635–17643; (e) Ebara, Y., Itakura, K., and Okahata, Y. (1996) *Langmuir*, **12**, 5165–5170.

42 (a) Higashi, N., Koga, T., Fujii, Y., and Niwa, M. (2001) *Langmuir*, **17**, 4061–4066; (b) Huo, Q., Sui, G., Zheng, Y., Kele, P., Leblanc, R.M., Hasegawa, T., Nishijo, J., and Umemura, J. (2001) *Chem. Eur. J.*, **7**, 4796–4804.

43 (a) Huo, Q., Dziri, L., Desbat, B., Russell, K.C., and Leblanc, R.M. (1999) *J. Phys. Chem.*, **103**, 2929–2934; (b) Miao, W., Du, X., and Liang, Y. (2003) *J. Phys. Chem.*, **107**, 13636–13642. Kim, Y.S., Chase, B., Kiick, K.L., and Rabolt, J.F. (2010) *Langmuir*, **26**, 336–343.

44 Oishi, Y., Torii, Y., Kato, T., Kuramori, M., Suehiro, K., Ariga, K., Taguchi, K., Kamino, A., Koyano, H., and Kunitake, T. (1997) *Langmuir*, **13**, 529–524.

45 (a) Nandi, N., and Vollhardt, D. (2003) *Chem. Rev.*, **103**, 4033–4075; (b) Vollhardt, D., Nandi, N., and Dutta Banik, S. (2011) *Phys. Chem. Chem. Phys.*, **13**, 4812–4829.

46 (a) Böhm, C., Möhwald, H., and Leiserowitz, L. (1993) *J. Als-Nielsen, K. Kjear, Biophys. J.*, **64**, 553–559; (b) Möhwald, H., Dietric, A., Böhm, C., Brezesinski, G., and Thoma, M. (1995) *Mol. Membr. Biol.*, **12**, 29–38.

47 (a) Stine, K.J., Uang, J.Y.-J., and Dingman, S.D. (1993) *Langmuir*, **9**, 2112–2118; (b) Parazak, D.P., Uang, J.Y.-J., Turner, B., and Stine, K.J. (1994) *Langmuir*, **10**, 3787–3793; (c) Huehnerfuss, H., Neumann, V., and Stine, K.J. (1996) *Langmuir*, **12**, 2561–2569; (d) Hoffmann, F., Huehnerfuss, H., and Stine, K.J. (1998) *Langmuir*, **14**, 4525–4534.

48 Melzer, V., Weidemann, G., Vollhardt, D., Brezesinski, G., Wagner, R., Struth, B., and Möhwald, H. (1997) *J. Phys. Chem. B*, **101**, 4752–4758.

49 Vollhardt, D., Gutberlet, T., Emrich, G., and Fuhrhop, J.-H. (1995) *Langmuir*, **11**, 2661–2668.

50 (a) Diamond, D., and McKervey, M.A. (1996) *Chem. Soc. Rev.*, **25**, 15–24; (b) Wallimann, P., Marti, T., Fürer, A., and Diederich, F. (1997) *Chem. Rev.*, **97**, 1567–1608.

51 (a) Baldini, L., Casnati, A., Sansone, F., and Ungaro, R. (2007) *Chem. Soc. Rev.*, **36**, 254–266; (b) Jose, P., and Menon, S. (2007) *Bioinorg. Chem. Appl.*, **2007**, 1–16.

52 Taneva, S., Ariga, K., Okahata, Y., and Takagi, W. (1989) *Langmuir*, **5**, 111–113.

53 (a) Chmurski, K., Bilewicz, R., and Jurczak, J. (1996) *Langmuir*, **12**, 6114–6118; (b) Parazak, D.P., Khan, A.R., D'Souza, V.T., and Stine, K.J. (1996) *Langmuir*, **12**, 4046–4049; (c) Kobayashi, K., Kajikawa, K., Sasabe, H., and Knoll, W. (1999) *Thin Solid Films*, **349**, 244–249; (d) Vico, R.V., Silvs, O.F., De Rossi, R.H., and Maggio, B. (2008) *Langmuir*, **24**, 7867–7874; (e) Flasinski, M., Broniatowski, M., Romeu, N.V., Dynarowicz-Latka, P., Moreno, A.G., Vilas, A.M., and Martin, M.L.G. (2008) *J. Phys. Chem. B*, **112**, 4620–4628; (f) Sallas, F., and Darcy, R. (2008) *Eur. J. Org. Chem.*, **2008**, 957–969.

54 (a) Kazantseva, Z.I., Lavrik, N.V., Nabok, A.V., Dimitriev, O.P., Nesterenko, B.A., Kalchenko, V.I., Vysotsky, S.V., Markovskiy, L.N., and Marchenko, A.A. (1997) *Supramol. Sci.*, **4**, 341–347; (b) Shinkai, S., and Ikeda, A. (1999) *Pure Appl. Chem.*, **71**, 275–280.

55 (a) Pietraszkiewicz, M., Prus, P., and Pietraszkiewicz, O. (2004) *Tetrahedron*, **60**, 10747–10752; (b) Guo, X., Lu, G.-Y., and Li, Y. (2004) *Thin Solid Films*, **460**, 264–268; (c) Weis, M., Janíček, R., Cirák, J., and Hianik, T. (2007) *J. Phys. Chem. B*, **111**, 10626–10631; (d) Liu, F., Lu, G.-Y., He, W.-J., Liu, M.-H., and Zhu, L.-G. (2004) *Thin Solid Films*, **468**, 244–249; (e) Shahgaldian, P., Pieles, U., and Hegner, M. (2005) *Langmuir*, **21**, 6503–6507.

56 Lednev, I.K., and Petty, M.C. (1996) *Adv. Mater.*, **8**, 615–630.

57 (a) Tsukanov, A.V., Dubonosov, A.D., Bren, V.A., and Minkin, V.I. (2008) *Chem. Heterocyclic Compounds*, **44**, 899–923; (b) Faridbod, F., Ganjali, M.R., Dinarvand, R., Norouzi, P., and Riahi, S. (2008) *Sensors*, **8**, 1645–1703.

58 Kimura, E., Kodama, M., and Yatsunami, T. (1982) *J. Am. Chem. Soc.*, **104**, 3182–3187.

59 (a) Ariga, K., Terasaka, Y., Sakai, D., Tsuji, H., and Kikuchi, J. (2000) *J. Am. Chem. Soc.*, **122**, 7835–7836; (b) Ariga, K., Nakanishi, T., Terasaka, Y., Tsuji, H., Sakai, D., and Kukuchi, J. (2005) *Langmuir*, **21**, 976–981; (c) Ariga, K., Nakanishi, T., Hill, J., Terasaka, Y., Sakai, D., and Kikuchi, J. (2005) *Soft Matter*, **1**, 132–137.

60 Aoyagi, N., Minamikawa, H., and Shimizu, T. (2004) *Chem. Lett.*, **33**, 860–861.

61 Azov, V.A., Beeby, A., Cacciarini, M., Cheetham, A.G., Diederich, F., Frei, M., Gimzewski, J.K., Gramlich, V., Hecht, B., Jaun, B., Latychevskaia, T., Lieb, A., Lill, Y., Marotti1, F., Schlegel, A., Schlittler, R.R., Skinner, P.J., Seiler, P., and Yamakoshi, Y. (2006) *Adv. Funct. Mater.*, **16**, 147–156.

62 Michinobu, T., Shinoda, S., Nakanishi, T., Hill, J.P., Fujii, K., Player, T.N., Tsukube, H., and Ariga, K. (2006) *J. Am. Chem. Soc.*, **128**, 14478–14479.

63 (a) Kuhn, H. (1972) *Chem. Phys. Lipids*, **8**, 401–404; (b) Möbius, D. (1978) *Ber. Bunsenges. Phys. Chem.*, **81**, 848–858; (c) Kuhn, H. (1979) *J. Photochem.*, **10**, 111–132; (d) Kuhn, H. (1979) *Pure Appl. Chem.*, **51**, 341–352.

64 Taiz, L., and Zeiger, E. (2010) *Plant Physiology, Fifth Edition, Photosynthesis: The Light Reactions*, Sinauer Associates, pp. 111–192.

65 (a) Fujihira, M., Nishiyama, K., and Yamada, H. (1985) *Thin Solid Films*, **132**, 77–82; (b) Fujihira, M., and Sakomura, M. (1989) *Thin Solid Films*, **179**, 471–476; (c) Fujihira, M. (1995) Photoinduced electron transfer in monolayer assemblies and its application to artificial photosynthesis and molecular devices, in *Organic Thin Films and Surfaces: Directions for the Nineties* (ed. A. Ulman), Academic Press, pp. 239–277.

66 (a) Echegoyen, L., and Echegoyen, K.L. (1998) *Acc. Chem. Res.*, **31**, 593–601; (b) Guldi, D.M. (2002) *Chem. Soc. Rev.*, **31**, 22–36; (c) Gust, D., Moore, T.A., and Moore, A.L. (2001) *Acc. Chem. Res.*, **34**, 40–48.

67 (a) Fujii, S., Moria, T., and Kimura, S. (2007) *Bioconjug. Chem.*, **18**, 1855–1859; (b) Fujii, S., Morita, T., and Kimura, S. (2008) *Langmuir*, **24**, 5608–5614.

68 (a) Chukharev, V., Tkachenko, N.V., Efimov, A., Guldi, D.M., Hirsch, A., Scheloske, M., and Lemmetyinen, H. (2004) *J. Phys. Chem. B*, **108**, 16377–16385; (b) Vuorinen, T., Kaunisto, K., Tkachenko, N.V., Efimov, A., Lemmetyinen, H., Alekseev, A.S., Hosomizu, K., and Imahori, H. (2005) *Langmuir*, **21**, 5383–5390; (c) Vivo, P., Vuorinen, T., Chukharev, V., Tolkki, A., Kaunisto, K., Ihalainen, P., Peltonen, J., and Lemmetyinen, H. (2010) *J. Phys. Chem., C* (114), 8559–8567; (d) Tkachenko, N.V., Vehmanen, V., Efimov, A., Alekseev, A.S., and Lemmetyinen, H. (2003) *J. Porphyrins Phthalocyanines*, **7**, 255–263; (e) Aleksee, A.S.A.S., Tkachenko, N.V., Efimov, A.V., and Lemmetyinen, H. (2010) *Russ. J. Phys. Chem. A*, **84**, 1230–1241.

69 Marczak, R., Sgobba, V., Kutner, W., Gadde, S., D'Souza, F., and Guldi, D.M. (2007) *Langmuir*, **23**, 1917–1923.

70 (a) Matsui, J., Mitsuishi, M., Aoki, A., and Miyashita, T. (2003) *Angew. Chem. Int. End.*, **42**, 2272–2275; (b) Matsui, J., Mitsuishi, M., Aoki, A., and Miyashita, T. (2004) *J. Am. Chem. Soc.*, **126**, 3708–3709; (c) Mitsuishi, M., Matsui, J., and Miyashita, T. (2009) *J. Mater. Chem.*, **19**, 325–329.

71 Voet, D., Voet, J.G., and Pratt, C.W. (2006) *Fundamentals of Biochemistry: Life at the Molecular Level*, 2nd edn, John Wiley & Sons, Inc.

72 (a) Ramstedt, B., and Slotte, J.P. (1999) *Biophys. J.*, **77**, 1498–1506; (b) Ramstedt, B., and Slotte, J.P. (2002) *FEBS Lett.*, **531**, 33–37.

73 (a) Knobler, C.M. (1990) *Adv. Chem. Phys.*, **77**, 397–449; (b) Knobler, C.M., and Desai, R.C. (1992) *Annu. Rev. Phys. Chem.*, **43**, 207–237.

74 (a) Weis, R.M., and McConnell, H.M. (1985) *J. Phys. Chem.*, **89**, 4453–4459; (b) Keller, D.J., Korb, J.P., and McConnell, H.M. (1987) *J. Phys. Chem.*, **91**, 6417–6422; (c) McConnell, H.M., and Moy, V.T. (1988) *J. Phys. Chem.*, **92**, 4520–4525; (d) McConnell, H.M. (1991) *Annu. Rev. Phys. Chem.*, **42**, 171–195; (e) Benvegnu, D.J., and McConnell, H.M. (1992) *J. Phys. Chem.*, **96**, 6820–6824; (f) Benvegnu, D.J., and McConnell, H.M. (1993) *J. Phys. Chem.*, **97**, 6686–6691; (g) Lee, K.Y.C., and McConnell, H.M. (1993) *J. Phys. Chem.*, **97**, 9532–9539.

75 (a) Heckel, W.M., and Möhwald, H. (1986) *Ber. Bunsenges. Phys. Chem.*, **90**, 1159–1167; (b) Helm, C.A., and Möhwald, H. (1988) *J. Phys. Chem.*, **92**, 1262–1266; (c) Möhwald, H. (1995) Phospholipid monolayers, Chapter 4, in *Handbook of Biological Physics*, vol. 1 (eds R. Lipowsky and E. Sackmann), Elsevier Science B.V., pp. 161–211.

76 (a) Mayer, M.A., and Vanderlick, T.K. (1994) *J. Chem. Phys.*, **100**, 8399–8407; (b) Mayer, M.A., and Vanderlick, T.K. (1995) *J. Chem. Phys.*, **103**, 9788–9794.

77 (a) Simons, K., and Ikonen, E. (2000) *Science*, **290**, 1721–1726; (b) Silvius, J.R. (2003) *Biochim. Biophys. Acta*, **1610**, 174–183; (c) Chong, P.L.-G., Zhu, W., and Venegas, B. (2009) *Biochim. Biophys. Acta*, **1788**, 2–11; (d) Róg, T., Pasenkiewicz-Gierula, M., Vattulainen, I., and Karttunen, M. (2009) *Biochim. Biophys. Acta*, **1788**, 97–121; (e) Harris, J.R. (ed.) (2010) *Cholesterol Binding and Cholesterol Transport Proteins: Structure and Function in Health and Disease*, vol. 51, Subcellular Biochemistry.

78 (a) Radhakrishnan, A., and McConnell, H.M. (1999) *J. Am. Chem. Soc.*, **121**, 486–487; (b) Keller, S.L., Radhakrishnan, A., and McConnell, H.M. (2000) *J. Phys. Chem.B*, **104**, 7522–7527.

79 (a) McConnell, H.M., and Radhakrishnan, A. (2003) *Biochim. Biophys. Acta*, **1610**, 159–173; (b) McConnell, H.M., and Vrljic, M. (2003) *Ann. Rev. Biophys. Biomol. Struct.*, **32**, 469–492; (c) Hagen, J.P., and McConnell, H.M. (1997) *Biochim. Biophys. Acta*, **1329**, 7–11; (d) Radhakrishnan, A., Li, X.-M., Brown, R.E., and McConnell, H.M. (2001) *Biochim. Biophys. Acta*, **1511**, 1–6; (e) Radhakrishnan, A., and McConnell, H.M. (2000) *Biochemistry*, **39**, 8119–8124.

80 (a) Simons, K., and Ikonen, E. (1997) *Nature*, **387**, 569–572; (b) Simons, K., and Toomre, D. (2000) *Nature Rev. Mol. Cell Biol.*, **1**, 31–39; (c) Binder, W.H., Barragan, V., and Menger, F.M. (2003) *Angew. Chem. Int. Ed.*, **42**, 5802–5827; (d) Hancock, J.F. (2006) *Nature Rev. Mol. Cell Biol.*, **7**, 456–462; (e) Lingwood, D., and Simons, K. (2010) *Science*, **327**, 46–50.

81 Stottrup, B.L., Stevens, D.S., and Keller, S.L. (2005) *Biophys. J.*, **88**, 269–276.

82 (a) Maget-Dana, R. (1999) *Biochim. Biophys. Acta*, **1462**, 109–140; (b) Zhang, L., Rozek, A., and Hancock, R.E.W. (2001) *J. Biol. Chem.*, **276**, 35814–35722; (c) Volinsky, R., Kolusheva, S., Berman, A., and Jelinek, R. (2006) *Biochem. Biophys. Acta*, **1758**, 1393–1407; (d) Bringezu, F., Majerowicz, M., Maltseva, E., Wen, S., Brezesinski, G., and Waring, A.J. (2007) *ChemBioChem*, **8**, 1038–1047.

83 (a) Maltseva, E., and Brezesinski, G. (2004) *ChemPhysChem*, **5**, 1185–1190; (b) Maltseva, E., Kerth, A., Blume, A., Möhwald, H., and Brezesinski, G. (2005) *ChemBioChem*, **6**, 1817–1824.

84 (a) Möhwald, H., Baltes, H., Schwendler, M., Helm, C.A., Brezesinski, G., and Haas, H. (1995) *Jpn. J. Appl. Phys.*, **34**, 3906–3913; (b) Huo, Q., and Leblanc, R.M. (2002) Langmuir and Langmuir-Blodgett films of proteins and enzymes, in *Encylopedia of Surface and Colloid Science* (ed. A. Hubbard), Marcel Dekker, Inc., pp. 2967–2996.

85 (a) Dahmen-Levison, U., Brezesinski, G., and Möhwald, H. (1998) *Progr. Colloid Polym. Sci.*, **110**, 269–274; (b) Balashev, K., Jensen, T.R., Kjear, K., and Bjørnholm, T. (2001) *Biochimie*, **83**, 387–397.

86 (a) Ronzon, F., Desbat, B., Chauvet, J.-P., and Roux, B. (2002) *Colloids Surf. B*, **23**, 365–373; (b) Caseli, L., Zaniquelli, M.E.D., Furriel, R.P.M., and Leone, F.A. (2002) *Colloids Surf. B*, **25**, 119–128; (c) Ronzon, F., Desbat, B., Chauvet, J.-P., and Roux, B. (2002) *Biochim. Biophys. Acta*, **1560**, 1–13; (d) Caseli, L.,

Furriel, R.P.M., de Andrade, J.F., Leone, F.A., and Zaniquelli, M.E.D. (2004) *J. Colloid Interface Sci.*, **275**, 123–130; (e) Caseli, L., Oliveira, R.G., Masui, D.C., Furriel, R.P.M., Leone, F.A., Maggio, B., and Zaniquelli, M.E.D. (2005) *Langmuir*, **21**, 4090–4095; (f) Caseli, L., Masui, D.C., Furriel, R.P.M., Leone, F.A., and Zaniquelli, M.E.D. (2007) *Thin Solid Films*, **515**, 4801–4807; (g) Nobre, T.M., de Sousa e Silva, H., Furriel, R.P.M., Leone, F.A., Miranda, P.B., and Zaniquelli, M.E.D. (2009) *J. Phys. Chem. B*, **113**, 7491–7497; (h) Clop, E.M., and Perillo, M.A. (2010) *Cell Biochem. Biophys*, **56**, 91–107.

87 (a) Mello, S.V., Mabrouki, M., Cao, X., Leblanc, R.M., Cheng, T.-C., and DeFrank, J.J. (2003) *Biomacromolecules*, **4**, 968–973; (b) Narayanan, R., Stottrup, B.L., and Wang, P. (2009) *Langmuir*, **25**, 10660–10665; (c) Goto, T.E., Lopez, R.F., Oliveira, O.N., Jr., and Caseli, L. (2010) *Langmuir*, **26**, 11135–11139; (d) Monteiro, D.S., Nobre, T.M., and Zaniquelli, M.E.D. (2011) *J. Phys. Chem. B*, **115**, 4801–4809.

88 (a) Notter, R.H. (ed.) (2000) *Lung Surfactants: Basic Science and Clinical Applications (Lung Biology in Health and Disease)*, vol. 149, Marcel Dekker, Inc, New York; (b) Wüstneck, R., Perez-Gil, J., Wüstneck, N., Cruz, A., Fainerman, V.B., and Pison, U. (2005) *Adv. Colloid Interface Sci.*, **117**, 33–58; (c) Serrano, A.G., and Pérez-Gil, J. (2006) *Chem. Phys. Lipids*, **141**, 105–118.

89 Veldhuizen, E.J., and Haagsman, H. (2000) *Biochim Biophys Acta*, **1467**, 255–270.

90 (a) Alonso, C., Waring, A., and Zasadzinski, J.A. (2005) *Biophys. J.*, **89**, 266–273; (b) Discher, B.M., Schief, W.R., Vogel, V., and Hall, S.B. (1999) *Biophys. J.*, **77**, 2051–2061; (c) Bringezu, F., Ding, J., Brezesinksi, G., and Zasadzinski, J.A. (2001) *Langmuir*, **17**, 4641–4648; (d) Takamoto, D.Y., Lipp, M.M., Von Nahmen, A., Lee, K.Y.C., Waring, A.J., and Zasadinski, J.A. (2001) *Biophys. J.*, **81**, 153–169; (e) Bringezu, F., Ding, J., Brezesinksi, G., Waring, A.J., and Zasadzinski, J.A. (2002) *Langmuir*, **18**, 2319–2325.

91 (a) Ding, J., Takamoto, D.T., Nahmen, A.J., Lipp, M.M., Lee, K.Y.C., Waring, A.J., and Zasadzinski, J.A. (2001) *Biophys. J.*, **80**, 2262–2272; (b) Yu, L.M.Y., Lu, J.J., Chiu, I.W.Y., Leung, K.S., Chan, Y.W., Zhang, L., Policova, Z., Hair, M.L., and Neumann, A.W. (2004) *Colloids Surf. B*, **36**, 167–176; (c) Alonso, C., Alig, T., Yoon, J., Bringezu, F., Warriner, H., and Zasadzinski, J.A. (2004) *Biophys. J.*, **87**, 4188–4202; (d) Wang, L., Cai, P., Galla, H.-J., He, H., Flash, C.R., and Mendelsohn, R. (2005) *Eur. Biophys. J.*, **34**, 243–254; (e) Ma, G., and Allen, H.C. (2006) *Langmuir*, **22**, 11267–11274; (f) Ma, G., and Allen, H.C. (2006) *Photochem Photobiol.*, **82**, 1517–1529; (g) Baoukina, S., Monticelli, L., Amrein, M., and Tieleman, D. (2007) *Biophys. J.*, **93**, 3775–3782.

92 Fricke, M., and Volkmer, D. (2007) *Top. Curr. Chem.*, **270**, 1–41.

93 Spann, N., Harper, E.M., and Aldridge, D.C. (2010) *Naturwissenschaften*, **97**, 743–751.

94 (a) Mann, S., Heywood, B.R., Rajam, S., and Birchal, J.D. (1988) *Nature*, **334**, 692; (b) Duffy, D.M., and Harding, J.H. (2002) *J. Mater. Chem.*, **12**, 3419–3425; (c) Loste, E., Díaz-Martí, E., Zarbakhsh, A., and Meldrum, F.C. (2003) *Langmuir*, **19**, 2830–2837; (d) Zhang, L.-J., Liu, H.-G., Feng, X.-S., Zhang, R.-J., Zhang, L., Mu, Y.-D., Hao, J.-C., Qian, D.-J., and Lou, Y.-F. (2004) *Langmuir*, **20**, 2243–2249; (e) Maas, M., Rehage, H., Nebel, H., and Epple, M. (2007) *Colloid Polym. Sci.*, **285**, 1301–1311; (f) Chen, Y., Xiao, J., Wang, Z., and Yang, S. (2009) *Langmuir*, **25**, 1054–1059; (g) Lendrum, C., and McGrath, K.M. (2009) *Cryst. Growth. Des.*, **9**, 4391–4400; (h) Lendrum, C., and McGrath, K.M. (2010) *Cryst. Growth. Des.*, **10**, 4463–4470.

95 Heywood, B.R., and Mann, S. (1994) *Chem. Mater.*, **6**, 311.

96 Mann, S., Heywood, B.R., Rajam, S., Walker, J.B.A., Davey, R.J., and Birchall, J.D. (1990) *Adv. Mater.*, **2**, 257–261.

97 (a) Huang, F., Shen, Y., Xie, A., Yu, S., Chen, L., Zhang, B., and Chang, W. (2009) *Cryst. Growth Des.*, **9**, 722–727; (b) Xiao, J., Wang, Z., Tang, Y., and Yang, S. (2010) *Langmuir*, **26**, 4977–4982.

98 (a) Volkmer, D., Fricke, M., Vollhardt, D., and Siegel, S. (2002) *J. Chem. Soc. Dalton Trans.*, 4547–4554; (b) Volkmer, D., Fricke, M., Agena, C., and Mattay, J. (2002) *Cryst. Eng. Commun.*, **4**, 288–295; (c) Volkmer, D., and Fricke, M. (2003) *Z. Anorg. Allg. Chem.*, **629**, 2381–2390; (d) Volkmer, D., Fricke, M., Agena, C., and Mattay, J. (2004) *J. Mater. Chem.*, **14**, 2249–2259; (e) Fricke, M., Volkmer, D., Krill, C.E., Kellermann, M., and Hirsch, A. (2006) *Cryst. Growth Des.*, **6**, 1120–1123.

99 Lendrum, C., and McGrath, K.M. (2009) *Cryst. Growth Des.*, **9**, 4391–4400.

100 Nagase, S., Kataoka, M., Naganawa, R., Komatsu, R., Odashima, K., and Umezawa, Y. (1990) *Anal. Chem.*, **62**, 1252–1259.

101 (a) Petty, M.C. (1991) *J. Biomed Eng*, **13**, 209–214; (b) Arisawa, S., Arise, T., and Yamamoto, R. (1992) *Thin Solid Films*, **209**, 259–263; (c) Shunichi, A., and Ryoichi, Y. (1992) *Thin Solid Films*, **210–211**, 443–445; (d) Zhu, D.-G., Cui, D.-F., and Petty, M.C. (1993) *Sens. Actuators B Chem.*, **12**, 111–114; (e) Dubrovsky, T., Vakula, S., and Nicolini, C. (1994) *Sens. Actuators B Chem.*, **22**, 69–73; (f) Wan, K., Chovelon, J.M., and Jaffrezic-Renault, N. (2000) *Talanta*, **52**, 663–670.

102. (a) Okahata, Y., Tsuruta, T., Ijiro, K., and Ariga, K. (1988) *Langmuir*, **4**, 1373–1374; (b) Okahata, Y., Tsuruta, T., Ijiro, K., and Ariga, K. (1989) *Thin Solid Films*, **180**, 65–72.

103 (a) Zheng, Y., Orbulescu, J., Ji, X., Andreopoulos, F.M., Pham, S.M., and Leblanc, R.M. (2003) *J. Am. Chem. Soc.*, **125**, 2680–2686; (b) Misawa, Y., Kubo, Y., Tokita, S., Ohkuma, H., and Nakahara, H. (2004) *Chem. Lett.*, **33**, 1118–1119; (c) de Miguel, G., Martín-Romero, M.T., Pedrosa, J.M., Muñoz, E., Pérez-Morales, M., Richardson, T.H., and Camacho, L. (2007) *J. Mater. Chem.*, **17**, 2914–2920; (d) Caseli, L., Moraes, M.L., Zucolotto, V., Ferreira, M., Nobre, T.M., Zaniquelli, M.E.D., Filho, U.P.R., and Oliveira, O.N., Jr. (2006) *Langmuir*, **22**, 8501–8508.

104 Miyahara, T., and Kurihara, K. (2004) *J. Am. Chem. Soc.*, **126**, 5684–5685.

105 (a) Seki, T., and Ichimura, K. (1989) *Thin Solid Films*, **179**, 77–83; (b) Seki, T., Tamaki, T., Suzuki, Y., Kawanishi, Y., Ichimura, K., and Aoki, K. (1989) *Macromolecules*, **22**, 3505–3506; (c) Seki, T., Sakuragi, M., Kawanishi, Y., Tamaki, T., Fukuda, R., Ichimura, K., and Suzuki, Y. (1993) *Langmuir*, **9**, 211–218; (d) Sekkata, Z., Büchela, M., Orendia, H., Knoblocha, H., Seki, T., Itoc, S., Kobersteina, J., and Knoll, W. (1994) *Opt. Commun.*, **111**, 324–330.

106 Tachibana, H., Nakamura, T., Matsumoto, M., Komizu, H., Manda, E., Niino, H., Yabe, A., and Kawabata, Y. (1989) *J. Am. Chem. Soc.*, **111**, 300–3081.

107 Valli, L., Giancane, G., Mazzaglia, A., Scolaro, L.M., Conoci, S., and Sortino, S. (2007) *J. Mater. Chem.*, **17**, 1660–1663.

4
Layer-by-Layer (LbL) Assembly
Katsuhiko Ariga

4.1
Concept and Mechanism

In the previous chapters, two representative methods for preparation of organized thin films were introduced. The self-assembled monolayer (SAM) technique is an excellent method to provide organized monolayers that can be immobilized on a surface. However, it is not basically a technique to prepare well-organized multilayer films. The Langmuir–Blodgett (LB) method is powerful method to construct multilayers films. Unfortunately, it is not applicable to a wide range of materials, although the number of examples of LB films with nonlipid components is increasing. In this chapter, a layer-by-layer (LbL) technique is introduced as a versatile method for multilayer preparation. Unlike films obtained from the former two techniques, films prepared by the LbL method are rather disorganized. However, the LbL technique provides multilayer films with a very simple and inexpensive procedure where many kinds of materials can be used. Therefore, the LbL method is regarded as a much more versatile technique for multilayer preparation, as compared with SAM and LB techniques.

Now, various kind of interactions have been used for multilayer assembly by the LbL adsorption technique. In the early history of this technique, electrostatic interactions were mainly used for layer adsorption. The concept of LbL adsorption was first proposed by Iler. Somewhat later, Decher and coworkers demonstrated layer-by-layer assemblies of polyelectrolytes and/or bola-type amphiphiles through electrostatic assembly. This concept was spread to many researchers who developed LbL films using various components including biological and inorganic materials. Similarly, the other molecular interactions such as metal coordination and hydrogen bonding started to be utilized for the LbL technique. Use of the LbL method is not limited to thin-film preparation on flat solid surfaces. Coated colloids and hollow capsules are fabricated through LbL assembly on invisible colloidal particles.

The details of various LbL methods will be described later in this chapter. Fundamental aspects and typical examples are shown here. Figure 4.1 illustrates layer-by-layer assembly between cationic polyelectrolytes and anionic particles

Organized Organic Ultrathin Films: Fundamentals and Applications, First Edition. Edited by Katsuhiko Ariga.
© 2013 Wiley-VCH Verlag GmbH & Co. KGaA. Published 2013 by Wiley-VCH Verlag GmbH & Co. KGaA.

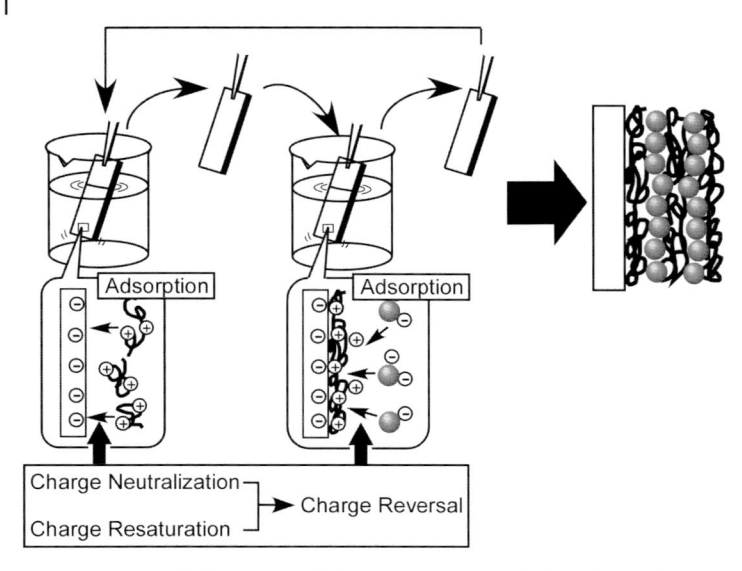

Figure 4.1 Layer-by-layer assembly between cationic polyelectrolyte and anionic particle.

through electrostatic interaction. In this example, the solid substrate has anionic charges on its surface. When this solid substrate is immersed in a solution containing cationic polyelectrolyte, the cationic polyelectrolyte starts to adsorb on the solid surface through electrostatic interaction. Negative charges on the solid surface are neutralized through adsorption of the cationic polyelectrolytes, but the adsorption does not stop at the neutral point. In most cases, overadsorption occurs, which lead to reversion of charges. Therefore, the surface charge of the solid substrate changes to positive. This overadsorption is spontaneously diminished via unfavorable electrostatic repulsion. Therefore, adsorption of cationic polyelectrolytes is basically equilibrated in a certain range. This reversion of surface charges was confirmed by surface force analyses. The surface-force measurement revealed that surface charges were reversed upon adsorption of polyelectrolytes onto countercharged surface above a certain level of polyelectrolyte concentrations.

According to this fundamental mechanism of LbL assembly, surface charges in adsorbates become essential. As a wide range of materials have certain surface charges, many kinds of materials can be immobilized on the surface in the layer-by-layer mode. In addition to electrostatic interaction, several interactions such as hydrogen bonding and metal coordination have been used for the LbL technique, as shown later, the materials range available for the LbL method becomes wider and wider. Therefore, this method is now used over a very wide research area including basic physics and practical biomedical fields. In this chapter, the versatile usage and high potential of the LbL technique is explained, including some recent advanced examples.

4.2
Preparation and Characterization

4.2.1
Applicable Materials and Interactions

Prior to an explanation of the detailed procedures of the LbL assembly and related characterization, materials applicable for the LbL method are overviewed here. Conventional and functional polyelectrolytes are the most frequently used materials in the LbL procedure. Figure 4.2 summarizes typical polyelectrolytes used in LbL assembly, poly(allylamine hydrochloride) (PAH), poly(diallyldimethylammonium chloride) (PDDA), and poly(ethyleneimine) (PEI) as cationic species and poly(sodium styrenesulfonate) (PSS), poly(sodium vinylsulfonate) (PVS), and poly(acrylic acid) (PAA) as anionic species. As described later, these polyelectrolytes were assembled to provide a precursor film on a solid substrate. The precursor film provides sufficient charge to the surface and helps further LbL assembly of target materials. Because the LbL method is an easy method to construct ultrathin films, functional polymers are often used to fabricate thin-film devices. For example, layered assemblies of conjugate polymers, poly(p-phenylenevinylene) (PPV), were used as a light-emitting diode. A cationic precursor of PPV was first assembled with polyanion, and then thermal treatment created conjugate chains of PPV. Nonlinear optical effect in alternately assembled films between azobenzene-containing polycation and polyanion was demonstrated. Light-sensitive polyelectrolytes can be assembled for preparation of micropatterned thin films. The shape of the polyelectrolyte is not limited to a linear one. For example, poly(amidoamine)

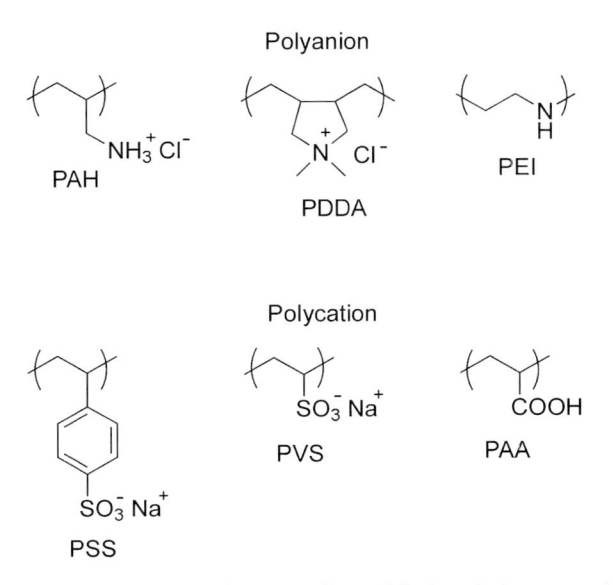

Figure 4.2 Polyelectrolytes typically used for layer-by-layer assembly.

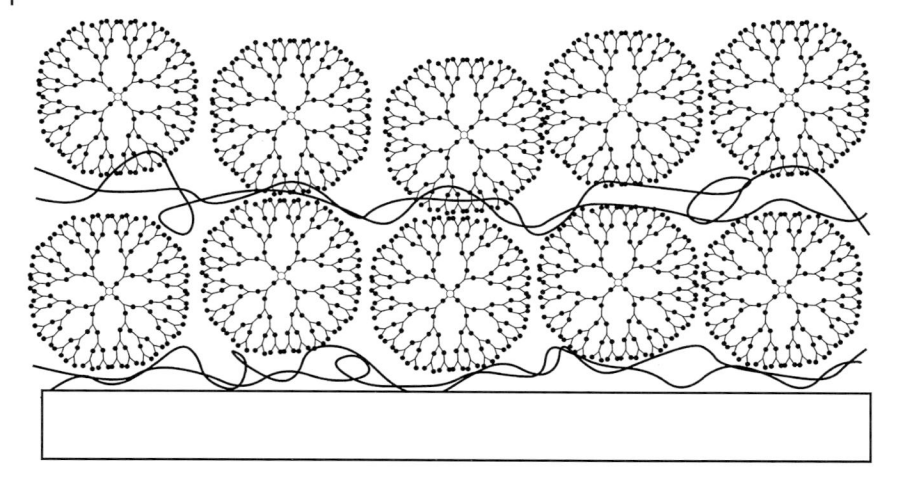

Figure 4.3 Layer-by-layer assembly between polyelectrolyte and dendrimer.

dendrimer (PAMAM) with positive charges at the correct pH, can be alternatively assembled with anionic PSS (Figure 4.3). Because dendrimers have interior space for inclusion of guest materials such as metal ions, this method is also effective to immobilize various substances in thin films.

Because charged biopolymers can be regarded as biological polyelectrolytes, many kinds of biomolecules including proteins, nucleic acids (deoxyribonucleic acid DNA and ribonucleic acid RNA), and charged polysaccharides are used in the LbL assembly. Most of the biopolymers are water soluble and/or have high affinity to water because of exposure of charged sites to external phase. This characteristic of the biopolymers is apparently appropriate for electrostatic LbL assembly. In addition, the biopolymers are simply adsorbed in the LbL assembly from their aqueous solution; that is, the assembling procedure is not conducted under severe physical conditions. Unnecessary disturbance should be minimized. According to these advantages, the LbL assemblies of proteins, especially enzymes, have been widely researched. For example, the wide applicability of the LbL assembly to aqueous proteins was confirmed by systematic studies based on a quartz crystal microbalance, which can precisely detect mass change based on the film assembly through shifts of resonant frequencies. In most of the cases tested, the assembly process of the proteins alternated with counterionic polyelectrolyte proceeded for unlimited numbers of cycles with high reproducibility. The estimated thickness of the protein assembly corresponds to the dimension of the adsorbed proteins. This is an important feature for optimizing the reactor structure, as demonstrated in a later section of this chapter. Other charged biological materials can also be assembled with suitable partner polyelectrolytes. Negatively charged DNA can be assembled with various cationic materials. Charged polysaccharides, chitosan and chondroitin sulfate, were assembled with polyanion and polycation, respectively. These materials are expected to be highly useful in biomedical fields. Therefore, biomedical applications of LbL assemblies have received much attention in recent researches. As a remarkable example, alternate assembly between a charged virus and polyelectrolyte was also demonstrated.

Charged inorganic materials are one of the best materials for electrostatic LbL assembly. The LbL assembly based on electrostatic interaction was originally proposed by Iler for negative and positive colloidal particles such as silica and alumina, and he estimated the layer thickness qualitatively by the interference color. Although direct assembly between negatively and positively charged silica particles was demonstrated by a quartz crystal microbalance (QCM), such direct assembly of conformationally rigid components such as colloidal particles and clays is basically difficult. Therefore, use of intermediate layers of flexible polyelectrolytes as an electrostatic glue is usually effective for successful assemblies of rigid inorganic materials. Spherical nanoparticles are frequently used as components for LbL assembly in order to construct defined structure of functional inorganic materials for fine device preparation. LbL assembly of nanoparticles with defined size always provides uniform films. However, the detailed assembling process doe not seem to be so simple. For example, Kotov and coworkers reported detailed observation of the surface image of assembled films of yttrium iron garnet (YIG) nanoparticles with PDDA, where they noticed the two kinds of assembly modes in nanoparticle deposition; that is, layer growth and lateral growth mode. The latter mode should be avoided for preparation of a sophisticated layered structure. Grafting of the charged polymer on the nanoparticles significantly increases particle–polyelectrolyte and particle–particle interaction, resulting in formation of a densely packed particle layer and avoidance of the unfavorable lateral growth deposition. Assembly of a large nonspherical inorganic substance by the LbL method was also reported. The LbL assembly of cationically modified zeolite crystals and PSS for assembling components was demonstrated. Successful assembly of the two-dimensional perovskite sheets with polyelectrolytes was also reported. The nanosheet-like clay objects can be good materials to be assembled by the LbL technique. Recently, Sasaki and coworkers extensively studied the LbL assembly of various nanosheet materials. For example, positively charged nanosheets with a lateral dimension of micrometers synthesized by directly delaminating a well-crystallized Mg–Al layered double hydroxide (LDH) could be assembled layer-by-layer with an anionic polymer, PSS, onto the solid surface to produce ultrathin nanocomposite films, demonstrating their usefulness as a positively charged lamellar nanoblock. The other inorganic materials such as nanorods and nanotubes are also used as components of LbL assembly as well as inorganic quantum materials including quantum dots and quantum. In addition, bulk materials with nanoregulated structures such as mesoporous materials are also assembled into a layer-by-layer structure.

Some kinds of assemblies of low molecular weight substances can behave like colloidal particles and can be therefore assembled into thin films through a LbL process. In particular, assemblies of functional elements such as supramolecular objects and nanomaterials have been investigated. Spherical molecular assemblies such as micelles and vesicles are also subjects for LbL assembly. For example, the LbL assembly between micelles with a polyanion and a polycation corona,

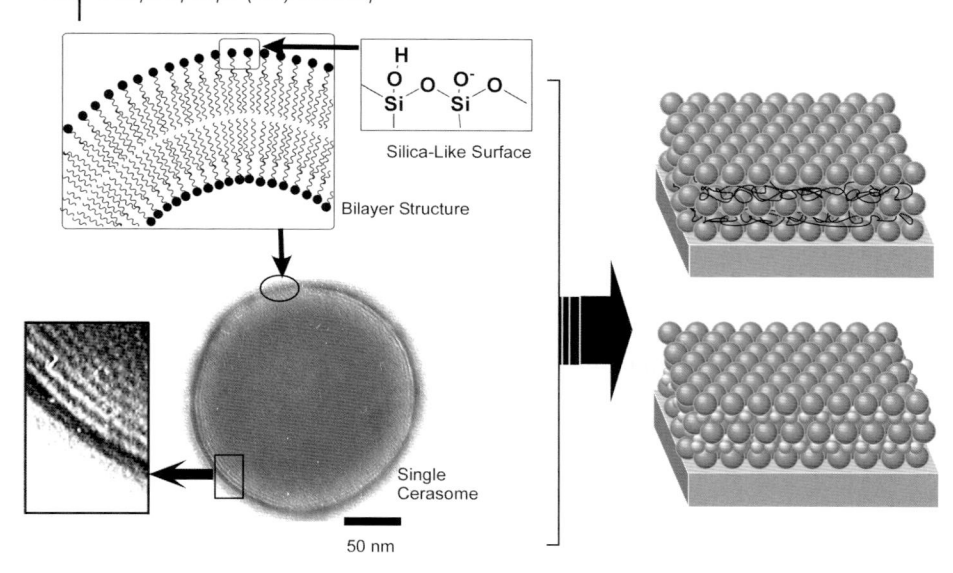

Figure 4.4 Layer-by-layer assembly of organic–inorganic hybrid vesicles.

poly(4-vinylpyridine)-b-poly(tert-butyl acrylate) and poly(acrylic acid)-b-poly(4-vinylpyridine), was realized without the aid of intermediate polyelectrolytes. The embedded layers can include hydrophobic functional molecules within their hydrophobic cores, affording great opportunities for immobilization of organic molecules in organized films. Vesicles, liposomes and their related structures of lipid-like components were also reported as components applicable to the LbL technique. This type of assembly was successfully demonstrated through the LbL assembly of lipid vesicles through stabilizing the assembly structures (Figure 4.4). This example used a superstable organic-inorganic vesicle, cerasome, in which silane-bearing amphiphiles form a silica network at the polar headgroup upon sol-gel reaction together with spontaneous formation of a bilayer structure resulted in cell-like vesicle structures. Anionic cerasome can be assembled by the LbL procedure with a cationic polyelectrolyte. In addition, successful LbL assembly between the anionic cerasome and the cationic cerasome without the aid of polyelectrolytes was demonstrated without rupture of the vesicular structures. The latter assemblies are expected to be used as multicellular mimics. The incorporation of phospholipid vesicles in the LbL films composed of poly (glutamic-acid) (PGA) and PAH was also demonstrated. These LbL structures were later used as a reactor for biomineralization.

Of course, flat molecular assemblies can be assembled into layered films through the LbL technique. One of the initial examples in the history of the LbL films was assembly between polyelectrolyte and bola-amphiphile films. Bola-amphiphiles have charged groups at both ends of the hydrophobic moiety such as a polymethylene chain and form a planar monolayer assembly upon exposure to aqueous phase (Figure 4.5). This planar film can be assembled alternately with counterionic

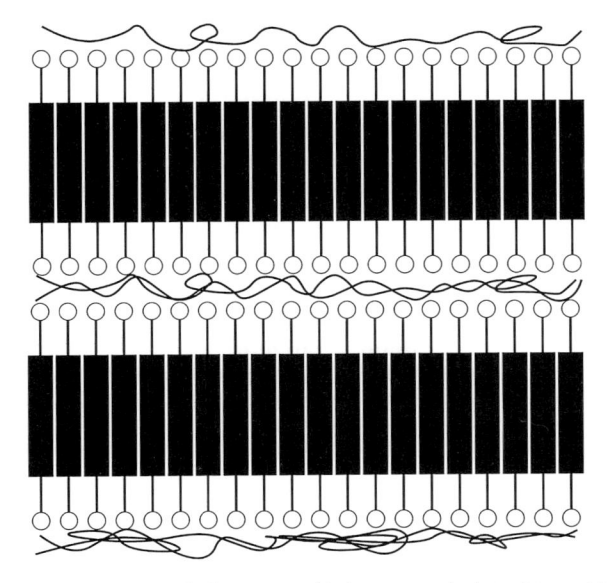

Figure 4.5 Layer-by-layer assembly between polyelectrolyte and bola amphiphile.

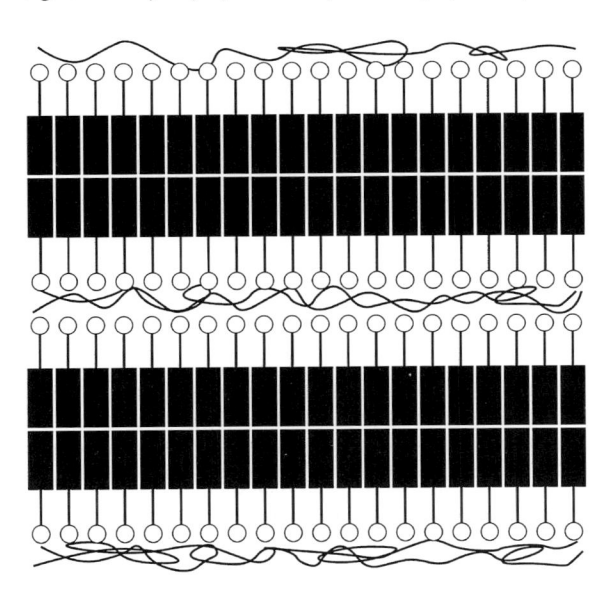

Figure 4.6 Layer-by-layer assembly between polyelectrolyte and lipid bilayer structure.

polyelectrolyte. Without covalent connection between two charged groups, amphiphilic compounds such as lipids and surfactants form a planar membrane by forming bilayer structures. The bilayer structures are formed through attachment of hydrophobic faces of the monolayer structures (Figure 4.6), where covalent bonding between hydrophobic parts does not exist. In spite of such unconnected structural motif, LbL assembly between lipid bilayers and polyelectrolyte were successfully demonstrated. The lipid molecule used for this example has a dye-like moiety in the middle of the structure. Stacking between the dye-like moieties is believed to stabilize nonbonded bilayer structures during the LbL process. Similarly, the Langmuir–Blodgett (LB) process was integrated with the LbL procedure. Bilayer LB films and polyelectrolytes can be assembled into thin films by alternate operation of LB and LbL processes. The LBL procedures for molecular assemblies are not limited to lipid-like structures. Successful LbL assemblies of dye molecules, whose structures are far from those of typical lipid structures, was also demonstrated. Because most of the dye molecules have an aromatic core with several charged groups, the dye molecules can adsorb on the surface of the polyelectrolyte layer and form stacking structures. Exposure of the charged groups after dye adsorption to the outer side is important for continuous electrostatic LbL assemblies.

4.2.2
Thin-Film Preparation: Fundamental Procedure and Characterization

The previous section describes the materials aspects of the LbL method. Here, its technical aspects are summarized. In the initial stage of the LbL history, LbL films were characterized by rather tedious techniques such as small-angle X-ray scattering (SAXS). Application of simpler apparatuses such as the quartz crystal microbalance (QCM) enables us to analyze the LbL process more conveniently and smoothly. For example, the resonant frequency of the QCM device sensitively changes in proportion to the mass adsorption on its electrodes, which enable us to evaluate molecular film adsorption with nanogram precision. Roughly speaking, a frequency change of 1 Hz corresponds to ca. 0.9 ng of adsorption over the entire electrode when we use the QCM system with 9 MHz of fundamental frequency.

In order to overview the LbL assembling process, several examples of LbL assembly of silica (SiO_2) nanoparticles alternately with polyelectrolyte are shown here. Prior to assembling SiO_2 particles, formation of three or four pairs of LbL assemblies between conventional cationic and anionic polyeletrolytes on the solid support (QCM plate in this case) is usually required. These preformed layers can be regarded as precursor layers that are supposed to form a well-charged uniform surface on the solid support. Next, the LbL adsorption processes between anionic SiO_2 nanoparticles and cationic polyelectrolyte (PDDA) were quantitatively investigated using the QCM technique. Figure 4.7 illustrates a practical procedure for this QCM measurement. Adsorption of SiO_2 nanoparticles or cationic polyelectrolyte (PDDA) was conducted on the QCM plates covered with the precursor layers. The frequency shifts of the QCM device were monitored in air after the assembled thin films were washed and dried. Figure 4.8 shows typical examples of QCM frequency shifts upon the LbL assemblies between nanoparticles and polyelectrolytes, where the QCM frequency shifts relative to the assembling processes of three sizes of SiO_2 particles (25, 45, and 78 nm in diameter) with polycation PDDA are exhibited. The QCM frequency changes showed two kinds of steps, larger and

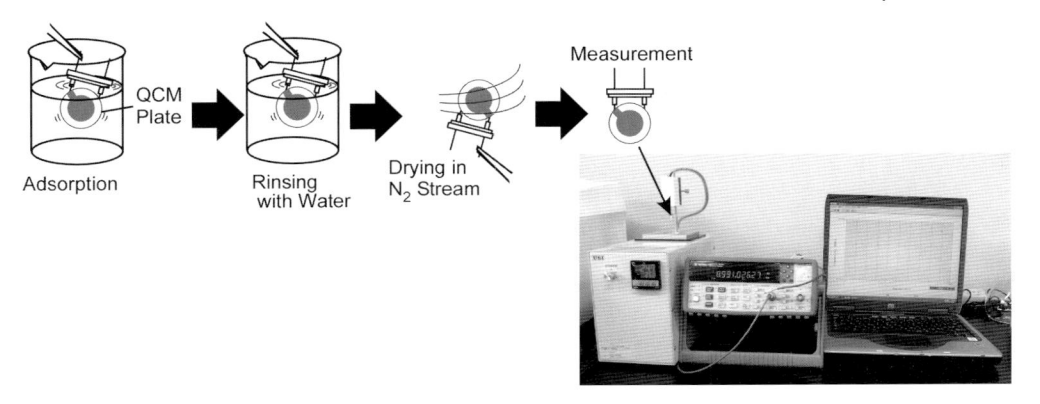

Figure 4.7 Evaluation of layer-by-layer assembly by QCM.

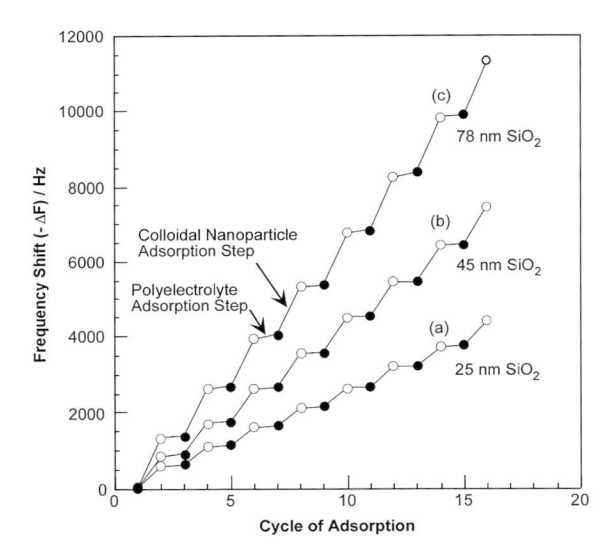

Figure 4.8 QCM frequency shifts on layer-by-layer assembly between polyelectrolyte and silica nanoparticle.

smaller steps, corresponding to adsorption of SiO_2 particles and PDAA, respectively. The sizes of growth steps clearly depends on the size of SiO_2 particles and the frequency shift of 520, 910, and 1350 Hz were assigned for adsorption of 25-, 45-, and 78-nm SiO_2 particles, respectively. These values are calculated to ca. 70% coverage of the surface on the assumption of hexagonal packing of the particles.

Figure 4.9 illustrates the less significant effect of adsorption time on SiO_2 adsorption process where the frequency shifts upon assembly between SiO_2 particles and PDDA through successively lessening to one-half of the previous adsorption time. Judging from the QCM frequency shifts, the film growth remained exactly the same as the adsorption time was changed from 20 min to 15 s. This

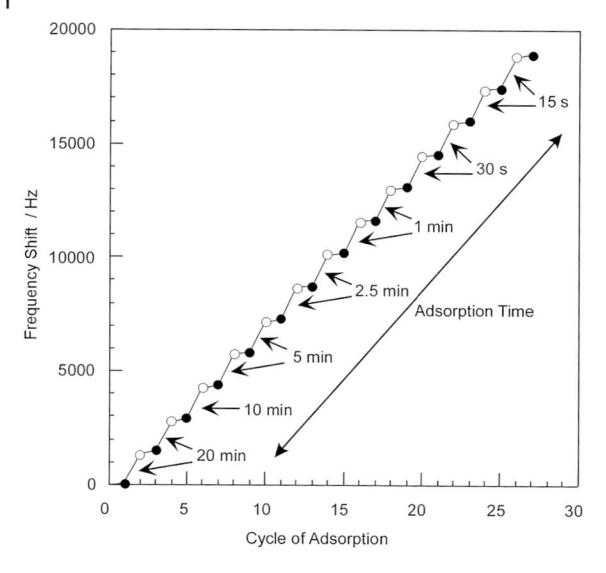

Figure 4.9 QCM frequency shifts on layer-by-layer assembly between polyelectrolyte and silica nanoparticle. with different adsorption times.

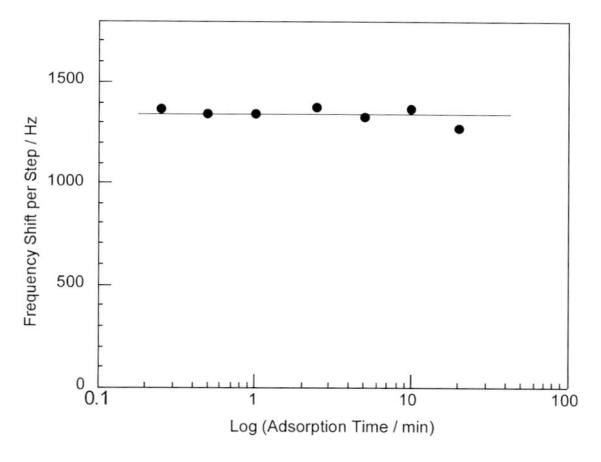

Figure 4.10 Effect of adsorption time on amount of adsorbed silica nanoparticle in layer-by-layer assembly.

characteristic is also indicated by Figure 4.10 that plots the adsorption amounts (frequency shifts) as a function of the adsorption time. This behavior is apparently different from those observed for adsorption of the other substances such as polyelectrolytes, proteins, dye assemblies, which usually needs 10 to 20 min to get saturation of the adsorption processes. These facts imply the necessity to investigate adsorption kinetics for efficient adsorption. Optimum adsorption times depend on the nature of substances for adsorption.

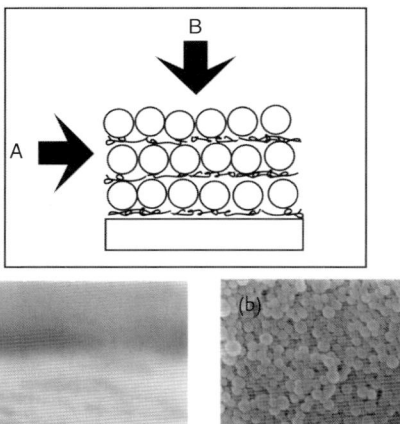

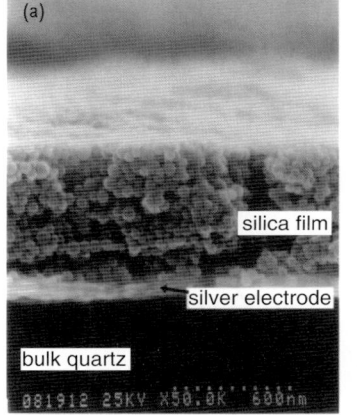

Figure 4.11 SEM images of layer-by-layer assembly between polyelectrolyte and silica nanoparticle.

The morphologies of the LbL films obtained from SiO_2 and PDDA were also investigated using scanning electron microscopy (SEM) and atomic force microscopy (AFM). Film morphology was sensitively tuned by the SiO_2 concentration and ionic strength of the SiO_2 solution, the latter can be adjusted by addition of appropriate inorganic salts. After many trials, the best quality film with a smooth surface and constant thickness was obtained from $10\,mg\,mL^{-1}$ of aqueous SiO_2 with 0.1-M NaCl. The morphologies of films prepared under this condition were closely analyzed by SEM observation. The cross-sectional image of the film (Figure 4.11a) indicated that the SiO_2 particles are closely packed in the layer, but long-range ordering is not found. The top view (Figure 4.11b) shows well-packed particles in a two-dimensional plane that is also confirmed by AFM in the noncontact mode. An AFM image of the PSS-PDDA precursor film (Figure 4.12a) demonstrates a highly flat morphology of the polyelectrolyte assembly. A height difference of only 1 nm was observed over 500 nm in the horizontal direction. In the AFM image of a SiO_2-PDDA film, where individual SiO_2 particles are clearly seen and where height profiling of the image shows a height difference of 10–20 nm (Figure 4.12b). The latter value is in reasonable agreement with the actual particle radius of 23 nm.

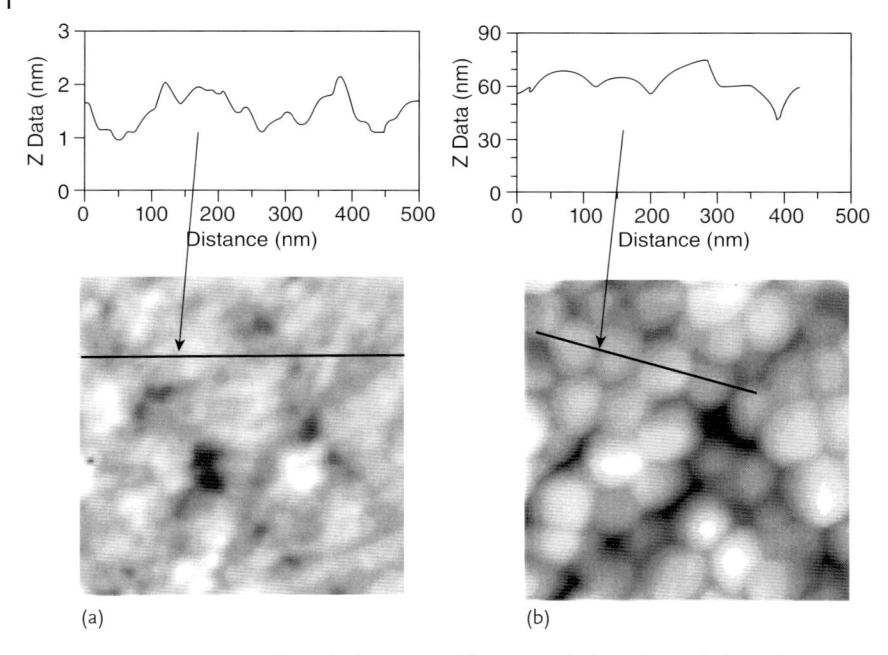

Figure 4.12 AFM images of layer-by-layer assemblies: (a) polyelectrolyte/polyelectrolyte; (b) polyelectrolyte/silica nanoparticle.

Next, modulation of film structures through the LbL assemblies is demonstrated using various SiO_2 particles. One of the most pronounced advantages of the LbL assembly is its excellent freedom of film construction. Figure 4.13 demonstrates the consecutive assembly of two kinds of layers in a separated layer mode, where the first four layers were made from 45-nm SiO_2 and the following six layers were assembled with 25-nm SiO_2. Growth steps for the two sizes of SiO_2 particles reasonably depend on the size of SiO_2 nanoparticles and both the growth profiles are highly constant. A neighboring layer mode was also demonstrated. In this assembly, a four-unit layer (45-nm SiO_2/PDDA/25-nm SiO_2/PDDA) was repeatedly assembled. As seen in Figure 4.14, individual growth steps in this heterogeneous film are identical to the corresponding adsorption steps in the assembly in the separated layer mode, which means that the adsorption steps are not affected by previous adsorption sequences. Hetero-type particle assembly was also conducted using both inorganic nanoparticles and biomaterial. A multilayer assembly with alternation of 45-nm SiO_2 and anionic glucose oxidase (GOD) with an intermediate PDDA layer, {SiO_2/PDDA/GOD/PDDA}n. Similarly to the previous examples, precise repetition of frequency shifts was confirmed. These demonstrations clearly indicated the basic independency of every adsorption process. This characteristic allows us to freely construct multilayer films with desired layering sequences and components through the LbL assembly.

The above-mentioned examples give the impression of huge freedom of film construction with rather simple procedure. Such characteristics are true in many

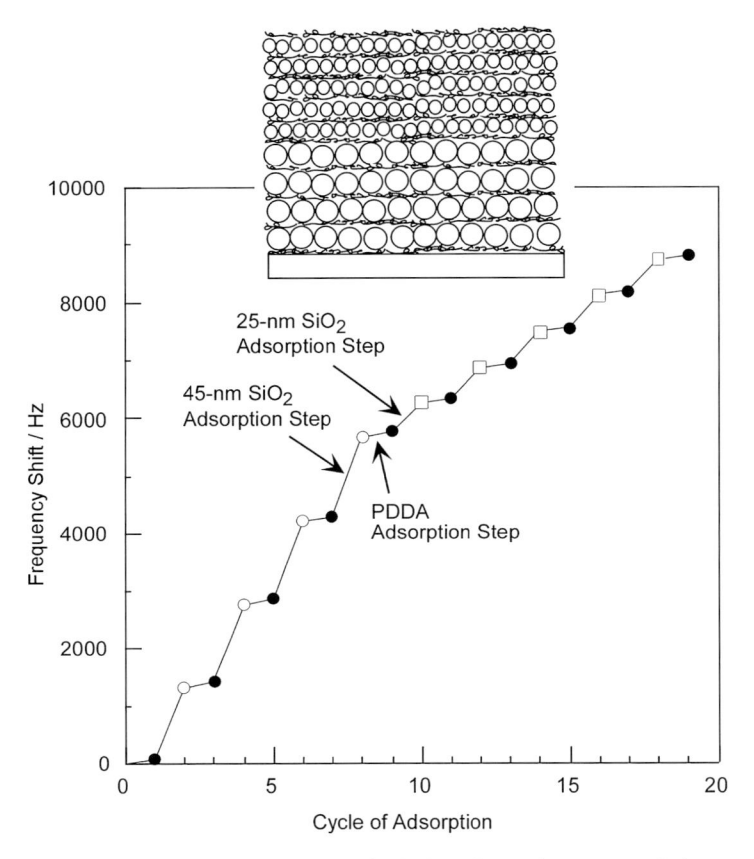

Figure 4.13 Layer-by-layer assembly of polyelectrolyte and two types of silica nanoparticle in a separated layer mode.

cases, not limited to assemblies of inorganic nanomaterials such as SiO_2 nanoparticles. There are also many modifications in assemblies that will be described in the next section, Prior to descriptions of modified methods, characterizations of the LbL assemblies are generally described here. As summarized in Figure 4.15, generally speaking, analyses on the LbL films are similar to those for the other thin films. In addition to the QCM technique described above, regular film growth can be detected by UV-Vis and FT-IR spectroscopies if the adsorbed materials have specific absorption bands. More generally, the thickness of the adsorbed layers can be evaluated by X-ray reflectivity, surface plasmon resonance (SPR), and scanning angle reflectometry (SAR). As exemplified previously, the morphology of the LbL films is evaluated by various microscopies including scanning electron microscopy (SEM), transmission electron microscopy (TEM), and scanning probe microscopies (SPMs).

In addition to these standard methods, various specialized analyses are available depending on assembling components. LbL thin films incorporating metallic

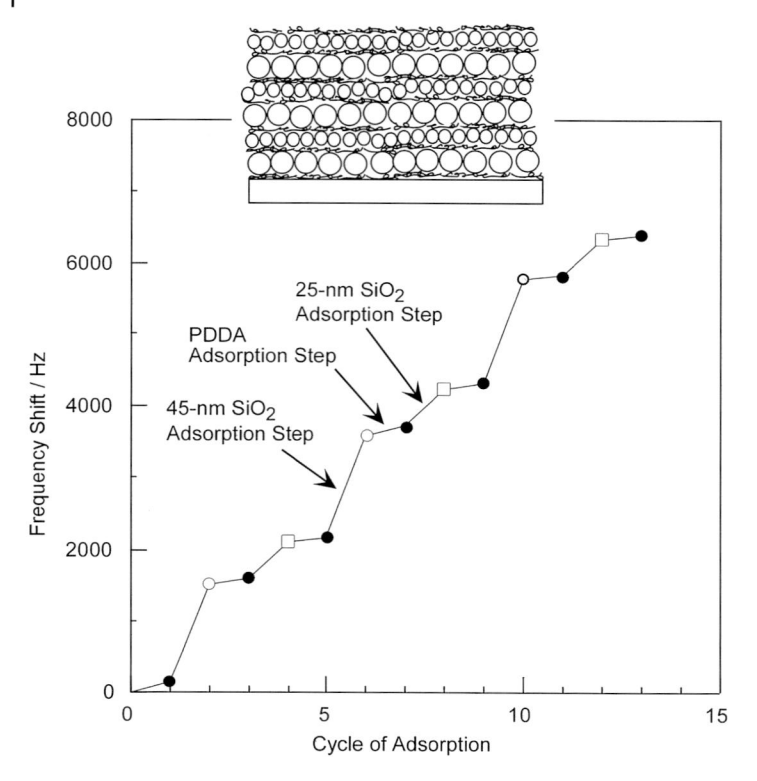

Figure 4.14 Layer-by-layer assembly of polyelectrolyte and two types of silica nanoparticle in an alternate layer mode.

nanoparticles can be examined for their use as substrates with surface-enhanced Raman scattering (SERS), which is a phenomenon involving large increases in Raman scattering cross sections of molecules adsorbed at the surfaces of nanometer-scale metallic particles. The surface plasmon resonance (SPR) peak is used for evaluation of the level of metallic nanoparticle aggregation. For *in situ* analyses on the LbL assembles, ellipsometry, quartz crystal microbalance with dissipation (QCM-D), and dual-polarization interferometry (DPI) are also used for the LbL thin films, the latter is a technique used to measure changes in both the thickness and the refractive index of adsorbed layers. The LbL films for sensor applications often include electroactive components. In such cases, cyclic voltammetry (CV) becomes a powerful method for evaluation of the films' redox behaviors.

4.2.3
Various Driving Forces and Techniques

Means to assemble LbL films have been widely developed in which various driving forces and film-preparation methods are used. These expansions of the assembly methodology are useful to prepare variously functionalized films.

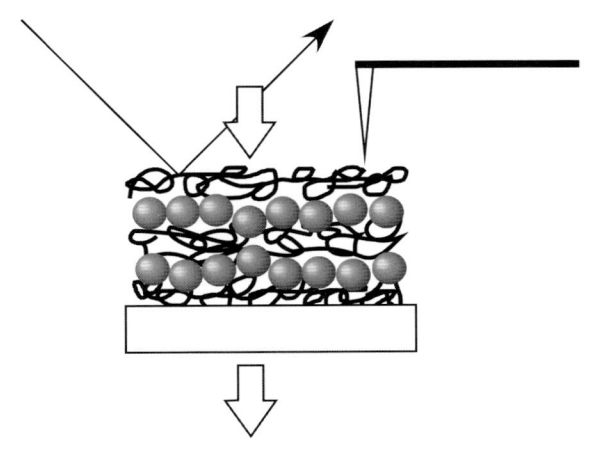

X-ray reflectivity
Surface plasmon resonance (SPR)
Scanning angle reflectometry (SAR)
Surface-enhanced Raman scattering (SERS)
Scanning electron microscopy (SEM)
Transmission electron microscopy (TEM)
Scanning probe microscopies (SPMs)

Figure 4.15 Analytical methods for layer-by-layer assemblies.

As described above, the most popular driving force for the LbL assemblies is electrostatic interaction. However, the driving force of the LbL assembly is not limited to electrostatic interactions. Many kinds of physicochemical interactions are actually available in the modified LbL techniques. For example, LbL assembly based on metal–ligand interaction and/or metal coordination has a long research history comparable with that of electrostatic LbL methods, which was initiated by Mallouk and coworkers based on metal–phosphate interaction (Figure 4.16). Various kinds of metal–ligand interactions such as palladium–pyridine complex have been used. This kind of interaction is often used to assemble rigid aromatic chromophore moieties in the LbL structures because many aromatic cores with heteroatoms are capable of forming coordination bonds. Hydrogen bonding has been also used as a driving force for the LbL assembly. Unlike electrostatic interaction, hydrogen bonding can be formed between selective functional groups where matched pair formation between hydrogen-bond donor and hydrogen-bond acceptor is crucial. This highly specific nature in interaction is advantageous for defining the direction and orientation of functional groups within the assembled films. For example, noncentrosymmetric layered structure was obtained through intermolecular hydrogen-bonding interactions (Figure 4.17), which align chromophore molecules head-to-tail and preferentially perpendicular to the substrate. The obtained thin films show second-harmonic generation (SHG) response.

Introduction of covalent bonding to the LbL technique would significantly strengthen the assembly structure, which is highly advantageous for practical

Figure 4.16 Layer-by-layer assembly based on metal–phosphate interaction.

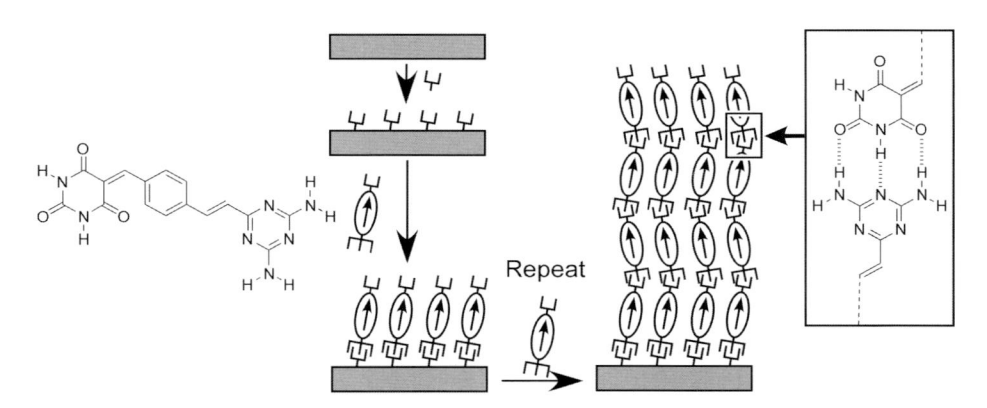

Figure 4.17 Layer-by-layer assembly based on hydrogen bonding for nonlinear optics.

applications. For example, photoreaction diazonium salt was used for stable LbL multilayer films that are more resistant to etching by organic solvents or solutions. Click chemistry was also used to form covalent for LbL assembly (Figure 4.18). The LbL assembly was performed by sequentially exposing the substrates to poly(acrylic acid) with either azide and alkyne functionality solutions containing copper sulfate, finally providing covalently linked ultrathin films with designed structures.

Layer-by-layer adsorption processes can be done in a step-wise mode of electrochemical coupling. For example, electrochemical coupling of N-alkyl carbazole was applied in a LbL process using electrochemical stimuli from an electrode surface

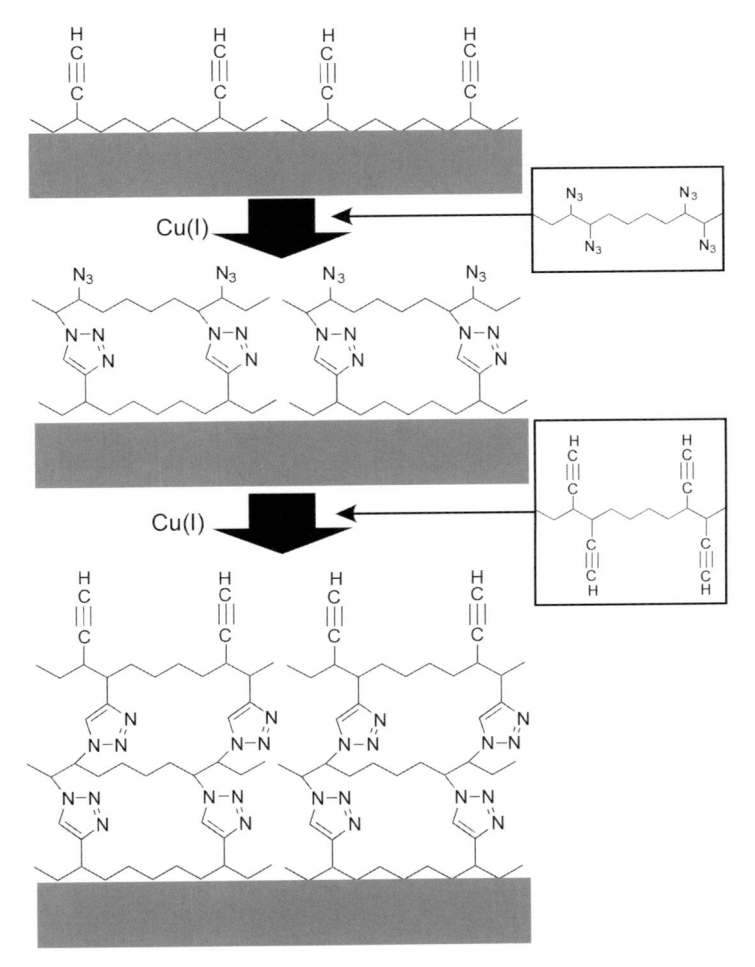

Figure 4.18 Layer-by-layer assembly through covalent bonds by a click reaction.

beneath the films without the use of additional reactants (Figure 4.19). Because N-alkyl carbazole and its dimer has a large hole transport mobility, electrochemical signals can be transmitted to the top layer in their films. This reagentless clean process is especially useful for constructing covalently linked layer-controlled thin films on a sensitive device surface. Several carbazole derivatives carrying distinctive donors such as porphyrins and acceptor moieties such as C_{60} are subjected to controlled layering of monomers in both homo- and heteroassemblies. Assembled films have fine flat structures and their photocurrent conversion device application was demonstrated as illustrated in Figure 4.20. The anodic photocurrent increased monotonically with increasing positive bias at the ITO electrode from $-0.5\,V$ to $2.0\,V$, whereas the dark current remained almost constantly zero, indicating photoconductive behavior.

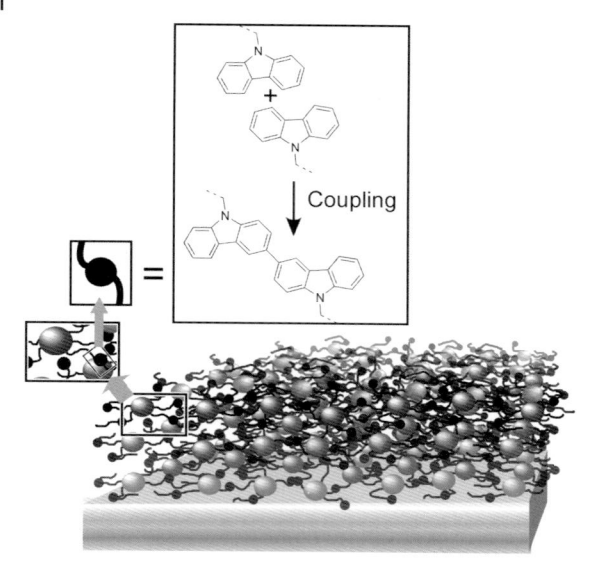

Figure 4.19 Electrochemical coupling layer-by-layer (ECC-LbL) assembly.

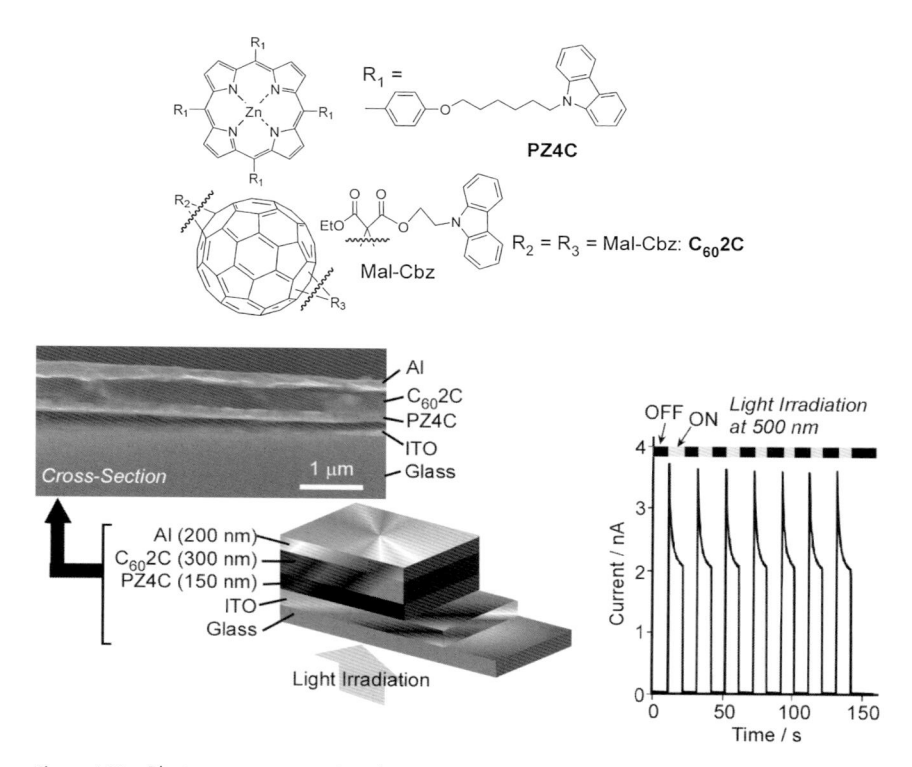

Figure 4.20 Photocurrent conversion device constructed by ECC-LbL assembly.

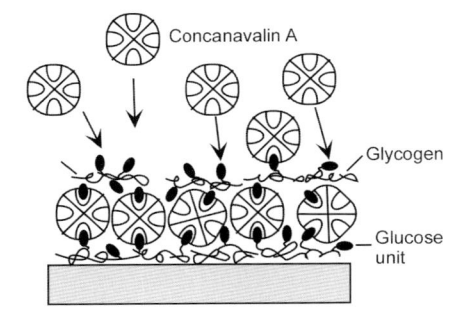

Figure 4.21 Layer-by-layer assembly based on biospecific recognition.

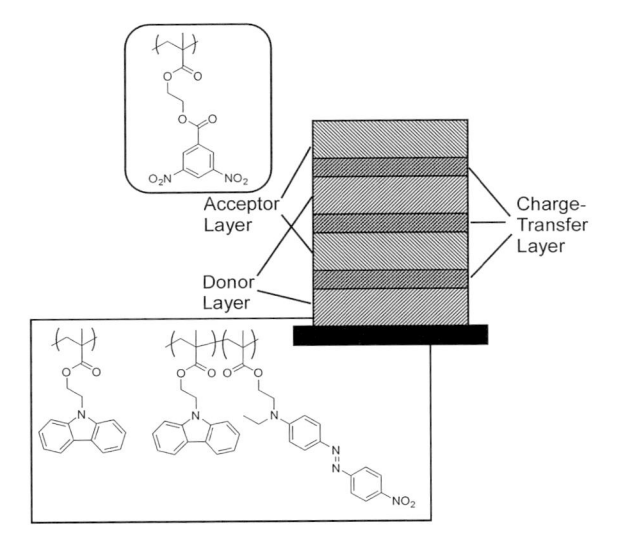

Figure 4.22 Layer-by-layer assembly based on charge-transfer interaction.

Biochemical interactions have been used widely for the LbL method. For example, LbL assemblies of rectin, concanavalin A with polysaccharides, glycogen (branched polyglucose) are possible (Figure 4.21). In addition, a well-investigated biospecific interaction, the biotin–avidin system, can also be applied to LbL assembly. Various physicochemical interactions can be good driving forces in the LbL process. Use of charge-transfer interactions for LbL assembly was demonstrating assemblies between an electron-accepting polymer [poly(2-(3,5-dinitrobenzoyl)oxy) ethylene methacrylate] and an electron-donating polymer [poly(2-(9-carbazoyl) ethyl) methacrylate] (Figure 4.22). Nonlinear optical properties of similarly assembled films by copolymerization of a nonlinear optical dye with donor polymer was reported. Alternate LbL assembly of syndiotactic hydrophobic poly(methyl methacrylate) (PMMA) and its isotactic counterpart was achieved through alternate repetition of physical adsorption and stereocomplexation. This film can be used as an enzyme-like reactor (Figure 4.23). Preparation of multilayer films comprised

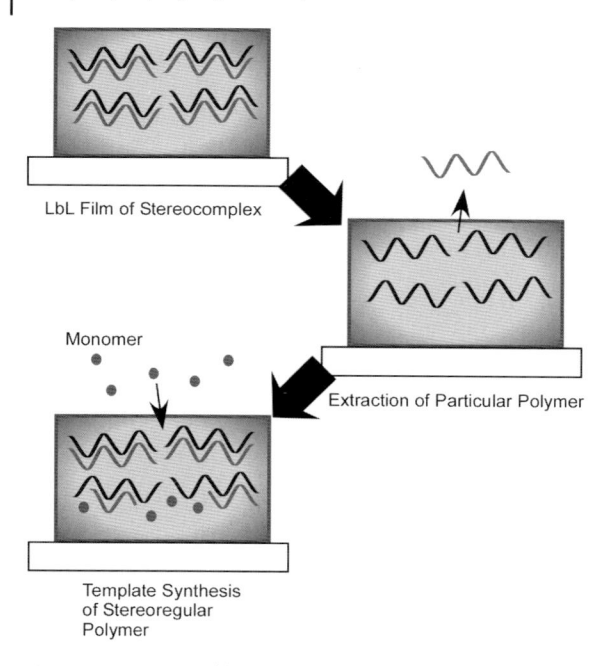

Figure 4.23 Enzyme-like reactor based on stereocomplex layer-by-layer assembly.

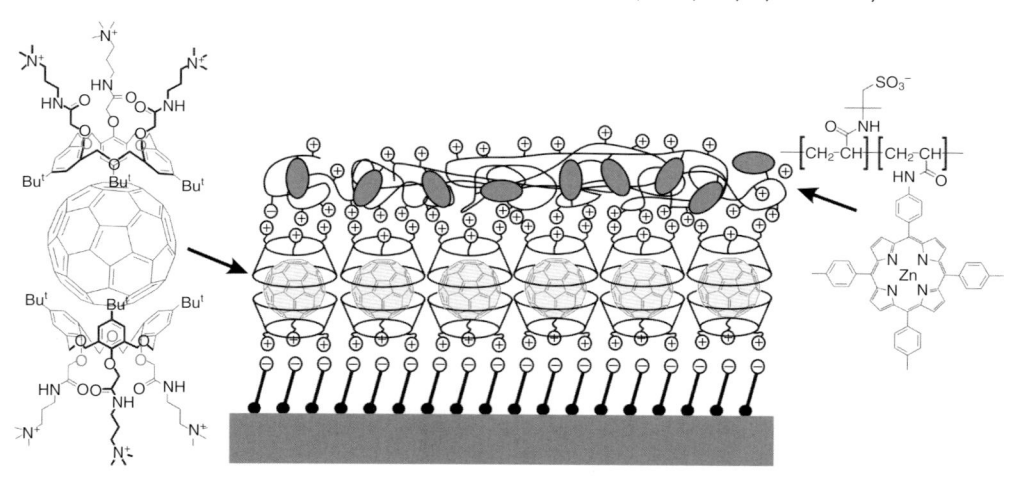

Figure 4.24 Layer-by-layer assembly based on supramolecular recognition.

of C_{60} fullerene and porphyrin layers was demonstrated using the LbL method based on a homohexacalix[3]arene-[60] fullerene 2:1 supramolecular complex formation (Figure 4.24). Step-wise sol-gel reaction at the surface of a solid support was also used for LbL assembly of metal oxide layers (Figure 4.25).

Modifications of assembling techniques are variously considered. Basically, procedures for LbL assembly are relatively simple, so that there is plenty of scope for

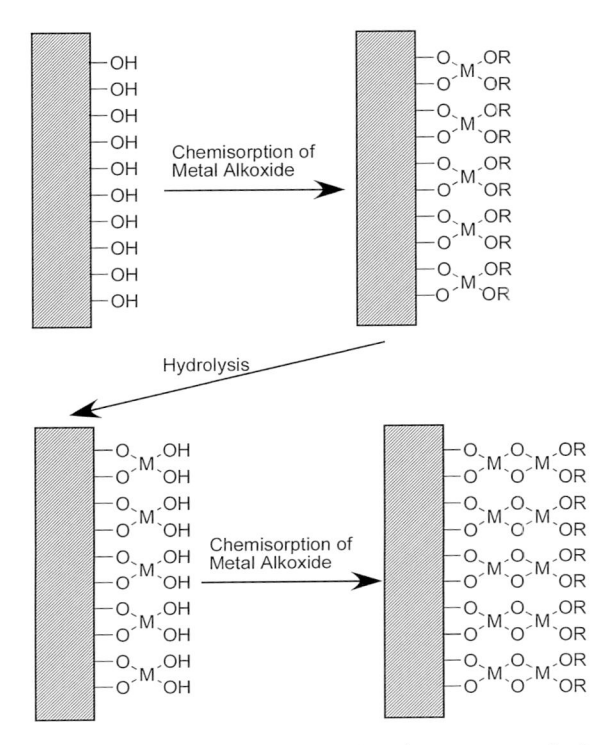

Figure 4.25 Layer-by-layer assembly based on stepwise sol-gel reaction.

technical improvement, for example, an automated LbL film assembler combined with a mass-controlled mechanism (Figure 4.26). A solid plate was attached to the arm of a robot together with QCM as a sensitive mass detector, which provided frequency shifts during adsorption of the materials. Feeding back the data acquired by the QCM from the deposition to the dipping time allowed preparation of a high-quality self-assembly film.

Combination of the LbL methods with other film-fabrication techniques resulted in various technical improvements. In the conventional LbL adsorption process, a self-diffusion process needs extra time, inducing significant adsorption of the weakly attached polyelectrolyte chains giving increased surface roughness and poor film quality. In order to avoid these problems, a spin-coating technique is combined with LbL assembly. The spin-coating LbL assembly method was demonstrated as an alternative for making well-organized multilayer films in a very short process time (Figure 4.27). The adsorption and rearrangement of adsorbed chains on the surface and the elimination of weakly bound polymer chains from the substrate are achieved almost simultaneously at a high rate of rotation over a short time period in the spin-coating process. This spin-coating LbL process yields a highly ordered internal structure far superior to the structure obtained with the conventional dipping method probably due to a mechanical effect upon the air shear force caused by the spinning process.

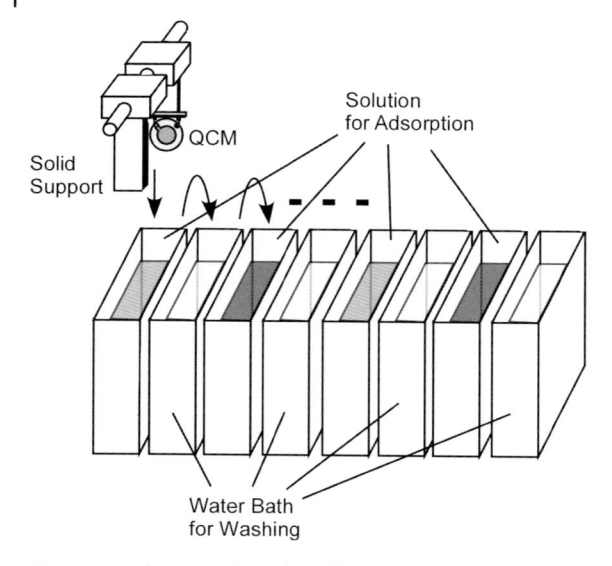

Figure 4.26 Automated machine for layer-by-layer assembly.

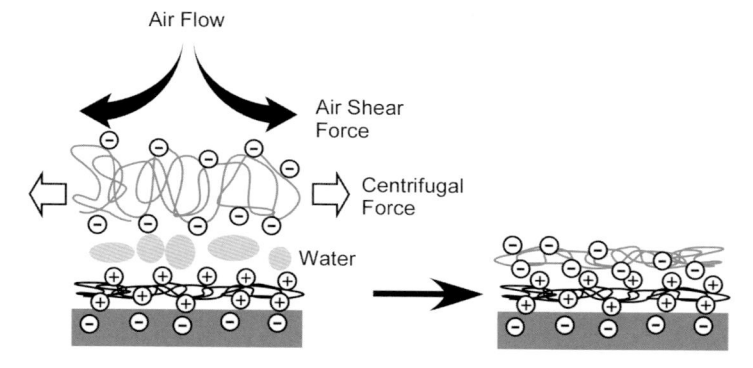

Figure 4.27 Spin-coating layer-by-layer assembly.

Combination of the LbL method and the spraying technique was also proposed. Figure 4.28 illustrates preparation of polyelectrolyte film by successive spraying of polycation and polyanion solutions. Various parameters such as spraying time, polyelectrolyte concentration, and the effect of film drying during multilayer construction are influential. With this technique, the LbL assembly process becomes very fast and leads to films with low surface roughness, as estimated by AFM and X-ray reflectometry. Spray deposition allows regular multilayer growth even under conditions for which dipping fails to produce homogeneous films, such as extremely short contact times. Drainage constantly removes a certain quantity of the excess material arriving at the surface, and the rinsing step can be eliminated, resulting in further improvements in preparation times for the whole film-construction process.

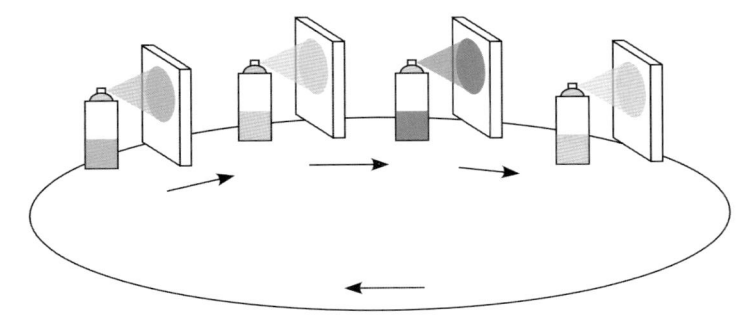

Figure 4.28 Spray layer-by-layer assembly.

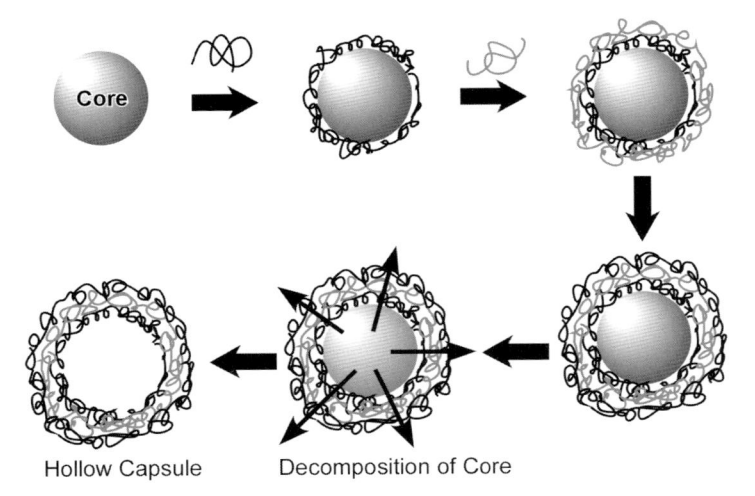

Figure 4.29 Preparation of capsule structure based on layer-by-layer assembly.

4.2.4
Three-Dimensional Assemblies

LbL films are not limited to fabrication as flat films. They are sometimes formulated into variously shaped objects such as microcapsules and tubes. A wide freedom in the LbL films' structure allows us to fabricate higher-dimensional structures from functional components. LbL assembly into hollow microcapsules is one of the most outstanding strategy modifications. As illustrated in Figure 4.29, the LbL films were assembled sequentially, similarly to the conventional assemblies, on a colloidal core. Dissolution of the central particle core upon exposure of the particles to appropriate solvents results in hollow capsules.

The prepared hollow microcapsules can be used for various purposes including encapsulation of guest substances and controlled release because permeability through the sphere skin is controllable using several factors. Proper selection of core materials leads to various functional nanosized systems. For example, the

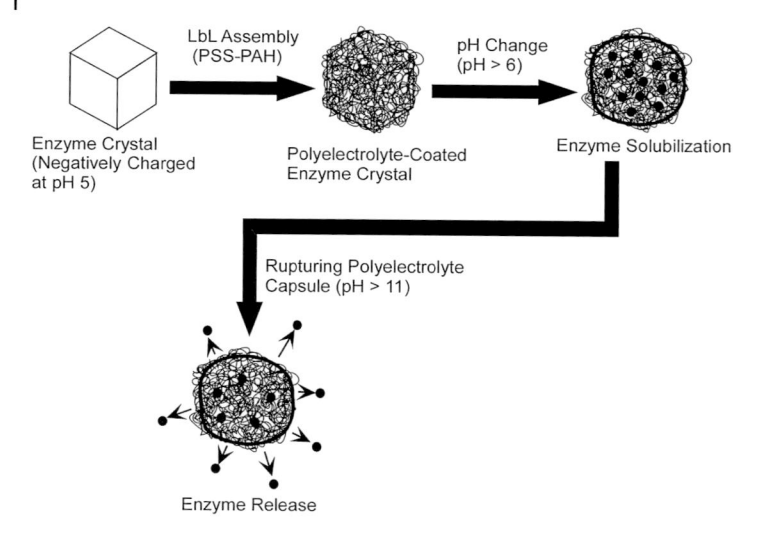

Figure 4.30 Layer-by-layer assembly using enzyme crystal as a core.

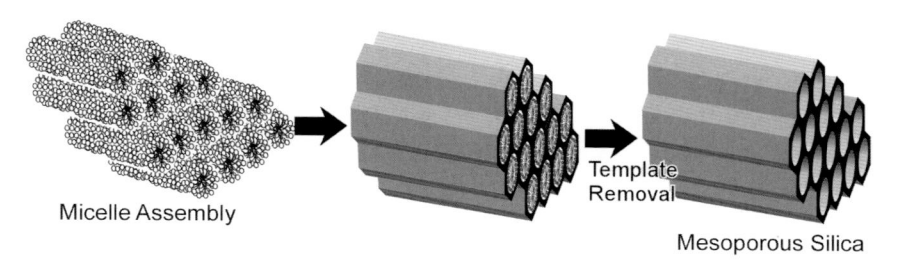

Figure 4.31 Synthesis of mesoporous silica.

LbL assembly on enzyme colloidal crystals provides inclusion of enzymes with an extremely high enzyme loading after dissolution of core crystals (Figure 4.30). As a successful demonstration of the LbL assembly at the mobile interfaces between liquid droplets and aqueous phases, formation of LbL multilayers on thermotropic liquid crystalline oil-in-water emulsions has influential effects on the orientation of the liquid crystals. A bipolar-to-radial ordering transition triggered by exposure of the LbL film-coated droplets to surfactant was observed to be slowed by 2 orders of magnitude relative to naked liquid-crystalline droplets.

The above-mentioned fabrication of LbL microcapsules can be classified as template synthesis that is frequently used for preparation of nanostructured materials such as mesoporous silica (Figure 4.31). Therefore, use of a well-designed template would resulted in three-dimensional LbL structures with more sophisticated structures and shapes. For example, coating of mesoporous microspheres by using LbL assembly for effective biocomponent immobilization was demonstrated. It can be a simple method for the preparation of densely enzyme-loaded particles through loading of the enzymes into mesoporous silica spheres followed

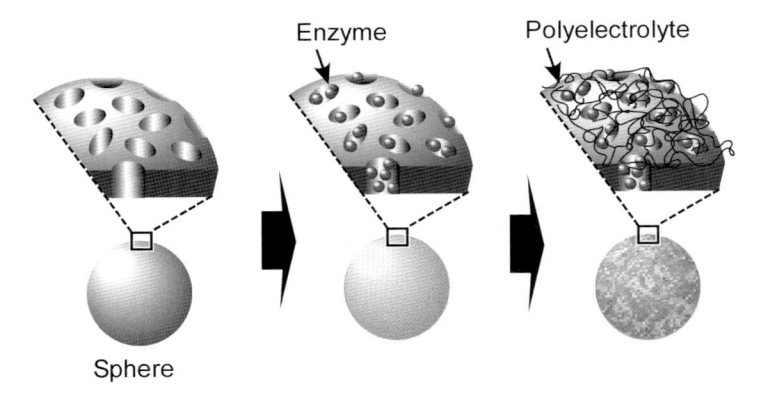

Figure 4.32 Loading of enzyme into mesoporous silica sphere.

by coating of the spheres with LbL multilayers (Figure 4.32). Following enzyme loading, a polyelectrolyte/nanoparticle LbL shell was assembled on the surface of porous spheres, thus preventing enzyme leakage, further resulting in high enzyme contents, enhanced enzyme activities, higher enzyme stabilities against pH, and protection of the encapsulated enzyme from proteolysis. According to the reported research, the encapsulated catalase can be recycled 25 times with an associated loss of activity of 30%, as compared to the 65% loss in activity for catalase immobilized in mesoporous silica spheres lacking the LbL coating. Even when the spheres with the LbL cover were subjected to proteolysis treatment, ca. 98% of the activity of the encapsulated catalase was maintained.

The above methodology can be combined with template (silica) removal for preparation of three-dimensional nanostructures without inorganic supports. It was demonstrated that the LbL coating of mesoporous silica spheres with polyelectrolytes and the subsequent removal of the silica templates results in formation of micrometer-sized nanoporous polyelectrolyte spheres (Figure 4.33). The sequential LbL adsorption of poly(acrylic acid) and poly(allylamine hydrochloride) in the silica sphere was followed by crosslinking by 1-ethyl-3-(3-dimethylaminopropyl) carbodiimide hydrochloride to add structural integrity to the nanoporous polyelectrolyte spheres. The subsequent removal of the silica component by exposure of the composite to hydrofluoric acid solution provided nanoporous polyelectrolyte spheres that could be collected after three centrifugation/water washing cycles.

This strategy can be used for preparation of nanoporous materials composed of biomaterials, as illustrated in Figure 4.34. In the initial step, proteins such as lysozymes were adsorbed into mesopores of silica spheres that was followed by several washing cycles to remove loosely adsorbed protein. The resulting protein-loaded mesoporous silica spheres were dispersed in an aqueous solution of poly(acrylic acid) at pH 4.5 where poly(acrylic acid) has the same charge as the silica substrate but is oppositely charged to the proteins. Further crosslinking using 1-ethyl-3-(3-dimethylaminopropyl) carbodiimide hydrochloride enhances the stability of the protein–polyelectrolyte complex. Exposure of the composite to

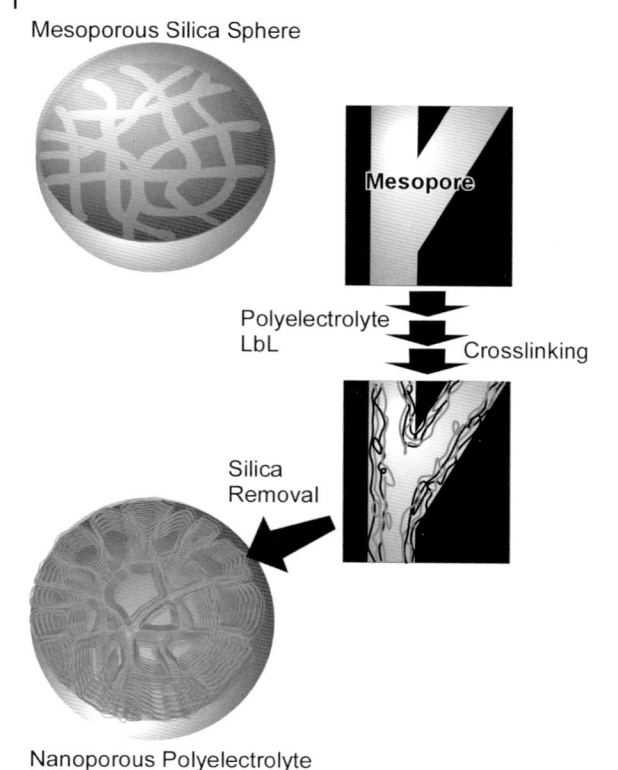

Figure 4.33 Preparation of nanoporous polyelectrolyte.

a HF/NH$_4$F buffer resulted in nanoporous protein spheres upon silica dissolution. The obtained high surface area material provides opportunities for easy approach of substrate molecules to proteins including enzymes, which would be useful in biosensing, catalysis, separations, and controlled drug delivery.

Use of the other porous template provides differently shaped LbL objects. If one selects a template anodic aluminum membranes with regular vertical pores, self-assembled microtubes can be fabricated by the LbL assembly technique (Figure 4.35). This template synthesis approach was used to fabricate microtubes and nanotubes through LbL assembly within pores and dissolution of the alumina template. These processes result in formation of tubular structures with outside diameters equivalent to the pore diameter of the template. For example, fabrication of microtubes based on hydrogen-bonding LbL self-assembly from poly(acrylic acid) and poly(4-vinylpyridine) was reported, where removal of the poly(acrylic acid) components from the tubes was also demonstrated. The porous-walled tubes resulting from the latter approach could also be useful as carriers in drug delivery or as catalyst supports.

Applying etching processes to the LbL techniques can be useful in the preparation of structure-fabricated materials. One of potential fruitful directions for the

Mesoporous Silica Sphere

Polyelectrolyte & Protein LbL

Mesopore

Cross-linking

Silica Removal

Nanoporous Protein

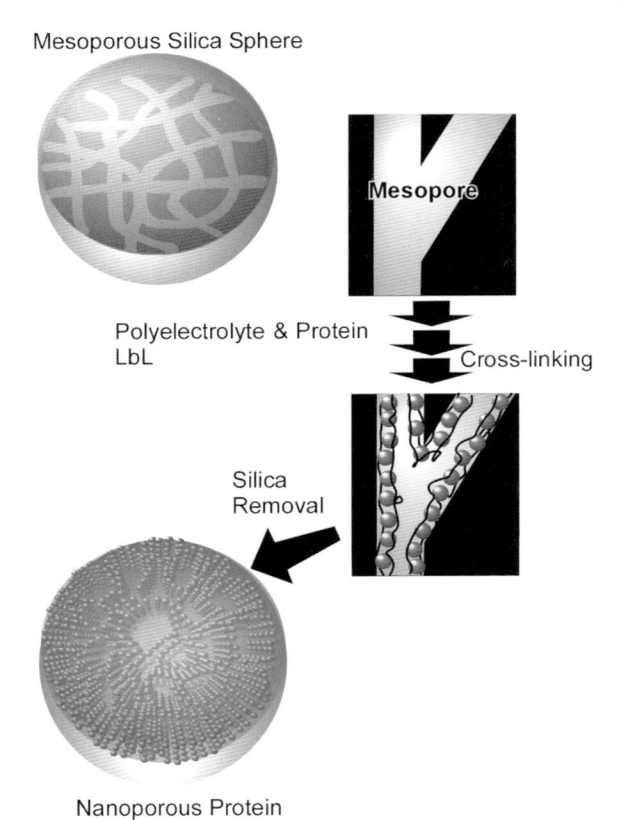

Figure 4.34 Preparation of nanoporous protein.

LbL assembly would be the combination of the LbL technique with currently existing top-down microfabrication and nanofabrication strategies. For example, use of poly(dimethylsiloxane) (PDMS) stamps prepared on a photolithographically prepared silicon master were used to prepare chemically modified patterned surfaces as templates for ionic multilayer assembly. Combining the LbL technique with ink-jet printing and photolithography was also demonstrated. An excellent demonstration of microfabrication of LbL films would be preparation of LbL-based free-standing microcantilever objects. As illustrated in Figure 4.36, sequenced procedures including patterning, phototreatment, etching, and LbL assembly were used to prepare this microscopic object. Preparation of free-standing LbL film was achieved by elective etching of the substrate surface. LbL films were first prepared on a cellulose acetate layer, which was selectively dissolved providing self-standing LbL films (Figure 4.37). In another approach, multilayered nanocomposite membranes containing gold nanoparticles were deposited alternately with oppositely charged polyelectrolytes on solid supports covered with a sacrificial layer of cellulose acetate, followed by coverage with additional assemblies of conventional polyelectrolytes. The multilayered polymer films with gold nanoparticles can be

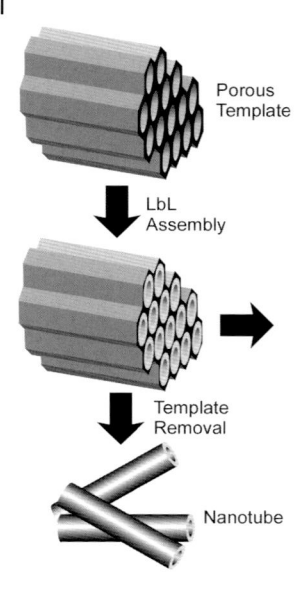

Figure 4.35 Preparation of nanotube through layer-by-layer assembly.

released from the supporting silicon wafer by dissolving the sacrificial cellulose acetate layer in acetone.

Many functional complexes in biology have highly hierarchic structures, as can be seen in various biological systems such as organelles, cells, tissues, and organs, which could be ultimate specimens for artificial functional structures. These hierarchic structures are constructed through spontaneous self-assembly that is achieved through molecular design and the appropriate selection of components. However, it would be very difficult to mimic all the available natural processes by using nonbiochemical approaches. One possible approach for constructing functional hierarchic structures would be a multistep process where first nanostructured materials are synthesized with their subsequent further assembly into organized structures of higher order. The LbL technique would be highly useful for the latter process. In the remaining parts of this section, preparation of hierarchic structures through LbL techniques is introduced.

One example shown here is a mimic of muliticellular assembly that was achieved by preparation of a lipid bilayer vesicle with covalent linkage of a siloxane framework and their LbL assembly. The transmission electron microscopic (TEM) images of the multilamellar cerasomes with a bilayer thickness of ca. 4 nm and vesicular diameter of 150 nm are clearly visible. Subjecting the cerasome structure to LbL techniques could result in multicellular mimics in a predesigned way. For example, the LbL assembly between cationic polyelectrolyte (poly(diallyldimethylammonium chloride), PDDA) and anionic vesicles was conducted (Figure 4.38). In addition, using both the anionic and cationic cerasomes, direct LbL assembly of cerasome structures in the absence of polyelectrolyte counterions became possible (Figure 4.39). The presence of closely packed cerasome particles in both

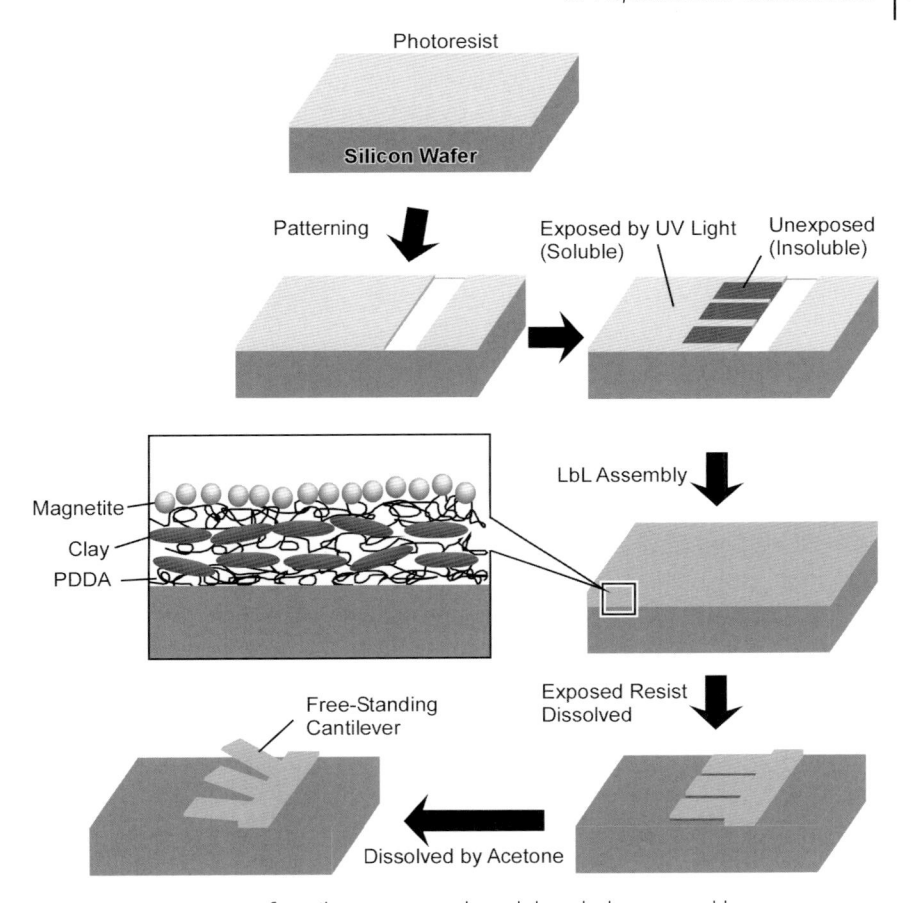

Figure 4.36 Preparation of cantilever structure through layer-by-layer assembly.

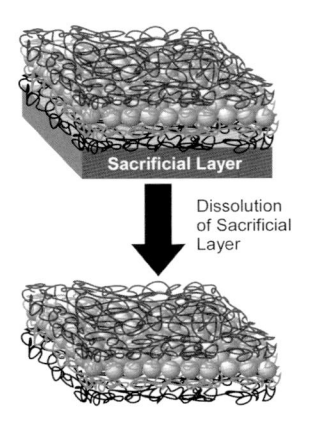

Figure 4.37 Preparation of self-standing layer-by-layer assembly.

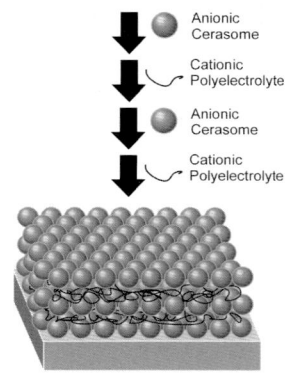

Figure 4.38 Layer-by-layer assembly between anionic cerasome and cationic polyelectrolyte.

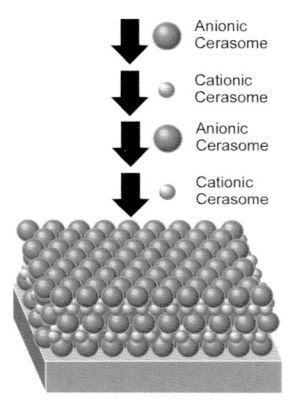

Figure 4.39 Layer-by-layer assembly between anionic cerasome and cationic cerasome.

layers was clearly confirmed by atomic force microscopic (AFM) observations of the surface of the assembled structures. The hierarchically structured assembly obtained can be regarded as a multicellular mimic and subsequently could be used as bioreactors or biosensors.

4.3
Functions and Applications

As one can understand from descriptions in the previous sections, the LbL technique has huge varieties in both applicable materials and assembly methods. Another particular feature of this method as a thin film fabrication is its surprisingly simple nature of the procedure. We can make nanosized films only using beakers and tweezers. These processes are short and costless. These characteristics of the LbL assembly result in huge possibilities of applications. In fact, a flood of research papers that propose application of the LbL assemblies can be currently found.

In the following section, various examples of applications of the LbL assemblies are introduced. Of course, all the topics cannot be included and balanced selections of the examples are indeed difficult. However, one can sense advantages of the LbL technologies in many applications from the examples presented below. The examples are roughly classified into (i) physicochemical and biological applications.

4.3.1
Physicochemical Applications of LbL Thin Films

The LbL assembly enables us to fabricate thin films in nanometer-scale layer interval and desired layering sequences, which are difficult to achieve by the conventional spin-coating process. Therefore, the LBL method offers a useful approach for fabrication of photonic devices with great accuracy in control over thickness. It is also possible to fabricate devices with alternating layers of hole transporter, photosensitizer, and electron-transporter materials. For example, photoactive LbL films from tellurium nanowires and polyelectrolyte were assembled. With the obtained films, a light-induced conductivity-switched the LbL thin films between on and off states and gave an optical gating phenomenon. Under ambient conditions, this light-on–light-off cycle is very stable without any signs of photodegradation in use for more than 100 repetitions. Photovoltaic cells using LbL films of poly(p-phenylene ethynylene)-based polyelectrolyte and a fullerene derivative were fabricated, The obtained cell structure showed efficient incident monochromic photon to current conversion efficiency response under low-intensity monochromatic light illumination (Figure 4.40). Because the LbL technique permits preparation of donor–acceptor films with relatively precise (molecular level) control over the structure and energies of the active layers of photovoltaic cells, the LbL structures becomes media appropriate for basic investigation of the effects of features such as energy-gradient-driven exciton and/or charge transfer on photoconversion efficiency in organic photovoltaic cells.

As is widely accepted, solar cells and fuel cells are important targets in current research. Therefore, various materials useful for these cell applications have been investigated using the LbL technique. For example, a solar cell was prepared from 50-cycle layers of TiO_2 large particle on 50-cycle layers of TiO_2 nanoparticles including photosensitive dye N719. The prepared films demonstrated advantages of the upper scattering layer for an increase in the current density. The methanol-crossover problem in direct methanol fuel cells (DMFC) was minimized by LbL assembly of composite polyelectrolyte multilayer thin films (Figure 4.41). Preparation of methanol-blocking multilayer thin film on a Nafion membrane using LbL assembly of oppositely charged polyelectrolytes induced a significant reduction in methanol crossover and on the enhancement of the performance of DMFCs. The LbL assembly approach also has a much smaller detrimental effect on the proton conductivity and chemical and thermal stabilities of the Nafion membrane, a significant advantage. In these demonstrations, advantages in the capability to design desired film structures using the LbL techniques was emphasized.

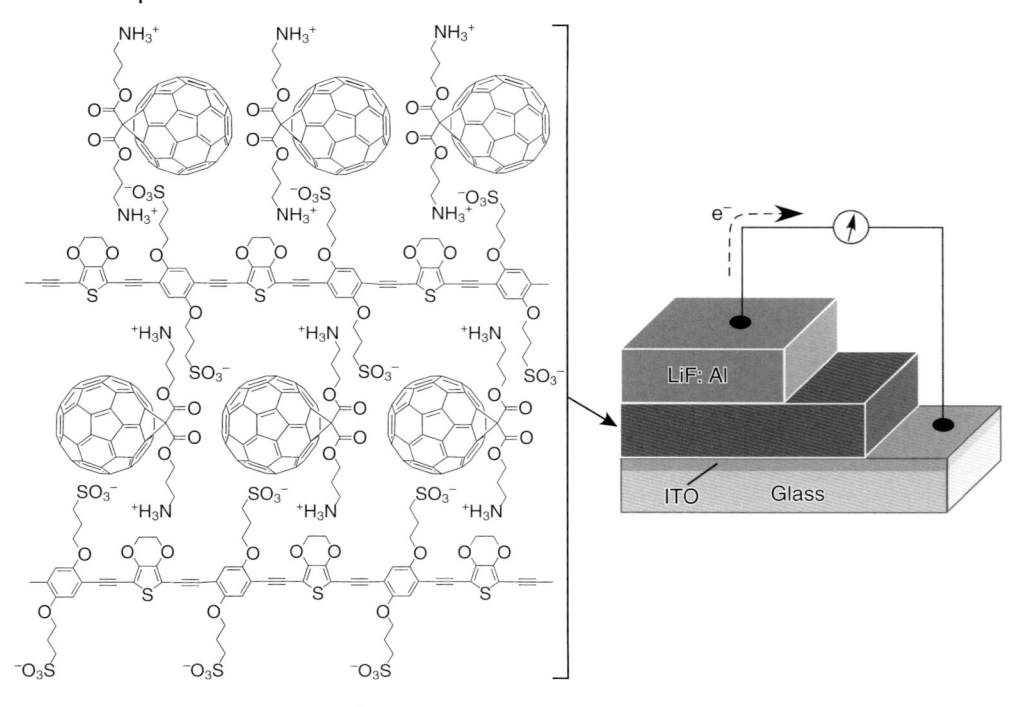

Figure 4.40 A device for monochromic photocurrent conversion based on layer-by-layer assembly.

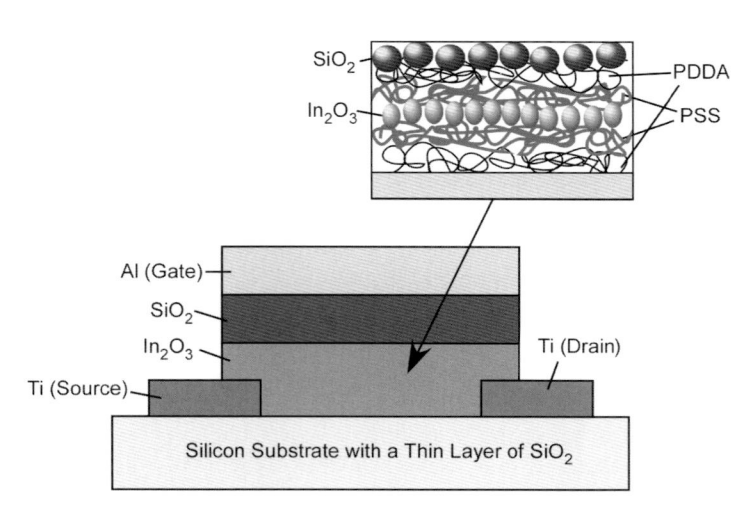

Figure 4.41 A direct methanol fuel cell structure based on layer-by-layer assembly.

As suggested in the above-mentioned examples, devices constructed with well-defined layer structures would be appropriate targets that can benefit from LbL technology. Aiming to construct an all-solid-state electrochromic device, the LbL films with poly(aniline-N-butylsulfonate)s as an electrochromic anionic polymer, and acid-doped polyaniline and vinylbenzyldimethyl-n-octadecylammonium salts as a polycation were fabricated. The obtained all-solid-state electrochromic device exhibited an electrochromic response at 3 V within 1 s, with a stable memory effect. In another example, metal-oxide-semiconductor field effect transistor (MOSFET) arrays were constructed using the LbL process on a silicon wafer. The on-off threshold voltage was 3 V, demonstrating an approach to fabricate low-cost MOSFETs and integrated circuits.

Design of a digitized rugate filter was also proposed. A special type of dielectric mirror possessing many thin layers that approximate a continuous, periodic refractive-index profile was fabricated through *in situ* growth of silver nanoparticles to selectively increase the refractive index of precisely defined regions of the LbL film. The presence of a reflection band grows in amplitude with increasing silver incorporation, reaching a peak reflectance of 75% after five silver loading and exchange cycles. Covering materials surfaces by the LbL films is an effective way to control materials wettability. Hydrophilic patterns were prepared on superhydrophobic surfaces using water/2-propanol solutions of a polyelectrolyte to produce surfaces with extreme hydrophobic contrast. Selective deposition of multilayer films onto the hydrophilic patterns introduces different properties to the area, including superhydrophilicity. The formed films can be potentially used for water-harvesting surfaces, controlled drug-release coatings, open-air microchannel devices, and lab-on-chip devices.

Permeation control through the films could be fundamental properties of thin-film materials. A sophisticated design of the LbL films should lead to advanced properties in the corresponding fields. For example, rather simple LbL films assembled with conventional polyelectrolytes containing amine and carboxylate groups can be used for removal of environmental undesired materials (Figure 4.42). After appropriate thermal treatment, the assembled LbL films were subjected to filter experiments for removal of environmentally unfriendly gases, ammonia gas (odorant) and sick-house-origin aldehyde (origin of sick-building syndrome). Basic ammonia can be effectively trapped through electrostatic interaction with carboxyl groups In addition, aldehyde species can be trapped by amino groups in the film through Schiff's base formation. In particular, suppression of aldehyde release is important for prevention of sick-building syndrome. Preparation of photoluminescent oxygen sensors based upon phosphorescent platinum and palladium porphyrins and ruthenium complexes was also reported; these dyes were adsorbed into the pores of mesoporous silica particles at submonolayer coverage on LbL films.

Well-defined layer structures are often useful for material detection. In the following parts, three types of chemical sensors based on the LbL structures of nanostructured carbons and polyelectrolytes are introduced. In the first example, sensor applications of the LbL structures prepared from mesoporous carbon

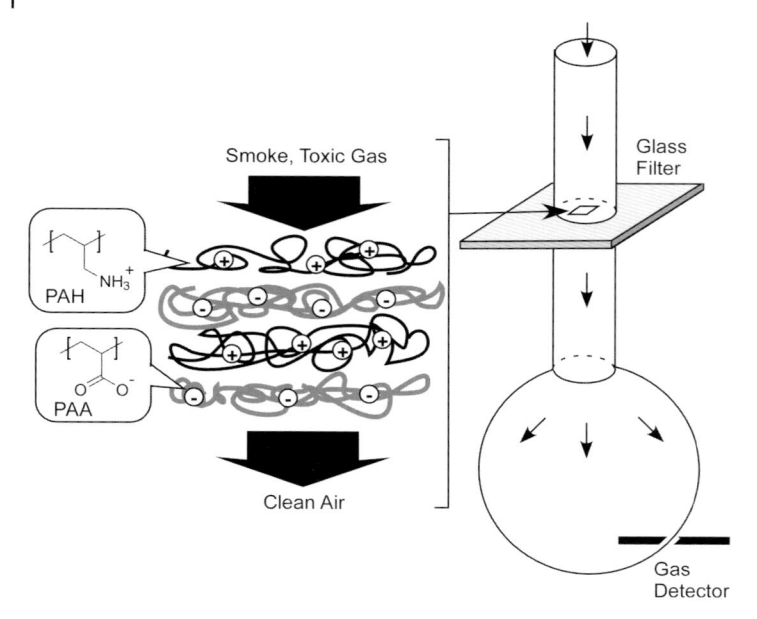

Figure 4.42 Removal of toxic components though layer-by-layer assembly.

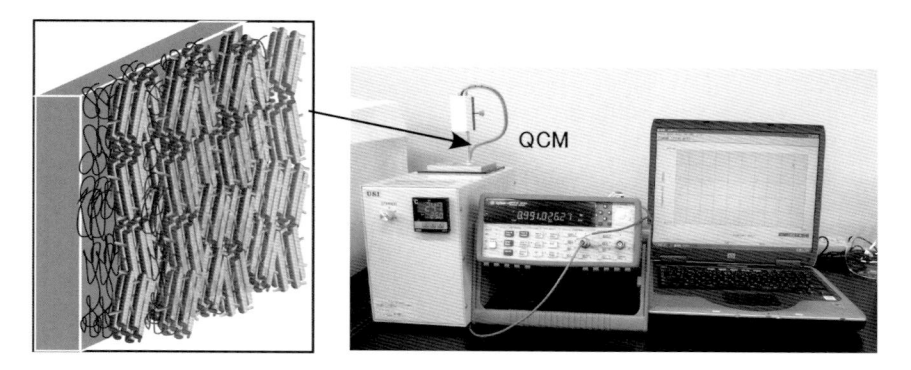

Figure 4.43 A sensor using a layer-by-layer assembly of mesoporous carbon and polyelectrolyte.

materials (CMK-3) and polyelectrolytes is described. Because mesoporous carbon materials do not possess surface charges sufficient for successful LbL assembly, surface oxidation of carbon to introduce negative carboxylate groups was conducted. LbL assembly of the oxidized CMK-3 was performed using polycation PDDA on a QCM plate (Figure 4.43). Sensing performances were investigated in aqueous solution where a QCM plate covered with the Hi-LbL film of CMK-3 was immersed and sensing target was injected. Frequency shifts upon adsorption of tannic acid greatly exceed those for catechin and caffeine. The superior adsorption capacity for tannic acid likely originates in its molecular structure, that is, multiple

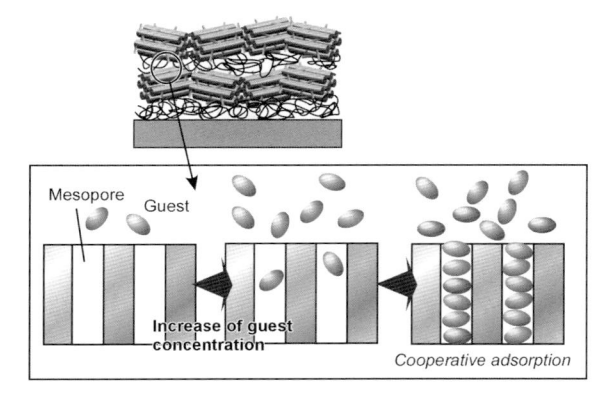

Figure 4.44 Cooperative binding of guest molecules within mesopores.

phenyl rings of the tannic acid molecule can interact with the carbon surface through π–π interactions and hydrophobic effects. In addition, size fitting of tannic acid (a roughly circular molecule with approximate diameter 3 nm) to the CMK-3 nanochannel may result in enhanced interactions between the guests themselves and/or the guest and carbon surface. Adsorption quantities of tannic acid to the CMK LbL film at equilibrium exhibited a sigmoidal profile at low concentrations, suggesting cooperative binding (Figure 4.44). This behavior might result from confinement effects during adsorption such as enhanced π–π and/or hydrophobic interactions. These observations will also promote our understanding of molecular interactions within nanospaces.

Carbon capsules were synthesized using zeolite crystals as templates with homogeneous dimensions ($1000 \times 700 \times 300\,nm^3$) and 35-nm thick mesoporous walls with a uniform pore-size distribution centered at 4.3 nm in diameter, and a specific surface area of $918\,m^2\,g^{-1}$. Surfactant covering enables us to assemble noncharged substances in the LbL process with aid of counterionic polyelectrolyte (Figure 4.45). Adsorption of various volatile substances onto the carbon capsule LbL films in vapor-saturated atmospheres was investigated by *in situ* frequency decrease of the QCM resonator used as the film support. Aromatic hydrocarbons such as benzene and toluene are better detected in this sensing system than aliphatic hydrocarbons such as cyclohexane. In particular, the amount of benzene adsorbed at equilibrium is ca. 5 times larger than that of cyclohexane, despite their very similar vapor pressures, molecular weights, and structures, indicating the crucial role of π–π interactions on volatiles' adsorption in the carbon capsule film.

Interesting modification of detection selectivity can be made to these films using interior spaces of the capsules. Detection selectivity can be easily tuned by impregnation with additional recognition components, that can be introduced after film preparation (Figure 4.46). High affinity to aromatic compounds of the caobon capsules can be altered. When the carbon capsule film was impregnated with lauric acid, the greatest affinities for nonaromatic amines and the second highest affinity for acetic acid were obserbed. Strong entrapment of amines through

Mesoporous Carbon Capsule

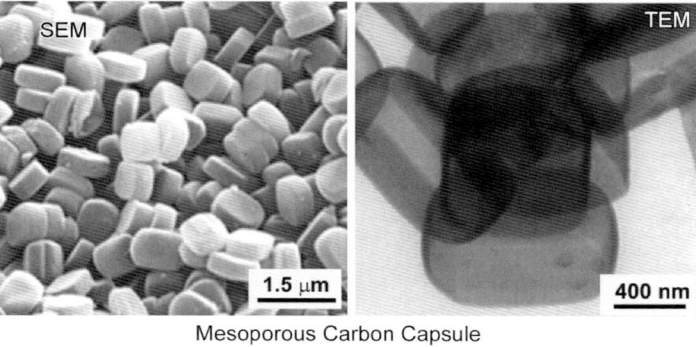

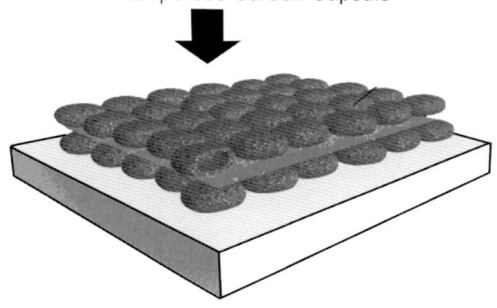

Figure 4.45 Layer-by-layer assembly of mesoporous carbon capsule.

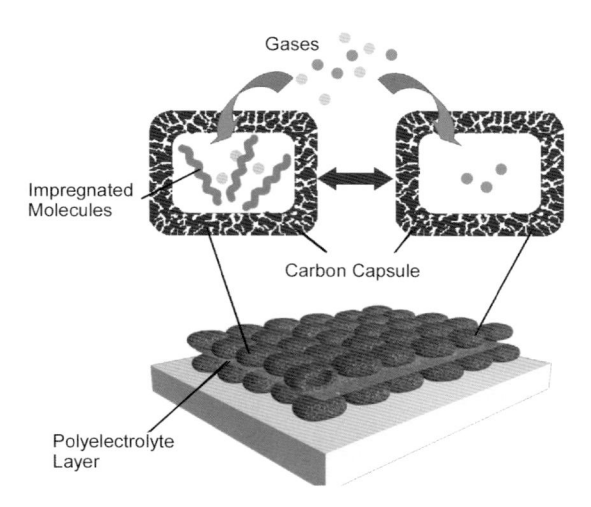

Figure 4.46 Layer-by-layer assembly of mesoporous carbon capsules with inside recognition component.

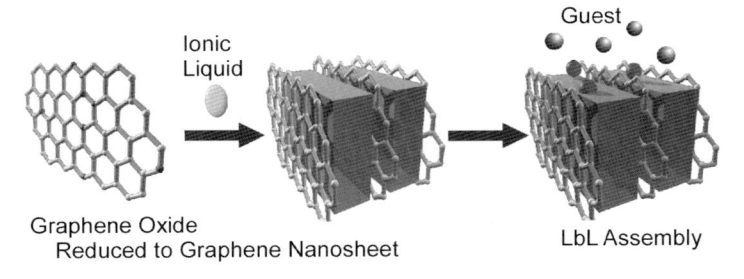

Figure 4.47 Layer-by-layer films of graphene nanosheet and ionic liquid.

acid–base interactions was suggested. In contrast, impregnation of dodecylamine into the carbon capsule films resulted in a strong preference for acetic acid. Such controls of detection selectivity of the sensor capability upon impregnation with second recognition sites to the capsule interiors within LbL films will find widespread applications as sensors or filters because of their designable guest selectivity. As the carbon materials used are stable in water, this system could also be used for removal of toxic materials from water.

Pieces of graphene can be disassembled from graphite and then reassembled into hierarchic structures through the LbL technique (Figure 4.47). Graphene oxide sheet was first prepared by oxidization of graphite under acidic conditions, followed by its reduction to graphene sheet in the presence of ionic liquids in water. Composites of graphene-sheet/ionic liquid (GS-IL) behave as charge-decorated nanosheets and were assembled alternately with poly(sodium styrenesulfonate) (PSS) by LbL adsorption on appropriate solid supports to provide layered assemblies of GS-IL composite with PSS on the surface of a QCM resonator for *in situ* detection due to gas adsorption. The large amount of adsorbed benzene clearly observed as the highly selective detection of aromatic guests within the well-defined π-electron-rich nanospace in the GS-IL films. Detection of vapors can be repeated through alternate exposure and removal of the subject solvents. Interestingly, gradual degradation of the ON/OFF response was noted for benzene detection, probably caused by the strong interactions between aromatic compounds and the graphene layer, while the response to cyclohexane was fully reversible. Responses to mixtures of benzene and cyclohexane at different mole ratios showed an approximately linear relation with small cooperative deviations, facilitating estimation of gas fractions in mixtures.

4.3.2
Biomedical Applications of LbL Thin Films

Important features of LbL assembly for biomaterials are its simplicity and mildness. LbL assembly can be performed in aqueous medium and does not require chemically harsh conditions. LbL films' structures are less densely packed than those of LB films and this is advantageous for material diffusion through the films.

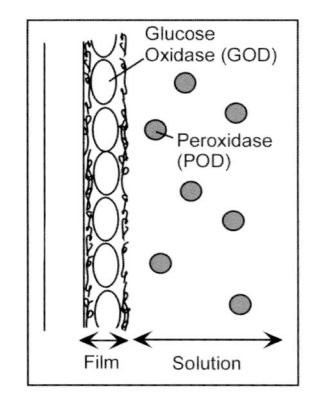

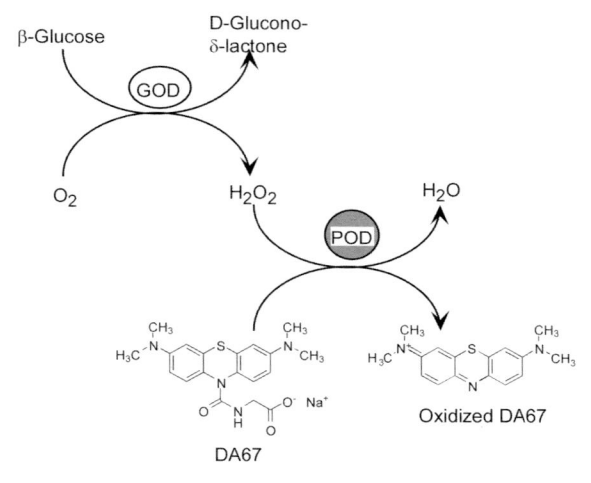

Figure 4.48 Layer-by-layer thin-film reactor.

In the initial part of this section, characteristics of the LbL films of biomaterials are described with several examples of the LbL enzyme reactors.

The first example of practical use of the LbL films containing enzymes is shown in Figure 4.48, where glucose oxidase (GOD) was assembled in the same film. The effect of the immobilization amount on the apparent activity was first investigated. Enzymatic activity was examined using films with various numbers of GOD layers. The initial rate of the reaction is plotted as a function of GOD immobilized in the film (Figure 4.49). Up to approximately 5 µg, regardless of the polycations used, the enzymatic activity is enhanced in proportion to the amount of GOD. However, suppression of the GOD activity at higher GOD contents apparently indicates limitation by substrate diffusion. Suppression of the reaction rate was observed for the GOD–PEI two-layer film (22.5 nm thick), which is, however, much larger than that for Langmuir–Blodgett (LB) films. The activity of GOD was sufficiently depressed in the case of a two-layer LB film (approximately 5 nm in thickness).

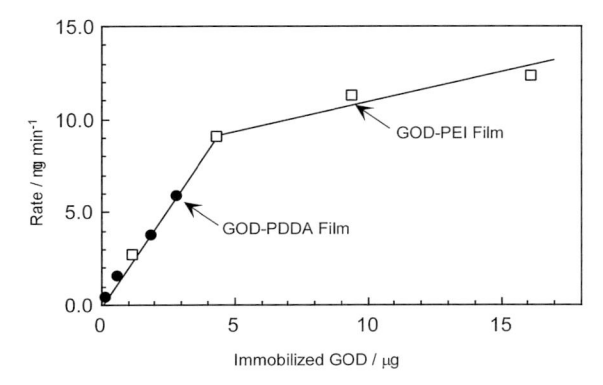

Figure 4.49 Reaction activity of glucose oxidase in layer-by-layer films with various thicknesses.

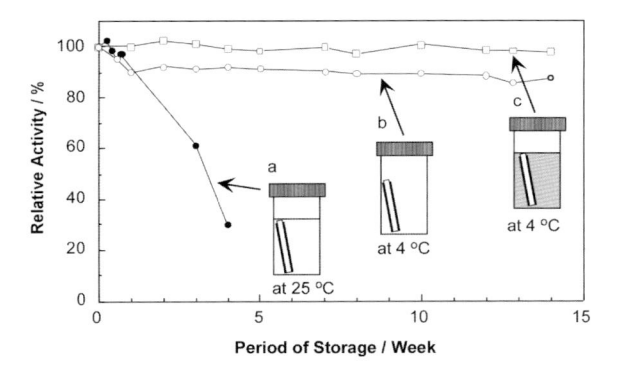

Figure 4.50 Stability of glucose oxidase in layer-by-layer films: (a) in water at 25 °C; (b) in air at 4 °C; (c) in water at 4 °C.

Therefore, the polyelectrolyte film is more permeable to water-soluble substrates than an LB film.

The storage stability of GOD LbL films was examined under three different conditions, stored in water at 25 °C, stored in 0.1 M PIPES buffer (pH 7) at 4 °C, and allowed to stand in air at 4 °C (Figure 4.50). Drastic decreases in GOD activity was observed for the film samples stored in water at 25 °C, probably due to bacterial growth. In contrast, the films kept in the buffer at 4 °C did not show a significant decrease in enzymatic activity over 14 weeks. A 10% decrease in the activity was observed in the first week for the films kept in air at 4 °C, but the activity was maintained during the following 13 weeks where the initial loss in activity was probably due to drying of the film. These examples demonstrated long-term maintenance of enzyme activities of the LbL films if they are kept under appropriate condition.

The thermostability of enzymatic activity for the GOD LbL films was also examined (Figure 4.51). A one-layer GOD film on a quartz plate was incubated in water at a given temperature for 10 min and their activities were evaluated at 25 °C.

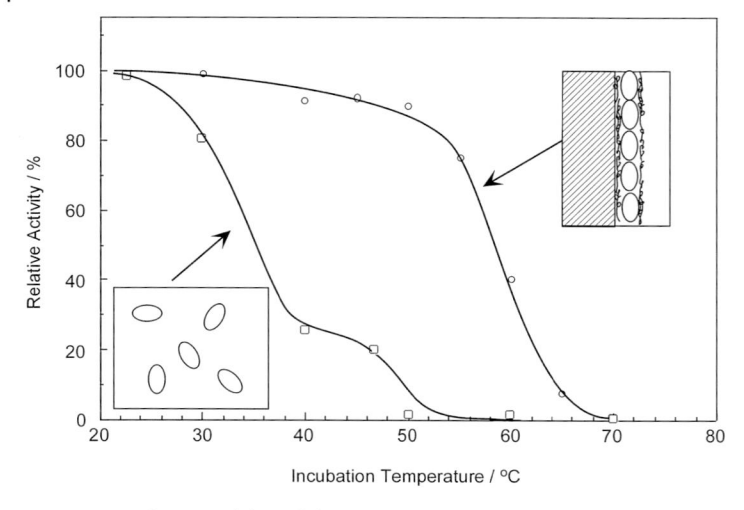

Figure 4.51 Thermostability of glucose oxidase in solution (left) and in layer-by-layer film (right).

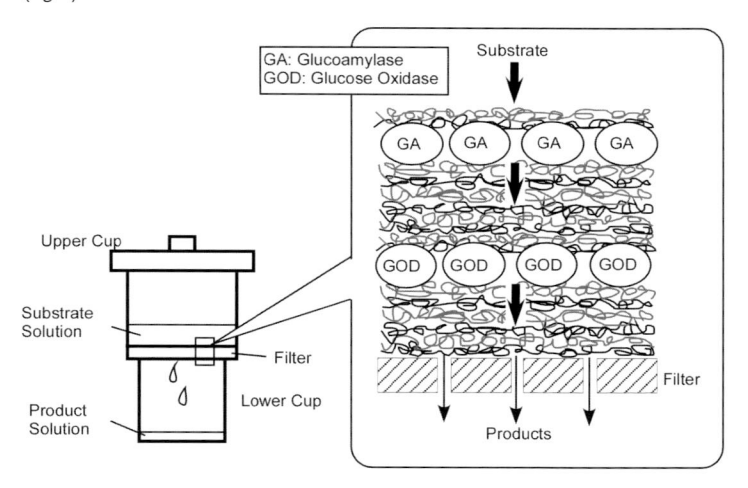

Figure 4.52 Layer-by-layer multienzyme reactor.

Activity measurement was similarly performed for GOD dissolved in water. The GOD dissolved in aqueous solution lost activity on incubation even at 30–40 °C and it became inactive at 50 °C. The thermostability of GOD assembled with PEI was remarkably improved, and a significant decrease in activity was not detected even after incubation at 50 °C. The improved enzymatic activity of GOD in the LbL film could be attributed to suppression of the conformational mobility of GOD upon complex formation with the surrounding polymer chains.

Freedom in film construction is one of the most pronounced advantages of the alternate layer-by-layer adsorption method. For demonstration of this feature, various LbL films of multienzyme reactors containing GOD and glucoamylase (GA) prepared on an ultrafilter (Figure 4.52). In these multienzyme reactors,

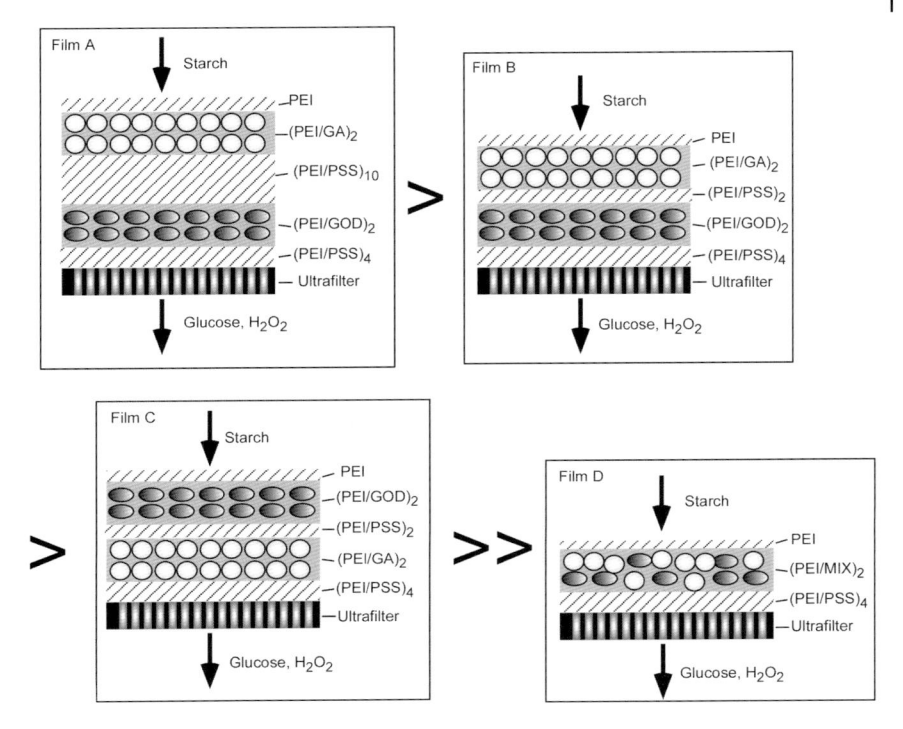

Figure 4.53 Activities of multienzyme reactors with various film constructions.

hydrolysis of the glycoside bond in starch by GA produces glucose, and glucose is then converted to gluconolactone by GOD with H_2O_2 as a coproduct. In order to investigate the effect of film organization on the reaction efficiency, the following enzyme films were prepared (Figure 4.53):

Film A: Filter + (PEI/PSS)$_4$ + (PEI/GOD)$_2$ + (PEI/PSS)$_{10}$ + (PEI/GA)$_2$ + PEI

Film B: Filter + (PEI/PSS)$_4$ + (PEI/GOD)$_2$ + (PEI/PSS)$_2$ + (PEI/GA)$_2$ + PEI

Film C: Filter + (PEI/PSS)$_4$ + (PEI/GA)$_2$ + (PEI/PSS)$_2$ + (PEI/GOD)$_2$ + PEI

Film D: Filter + (PEI/PSS)$_4$ + (PEI/MIX)$_2$ + PEI

In the film D, MIX represents adsorbed layers from an aqueous solution of an equimolar mixture of GOD and GA. The highest yield of the products was obtained with film A, and film B showed the second highest yield because of the layering order of the enzymes appropriate for the sequential reaction. When the disposition of the two enzyme layers is reversed (film C), the reaction efficiency remains low.

The difference observed between film A and film B suggests the importance of the separation between the two enzyme layers. A plausible mechanism is related to the inhibition of GA activity by gluconolactone. Larger separation between the GOD and GA layers would reduce the activity loss of GA. Film D showed the

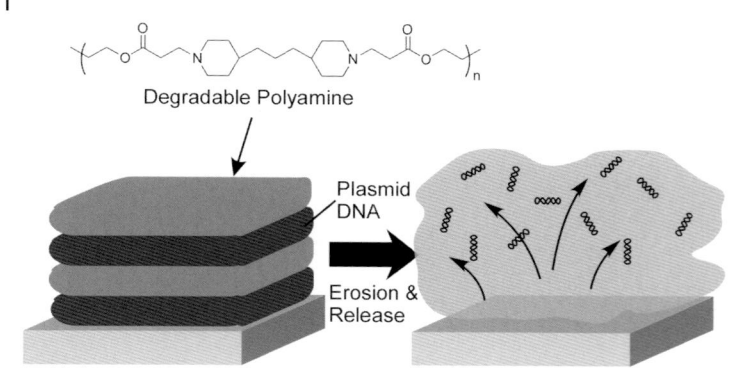

Figure 4.54 Release of DNA from layer-by-layer film.

lowest yield. The coexistence of GA and GOD might result in the lowest activity due to the inhibition of GA by gluconolactone produced by GOD and/or uncontrolled immobilization would result in an unbalanced ratio of the two enzymes. As demonstrated above, with the LbL method, various film structures are easily obtainable by quite simple procedures. The LbL film can be prepared on any solid support having some charges or a surface modified by charge-introduction treatment that activates our imagination to design protein-based nanodevices.

Not limited to the enzyme reactors shown above, various biochemical applications of the LbL films have been reported. Several examples are briefly explained below. Controlled DNA delivery was realized using a degradable polymer as a component of the LbL films (Figure 4.54). Multilayered LbL films up to 100 nm thick containing a synthetic degradable cationic polymer and plasmid DNA coded with green fluorescent protein were fabricated. Degradation of the former components induced release of the plasmid, which was confirmed by high levels of enhanced green fluorescent protein in the cell. The pH-sensitive bipolar ion permselective films of polyelectrolyte multilayers, prepared by LbL assembly and photocrosslinking of benzophenone-modified poly(acrylic acid) and poly(allylamine hydrochloride) with the aim of bipolar pH switching of permselectivity for both cationic and anionic molecules. The LbL film is permeable to the cationic probe but impermeable to the anionic probe at high pH because deprotonation of free carboxylic acid promotes and suppresses the permeability of cationic and anionic probes, respectively. At lower pH, the same LbL film becomes permeable to an anionic probe but less permeable to a cationic probe due to the protonation of free amine.

Covering electrodes with LbL films is a powerful method to prepare biosensors. For example, direct electrochemical luminescence measurement for a 8-oxoguanine detection method was proposed based on catalytic oxidation of guanine in DNA with $[Os(bpy)_2\text{-}(PVP)_{10}]$ leading to photoexcited OsII* sites and subsequently electrochemiluminescence signals (PVP represents poly(vinylpyridine)) (Figure 4.55). LbL films containing GOD and an Os-complex-derivatized PAH on an electrode were also reported in which the latter redox-active component can work as an

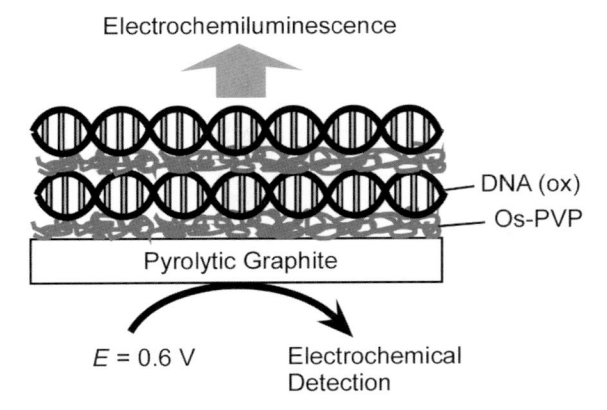

Figure 4.55 A layer-by-layer film for electrochemical luminescence sensor.

electric wire to mediate electron transfer between the enzyme and the electrode. Coating with LbL films has also been used for bioprotective purposes. Wrapped human pancreatic islets with polyelectrolyte LbL films has only a minor influence on the release of insulin but significantly suppressed antibody recognition. The suppression of antibody recognition may be caused by exclusion of the antibody due to the small cutoff of the multilayer structure and/or masking of the corresponding epitope on the cell surface.

Hierarchically assembled LbL films show unique materials-release profiles as exemplified in Figure 4.56 where anionic silica capsules were deposited on a QCM resonator using LbL assembly with the aid of anionic silica nanoparticles as a coadsorber. After immersing the compartment film on the QCM resonator into water and drying under a nitrogen flow, the net change in weight of the film after each cycle was measured in air by using QCM. Surprisingly, the frequency shifts upon water evaporation from the mesoporous nanocompartment films possess a stepwise profile even though no external stimulus was applied. A plausible mechanism for automodulated stepwise water release from the mesoporous naocompartment films is illustrated in Figure 4.57. Initially, water entrapped in mesopore channels evaporates to the exterior, which is observed as the first step of water release. After most of the water has evaporated from the mesopore channels, water enters that region from the capsule interior probably through rapid capillary penetration. Subsequently, water again evaporates from the mesopores to the exterior and is apparent as the second evaporation step. This release profile was used in a demonstration of controlled release of various fluid drugs such as fragrance molecules. Most of the currently available control release systems perform modulation in release using some external stimulus. However, we have presented here a rare example of *a stimulus-free controlled release medium*, which operates in a stepwise manner with prolonged release efficiency, a feature useful for controlled-release drug delivery. This new system has been shown to possess features of controlled loading/release, which are of great utility for development of energyless and clean *stimulus-free* controlled drug-release applications.

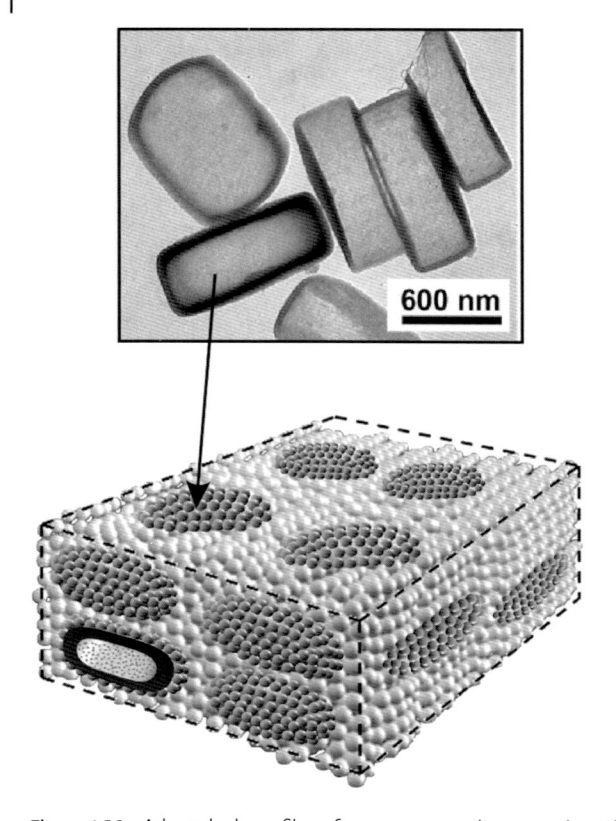

Figure 4.56 A layer-by-layer film of mesoporous silica capsule with silica nanoparticles.

The LbL films in capsule structures have received much attention in practical biomedical application fields, especially in controlled release of drugs. For example, glucose-sensitive polyelectrolyte capsules through LbL assembly of cationic polymer bearing phenylboronic acid moieties and PSS were fabricated. Upon addition of glucose, the phenylboronic acid moieties become negatively charged and start to interfere strongly with the polyelectrolyte multilayers to promote release of entrapped molecules. The permeability of polyelectrolyte LbL capsules composed of polyanions and polycations of poly(ferrocenylsilane) can be sensitively tuned via chemical oxidation, resulting in a fast capsule expansion accompanied by a drastic permeability increase in response to a very small trigger. DNA encapsulation inside a biocompatible polyelectrolyte was proposed (Figure 4.58), where $MnCO_3$ particles as template core materials were suspended in DNA solution, and the addition of spermidine solution into stirred $MnCO_3$/DNA solution caused precipitation of a water-insoluble DNA/spermidine complex onto the $MnCO_3$ particle. The $MnCO_3$ template particles were first dissolved, resulting in biocompatible capsules containing DNA/spermidine complex. Further decomposition of the DNA/spermidine complex led to selective release of low molecular

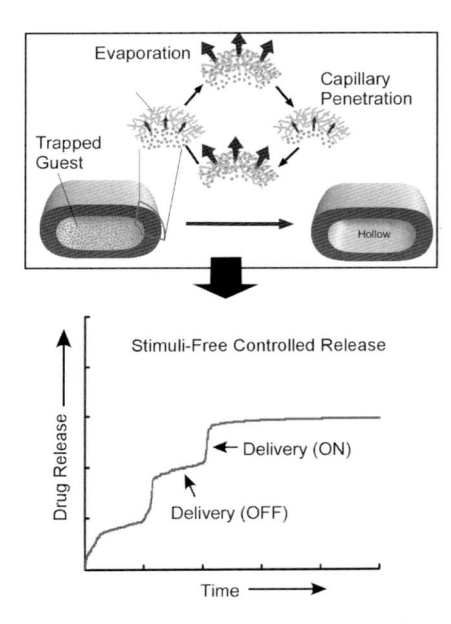

Figure 4.57 Automatic stepwise release of trapped materials from mesoporous silica capsule.

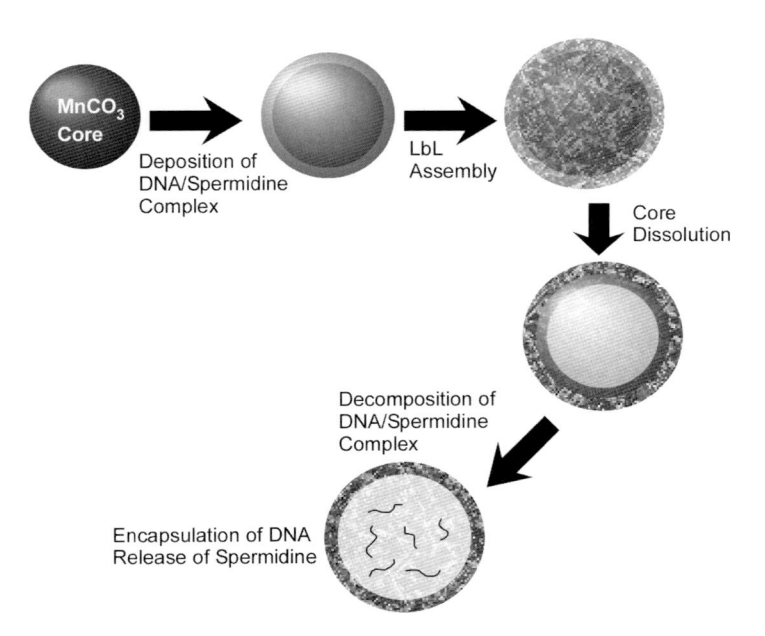

Figure 4.58 Encapsulation of DNA within layer-by-layer capsule.

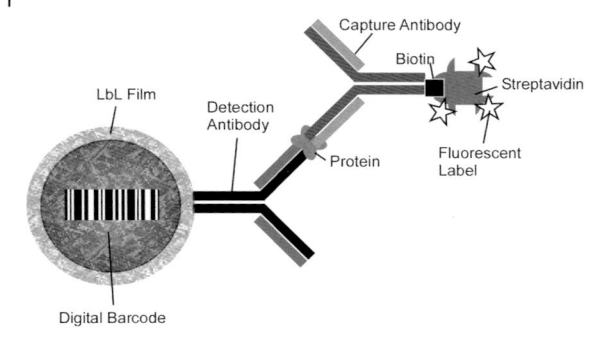

Figure 4.59 Antibody-carrying layer-by-layer capsule with digital barcode.

weight spermidine to complete DNA entrapment. As a more advanced analytical system, systems with use of fluorescent polystyrene microspheres encoded with a "barcode" and captured antibodies (Figure 4.59). This system enabled them to quantify proteins in serum and plasma.

Protein delivery is also an active research subject in the corresponding areas. Upon shell uptake by living cells, the walls of the microshells were actively degraded and digested by intracellular proteases. Wall degradation enabled intracellular proteases to reach the protein cargo in the cavity of the microshells. Enzymatic fragmentation of the fluorescence-labeled protein led to individual fluorescence-labeled peptides. In this case, only prodrugs inside the microshells reaching cells are activated and those in shells in an extracellular environment cannot be activated. Variously designed polymers have been used as components of the LbL microshells. For example, polymer hydrogel microshells based on disulfide crosslinked poly(methacrylic acid) were prepared through LbL assembly with poly(vinylpyrrolidone). The disulfide crosslinking provided a redox-active trigger for degradation that was initiated by a cellular concentration of glutathione. Combination of the LbL technique with molecular assemblies such as micelles and liposomes creates various unique shell systems. Preparation of LbL microshells constructed entirely from a cationic/zwitterionic pair of pH-responsive block-copolymer micelles was reported. Layers of anionic poly[2-(dimethylamino)ethyl methacrylate-block-poly(2-(diethylamino)ethyl methacrylate)] and cationic poly(2-(diethylamino)ethyl)methacrylate-block-poly(methacrylic acid) copolymer micelles were alternately assembled onto calcium carbonate colloidal templates. Addition of dilute ethylenediaminetetraacetic acid solution resulted in dissolution of the calcium carbonate and formation of hollow polymer microshells. These LbL microshells composed entirely of pH-responsive block-copolymer have potential applications in the encapsulation and triggered release of active therapeutic agents. Although LbL shells have been rapidly developed for potential applications, further technical and conceptual innovations are always required. Most the examples of LbL shells are categorized here as LbL microshells, because their diameters are in the micrometer to submicrometer range. However, injection of drug carriers requires very small diameters of drug encapsulants. There are currently only a

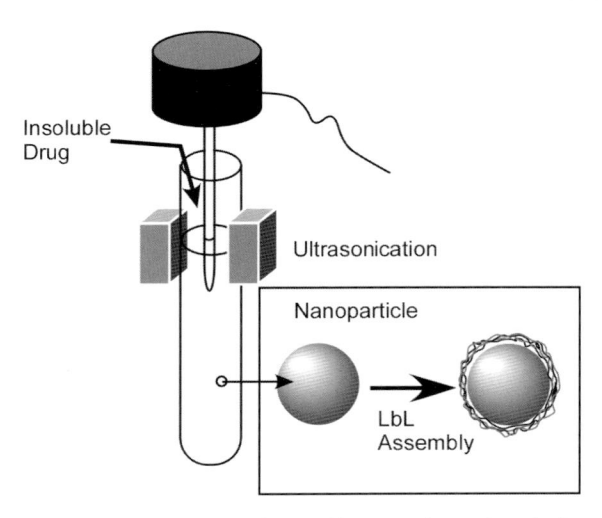

Figure 4.60 Entrapment of insoluble materials into layer-by-layer capsule with assistance of ultrasonics.

limited number of nanoscale delivery system established for the continuous sustained release of drugs. Nanoformulation of insoluble drugs, such as paclitaxel, tamoxifen, dexamethasone, and camptothecin is especially challenging. However, low solubility in water tends to be an intrinsic property of many high-potency drugs, including some powerful anticancer agents. An innovative method for LbL nanoshells for insoluble cancer drugs was recently proposed. Aqueous suspensions of insoluble drugs are subjected to powerful ultrasonic treatment in order to obtain nanosized cores (Figure 4.60). The resulting nanoparticles can be stabilized by maintaining the solution under sonication to prevent their fast reaggregation. Sequential addition of polycations and polyanions to the particle solution leads to the assembly of ultrathin polyelectrolyte shells onto the nanosized drug particles. This LbL strategy allows preparation of multilayer nanoshells with thicknesses of 5 to 50 nm and the requisite composition.

4.4
Brief Summary and Perspectives

In this chapter, preparations and functions of thin films through layer-by-layer assembly (LbL) are overviewed. Control of orientation and packing of components materials within the thin films by the LbL assembly is probably inferior to the other major thin-film preparation techniques such as the Langmuir–Blodgett (LB) method and the self-assembled monolayer (SAM) strategy. However, the LbL assembly is much simpler and versatile. This method allow us to prepare nanosized thin films in well-desired sequence and thickness with a huge variety of component selection. All these processes can be done by simple procedures of solution dipping, which can be, however, replaced by other technique such as spin coating, spraying, and even automated machines. The latter characteristic would be advantageous for physicochemical applications including device preparation. Another strong feature is the mildness of assembling process. The LbL assembly does not require any harsh conditions such as application of high temperature and contact with harmful chemicals. This point is highly advantageous for immobilization of biomaterials and their biorelated applications. Therefore, the LbL assembly has a huge potential in a wide range of applications. This excellent method for thin-film preparation must be further researched with the aim of practical applications.

Further Reading

1 Iler, R.K. (1996) Multilayer of colloidal particles. *J. Colloid Interface Sci.*, **21**, 569.
2 Decher, G., and Hong, J.D. (1991) Consecutively adsorption of anionic and cationic bipolar amphiphiles on charged surfaces. *Makromol. Chem. Macromol. Symp.*, **46**, 321.
3 Decher, G., and Hong, J.D. (1991) Consecutively adsorption of anionic and cationic bipolar amphiphiles and polyelectrolyte on charged surfaces. *Ber. Bunsen-Ges. Phys. Chem. Chem. Phys.*, **95**, 1430.
4 Decher, G., and Hong, J.D. (1992) Consecutively alternating adsorption of anionic and cationic polyelectrolytes on charged surfaces. *Thin Solid films*, **210**, 831.
5 Keller, S.W., Kim, H.N., and Mallouk, T.E. (1994) Layer-by-layer assembly of interaction compounds and heterostructures. Towards molecular beaker epitaxy. *J. Am. Chem. Soc.*, **116**, 8817.
6 Lvov, Y., Ariga, K., Ichinose, I., and Kunitake, T. (1995) Assembly of multicomponent protein films by means of electrostatic layer-by-layer adsorption. *J. Am. Chem. Soc.*, **117**, 6117.
7 Decher, G. (1997) Fuzzy nanoassemblies: toward layered polymeric multicomposites. *Science*, **277**, 1232.
8 Caruso, F., Caruso, R.A., and Möhwald, H. (1998) Nanoengineering of inorganic and hybrid hollow spheres by colloidal templating. *Science*, **282**, 1111.

9 Shiratori, S.S., and Rubner, M.F. (2000) pH-Dependent thickness behavior of sequentially adsorbed layers of weak polyelectrolytes. *Macromolecules*, **33**, 4213.
10 Caruso, F. (2000) Hollow capsule processing through colloidal templating and self-assembly. *Chem. Eur. J.*, **6**, 413.
11 Caruso, F. (2001) Nanoengineering of particle surfaces. *Adv. Mater.*, **13**, 11.
12 Bertrand, P., Jonas, A., Laschewsky, A., and Legras, R. (2000) Ultrathin polymer coatings by complexation of polyelectrolytes at interfaces: suitable materials, structure and properties. *Macromol. Rapid Commun.*, **21**, 319.
13 Hammond, P.T. (2004) Form and function in multilayer assembly: new applications at the nanoscale. *Adv. Mater.*, **16**, 1271.
14 Tang, Z., Wang, Y., Podsiadlo, P., and Kotov, N.A. (2006) Biomedical applications of layer-by-layer assembly: from biomimetics to tissue engineering. *Adv. Mater.*, **18**, 3203.
15 Jiang, C.Y., and Tsukruk, V.V. (2006) Freestanding nanostructures via layer-by-layer assembly. *Adv. Mater.*, **18**, 829.
16 Ariga, K., Hill, J.P., and Ji, Q. (2007) Layer-by-layer assembly as a versatile bottom-up nanofabrication technique for exploratory research and realistic application. *Phys. Chem. Chem. Phys.*, **9**, 2319.
17 De Geest, B.G., Sanders, N.N., Sukhorukov, G.B., Demeester, J., and De Smedt, S.C. (2007) Release mechanisms

for polyelectrolyte capsules. *Chem. Soc. Rev.*, **36**, 636.

18 Quinn, J.F., Johnston, A.P.R., Such, G.K., Zelikin, A.N., and Caruso, F. (2007) Next generation, sequentially assembled ultrathin films: beyond electrostatics. *Chem. Soc. Rev.*, **36**, 707.

19 Ariga, K., Hill, J., Lee, M.V., Vinu, A., Charvet, R., and Acharya, S. (2008) Challenges and breakthroughs in recent research on self-assembly. *Sci. Technol. Adv. Mater.*, **9**, 014109.

20 Wang, Y., Angelatos, A.S., and Caruso, F. (2008) Template synthesis of nanostructured materials via layer-by-layer assembly. *Chem. Mater.*, **20**, 848.

21 Srivastava, S., and Kotov, N.A. (2008) Composite layer-by-layer (LBL) assembly with inorganic nanoparticles and nanowires. *Acc. Chem. Res.*, **41**, 1831.

22 Jewell, C.M., and Lynn, D.M. (2008) Multilayered polyelectrolyte assemblies as platforms for the delivery of DNA and other nucleic acid-based therapeutics. *Adv. Drug. Deliv. Rev.*, **60**, 979.

23 Boudou, T., Crouzier, T., Ren, K., Blin, G., and Picart, C. (2010) Multiple functionalities of polyelectrolyte multilayer films: new biomedical applications. *Adv. Mater.*, **22**, 441.

24 Mora-Huertas, C.E., Fessi, H., and Elaissari, A. (2010) Polymer-based nanocapsules for drug delivery. *Int. J. Pharm.*, **385**, 113.

25 Ariga, K., Ji, Q., and Hill, J.P. (2010) Enzyme-encapsulated layer-by-layer assemblies: current status and challenges toward ultimate nanodevices. *Adv. Polym. Sci.*, **229**, 51.

26 Ariga, K., Lvov, Y.M., Kawakami, K., Ji, Q., and Hill, J.P. (2011) Layer-by-layer self-assembled shells for drug delivery. *Adv. Drug Deliv. Rev.*, **63**, 762.

5
Other Thin Films

Mineo Hashizume, Takeshi Serizawa, and Norihiro Yamada

This chapter consists of six sections including various topics except for those picked up in the proceeding chapters. Sections 5.1 and 5.2 discuss a bilayer membrane and relating molecular assemblages. The studies on these assemblages provide the conceptual background relevant to the main subject of this book. Sections 5.3–5.5 discuss unique techniques for preparing organized organic ultrathin films. The last section includes new techniques.

5.1
Bilayer Vesicle and Cast Film

5.1.1
Definition of a Bilayer Structure, a Bilayer Membrane, and a Bilayer Vesicle

In 1972, Singer and Nicolson proposed the fluid mosaic model as the structural basis of the biomembrane (Figure 5.1) [1]. According to this model, proteins are incorporated either on the surface or in the interior of the lipid bilayer membrane, which is composed of various kinds of lipids and steroids. The lipid bilayer membrane should be the very prototype of the bilayer membrane. Prior to the advent of the fluid mosaic model, Bangham and Horne had reported that the isolated lecithin or the lecithin-cholesterol of equal molar proportions formed "spherulites composed of concentric lamellae", when they were mechanically dispersed in water [2]. The spherulites are, of course, the multiwalled bilayer vesicles, and hence, it is the first finding of the bilayer vesicle. The spherulite was called the liposome before long, and its formation had been considered to be the intrinsic phenomenon of lipids, because such biological materials were so unstable that it was difficult to obtain without any help of living things. This consideration lacked the rigorous bases.

In 1977, Kunitake and Okahata demonstrated that the same spherical bilayer vesicle was formed when the man-made, unnatural surfactants were dispersed in water by irradiation of supersonic wave (sonication) [3]. They assumed that the unique structure of the one hydrophilic head group and two hydrophobic alkyl chains would be the essential framework to form a bilayer membrane. Thus,

Organized Organic Ultrathin Films: Fundamentals and Applications, First Edition. Edited by Katsuhiko Ariga.
© 2013 Wiley-VCH Verlag GmbH & Co. KGaA. Published 2013 by Wiley-VCH Verlag GmbH & Co. KGaA.

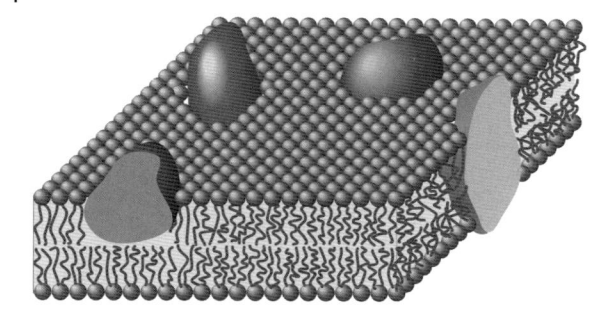

Figure 5.1 The fluid mosaic model of a biomembrane.

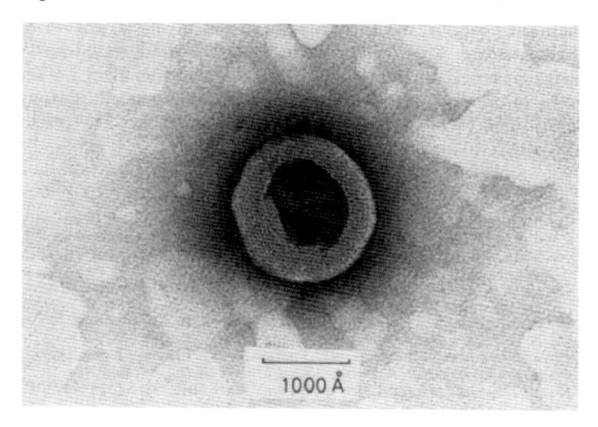

Figure 5.2 Synthetic bilayer membrane (taken from reference [3b]). Transmission electron microscope (TEM) picture of Didodecyldimethylammonium bromide dispersed in water by sonication. The sample solution was sonicated in the presence of uranyl acetate.

didodecyl dimethylammonium bromide, which has the simplest chemical structure, were synthesized as the first candidates [3b]. Figure 5.2 is a transmission electron microscope (TEM) picture of the multiwalled vesicle of didodecyl dimethylammonium bromide, which was enough to verify that the formation of a bilayer vesicle is not the intrinsic phenomenon of nature.

The finding of the totally synthetic bilayer membrane brought two important things, except for the generalization of the biological process described above. First, the synthetic bilayer membrane is helpful to study the lipid bilayer membrane, because the synthetic amphiphiles are easily prepared from simple precursors at low cost, and the resulting product is chemically and physically stable. Secondly, the finding and the subsequent studies break new ground, namely, molecular organization chemistry. Kunitake and coworkers have synthesized more than 600 amphiphiles, and studied their aggregation behavior [4]. These enormous accomplishments make it easy to prepare the molecule that is self-assembled into the desired assemblage.

Before continuing the explanation, it is necessary to define terminology, namely, a bilayer structure, a bilayer membrane and a bilayer vesicle. They are not synony-

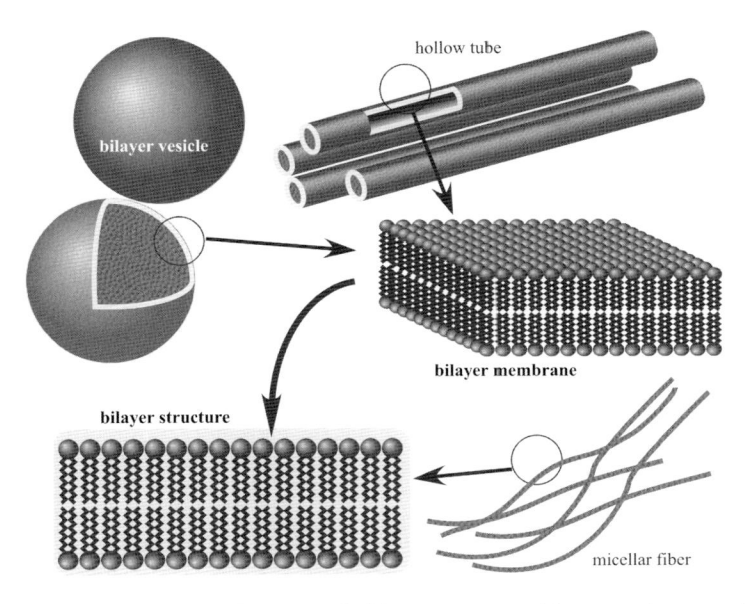

Figure 5.3 Schematic illustration of a bilayer structure, a bilayer membrane, and a bilayer vesicle.

mous. A bilayer membrane is a molecular assemblage of the simple amphiphilic molecule (amphiphile) in water, which is composed of a hydrophilic head group and hydrophobic alkyl chains (the other, new class of amphiphiles is described later). The molecular assembly was caused by the hydrophobic interaction. Due to hydrophobic repulsion, the alkyl chains are kept away from water, and form a double unimolecular layer, whose hydrophilic parts are kept in contact with bulk water as shown Figure 5.3. As a result, a two-dimensional membrane, in which the hydrophobic chains are sandwiched between hydrophilic parts, is formed. This is **the bilayer membrane**. A closed structure of the bilayer membrane is **a bilayer vesicle**, which is one of the aggregation structures in water. In the cross section of the bilayer membrane, we can see **a bilayer structure** as illustrated in Figure 5.3. The same bilayer structure can be seen in the longitudinal section of a micellar fiber (thread), although it is not a bilayer membrane. The micellar fiber is also one of the aggregation structures in water as is the bilayer vesicle. Furthermore, the bilayer structures are found in tapes, hollow tubes and helical superstructures, which will be discussed in Section 5.2. In the present section, the bilayer vesicle is mainly discussed.

5.1.2
Formation of a Bilayer Structure

5.1.2.1 Bilayer Forming Amphiphiles
Stated simply, the double-chain amphiphile, such as phospholipids and dialkyl ammonium salts, forms a bilayer membrane, whereas the single-chain amphiphile

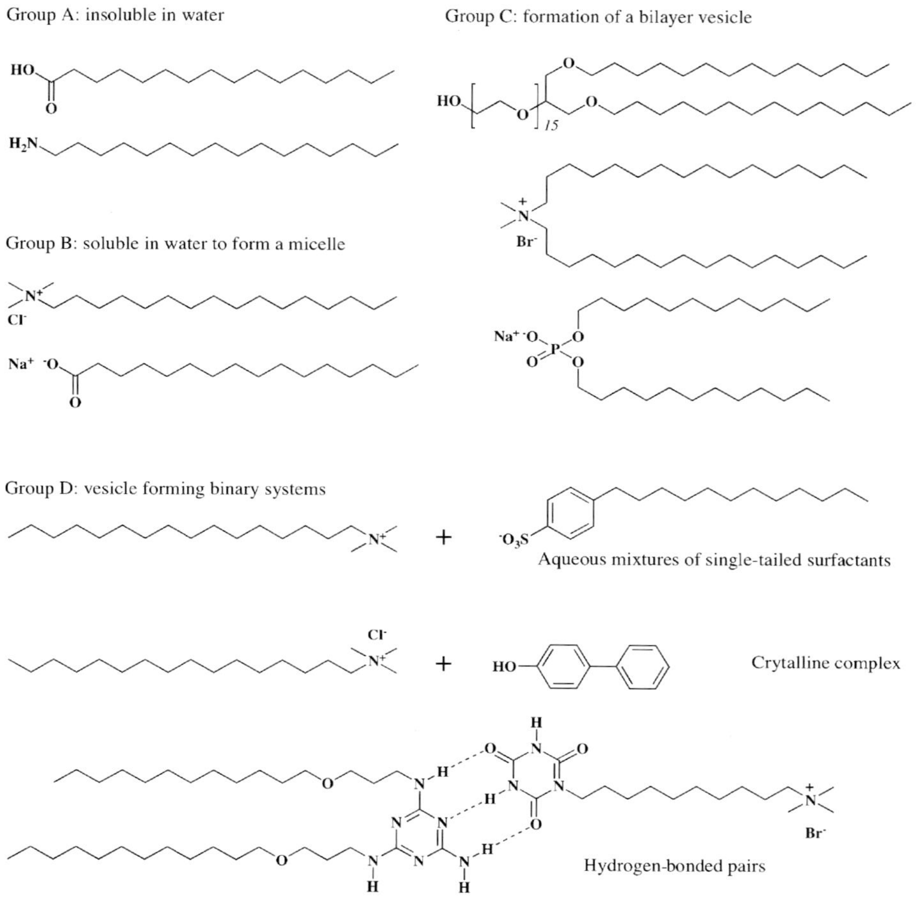

Figure 5.4 Several amphiphiles forming any assemblage.

forms a micelle. However, it is not so easy to set out the guideline to design new bilayer forming amphiphiles. A trustworthy method of the molecular design is to infer the needed structure inductively from structures of previously studied amphiphiles.

Thus, the amphiphiles were classified into four types (groups A–D) chiefly depending upon solubility into water (Figure 5.4). The amphiphiles belonging to group A are insoluble in water, which are unable to form a bilayer membrane but form a monolayer membrane at the air/water interface (Chapter 3). The amphiphiles belonging to group B are readily soluble in water to form a conventional spherical micelle. The amphiphiles belong to group C are slightly soluble in water, which can form a bilayer membrane. In order to avoid the confusion whether the aggregate is a micelle or a bilayer membrane, the following terms are sometimes useful.

1) A micelle solution is foamy, while an aqueous bilayer membrane is not foamy.
2) A micelle solution is transparent, while an aqueous bilayer membrane is translucent.
3) The micelle-forming amphiphiles are easily soluble in water, while the bilayer-forming amphiphiles are slightly soluble in water. Because of this property, the mechanical energy, such as supersonic waves, is often supplied to the latter system.

The last group D includes binary systems, namely, cationic and anionic surfactant complexes [5], crystalline complexes of monoalkyl ammonium surfactant and a certain aromatic compound [6], and complementary H-bond complexes of melamine and cyanuric acid derivatives [7]. These binary systems are useful to consider why the double-chain amphiphile prefers to form a bilayer membrane rather to form a micelle.

Using a "critical packing parameter", Israelachvili summarized the connection between the molecular shape and the aggregate morphology [8]. The critical packing parameter (P) consists of the volume of hydrocarbon chain (v), the optimal surface area (a_0), and the critical chain length (l_c).

$$P = v/a_0 l_c$$

Because all of the components, namely, v, a_0, and l_c, are estimated about the amphiphile whose alkyl chain melted, it is not so easy to calculate the parameter. However, when the amphiphile possesses a long alkyl chain, the term of v/l_c become constant (ca. 0.21 nm^2), because v and l_c for a saturated carbon chain possessing n carbon atoms are given below.

$$v \approx (27.4 + 26.9\,n) \times 10^{-3} \text{ nm}^3$$

$$l_c \approx l_{max} \approx (0.154 + 0.1265\,n) \text{ nm}$$

For example, the v/l_c values are 0.2095 nm^2 when $n = 10$, and 0.2099 nm^2 when $n = 12$. The component (a_0) is thermodynamically estimated. Finally, a bilayer membrane is formed, when $1/2 < v/a_0 l_c < 1$, whereas a spherical micelle is formed, when $v/a_0 l_c < 1/3$, as shown in Figure 5.5. This theory can explain the aggregation behavior for the amphiphiles possessing a simple molecular structure (Figure 5.4 groups A–C).

However, a new class of amphiphiles that recently appeared does not always obey this guideline. These new amphiphiles contain a third structural part other than the hydrophilic head group and the hydrophobic tail group in most cases. The third group is introduced into the amphiphile for various purposes, especially, of immobilizing the molecular alignment within the aggregate. In this case, the interaction between the third groups affects the aggregate structure. A typical example can be seen in the micellar fiber of the double-chain amphiphiles, which posses a tripeptide group [9]. For example, the critical packing parameter of the tripeptide-containing amphiphile is estimated at values above 1. Nevertheless, these amphiphiles did not form a bilayer vesicle at all, but formed a micellar fiber.

$v/a_0l_0 < 1/3$ $v/a_0l_0 = 1$

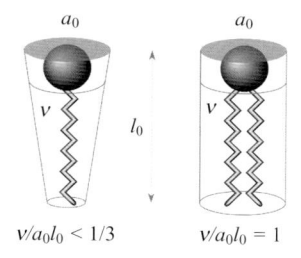

Figure 5.5 Relationship between aggregate morphology and molecular shape. Aggregate morphology of amphiphiles with a simple structure should be expected by a packing parameter; $P = v/a_0l_0$. Amphiphiles with $P < 1/3$ form a micelle and those with $1/2 < P < 1$ form a bilayer membrane [8].

On the other hand, a structurally similar dipeptide-containing amphiphiles did not form a micellar fiber, but formed a bilayer vesicle. Because a parallel chain β-sheet structure was formed in the direction of the long axis of the fiber, the H-bonding caused this exceptional instance.

The intermolecular interaction binds nearest-neighbor molecules, and promotes the stability of the molecular aggregate. Therefore, introducing the interactive group into a micelle-forming amphiphile, the spherical micelle should be transfigured into a superior molecular aggregate. In order to prepare the desired molecular assemblage, importance is placed on the controllability of the interactive group. Regrettably, the molecular interaction is not controlled in the present stage, and the obtained morphology is serendipitous in most cases.

5.1.2.2 Properties of Bilayer Membrane and Diagnostics of Bilayer Formation

Every molecular assemblage possesses its own peculiar aggregate structure, and the characteristic properties of the assemblage reflect the peculiar aggregate structure. Thus, representation of the transmission electron microscope (TEM) picture gives the strongest evidence to diagnose whether or not the bilayer membrane is formed. The scanning electron microscope (SEM) is also usable, but the SEM is inferior to the TEM in resolution. The TEM observation is carried out as briefly described below.

The sample amphiphile is dissolved into a solution containing a heavy-metal salt such as uranyl acetate (1–2%). A drop of the solution is put onto a cupper grid coated by amorphous carbon (TEM grid), and is held for 1–2 min. Aggregates adhere to the amorphous carbon in the short time, then, excess liquid is blotted off and the grid is dried *in vacuo*. The bilayer structure can be seen as a black line in the transparent background in a negative, or a white line in the black background on a positive, because the bilayer structure excludes heavy metal that obstructs the electron beam. This is the principal of the negative staining method of transmission electron microscopy. When the line width corresponds to the bimolecular length of the expanded molecule, it indicates the bilayer thickness. In general, a negative of 30 000 magnifications is required, at which the 7 nm thickness was magnified to 0.21 mm, and 1.05 mm on a 5 times magnified positive. The mem-

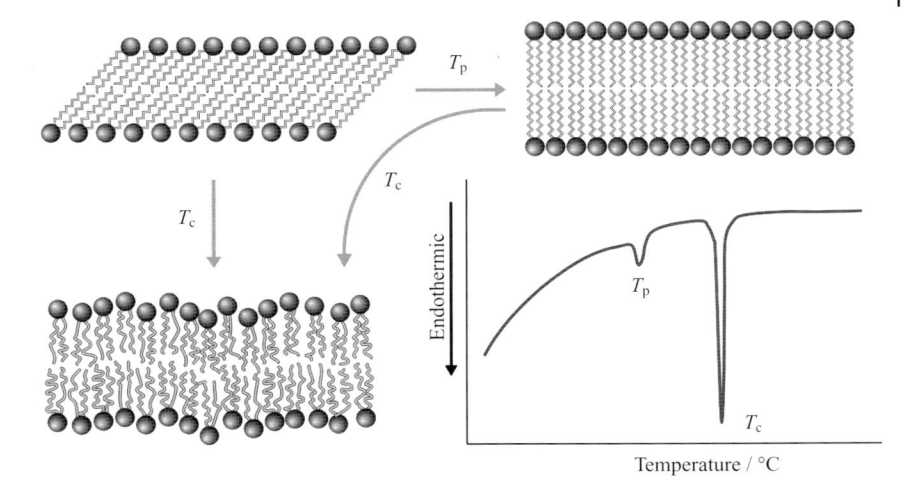

Figure 5.6 Gel to liquid crystal phase transition.

brane thickness is always less than that estimated from the expanded molecular length, because the tilt of the amphiphile reduces the membrane thickness.

As described above, an intrinsic aggregate structure induces its own characteristic phenomena. The thermal anomaly of a diluted aqueous aggregate is one of the characteristic phenomenon derived from a bilayer structure. The thermal anomaly of the bilayer membrane has been known as gel to liquid crystal phase transition. This critical transition temperature is denoted as T_c or T_m (suffixes c and m mean critical and main, respectively). All of the amphiphile maintains a *trans* zigzag structure at the alkyl chain part below T_c, which is the gel or the crystalline state. Above T_c, free rotation around the C–C axis occurs, and the alkyl chains become to fluid, which is the liquid crystal state (Figure 5.6).

The thermal behavior of the bilayer membrane is measured by means of differential scanning calorimetry (DSC), and the phase-transition temperature is observed as endothermic peaks on the DSC thermogram. The DSC thermogram sometimes accompanies a small endothermic peak as a pretransition temperature (T_p) or subtransition temperature (T_s) at the temperature below T_c. This pretransition peak should be attributed to a polymorphic conversion if the bilayer membrane adopted a different molecular packing at the crystalline state.

On the other hand, the alkyl chain in the micellar aggregate is fluid in the wide temperature range from 0 to 100 °C, and thereby, a diluted aqueous micelle solution does not show the gel to liquid crystal phase transition. However, a concentrated aqueous micelle solution sometimes shows a misleading thermal anomaly called the Krafft point (K_p). Below K_p, the micelle solution yields precipitates, whereas a solid amphiphile sharply dissolves and forms a micelle above K_p. In contrast, the aqueous bilayer membrane never produces precipitates below T_c. Consequently, the Krafft point is the critical temperature between the micelle and hydrated lamella phases, whereas the gel to liquid crystal phase transition is caused by the chain melting. Presence of T_c, thus, is the firm evidence for the formation of the bilayer structure. From this point of view, a micellar fiber is not a micelle, because the aqueous solution including the micellar fibers designates the gel to liquid crystal phase-transition phenomenon.

5.1.2.3 Mechanism and Preparation of Bilayer Formation

The bilayer membrane is easily prepared by stirring solids of amphiphiles in water or sometimes in buffer solution at temperatures above T_c. A probe-type or a bath-type sonicator is used, if the amphiphile is slightly soluble in water. However, irradiation of the supersonic wave makes the vesicle small. The bilayer vesicle with a larger size is prepared as follows; that is, the amphiphile is dissolved in chloroform, a small amount of which is injected into a vial, spread onto the inner surface of the vial, and dried in a stream of nitrogen. Water (or buffer) is added to the vial, and the coated amphiphile was dissolved by a moderate vortex at temperatures above T_c.

It has been considered that the micelle was formed by aggregation of monomerically dispersed surfactants in water at the concentration above CMC (critical micellar concentration). The CMC was determined by measuring the electrical conductivity of the ionic surfactants with a variety of concentration. The CMC values are, for example, 8.1×10^{-3} M for SDS (sodium dodecyl sulfate) and 9.2×10^{-4} M for CTAB (cethyl trimethylammmonium bromide). Thus, the CMC values of the micelle-forming, single-chain amphiphiles are about 1×10^{-3} M [10]. On the other hand, the CMC values of bilayer forming amphiphiles in several reports are $<10^{-6}$ M. However, the electrical conductivity of ultrapure water is about 1×10^{-7} S/cm, and 1 M of an aqueous solution of strong electrolyte is about 1×10^{-1} S/cm in general. These values suggests that measurement of an electrolyte solution whose concentration is less than 10^{-6} M is meaningless, because the obtained value is out of the allowable margin of error. Therefore, the CMC of the bilayer-forming amphiphile is too small ($<10^{-6}$ M) to measure by the electrical conductivity method. According to the definition of CMC, it seems that an aggregate should be formed by recombination (aggregation) of once dispersed molecules, without regard to whether the aggregate is a micelle or a bilayer membrane. This resolution-recombination process is, however, imaginary for the formation of a bilayer membrane. Perhaps, the bilayer-forming amphiphile directly forms a bilayer membrane as described below.

In contrast with the micelle, the bilayer membrane should be directly formed from the hydrated crystallite. Optical microscopic observation supports this idea, because the optical microscope revealed an initial event when the crystallite of the amphiphile comes into contact with water. Growth of Myelin figure should be the important initial event, which is a precursor of the bilayer vesicle. The Myelin figure is a relatively large object with birefringence, and hence, phase-contrast microscopy or polarized microscopy is especially suitable for the observation. Figure 5.7 is a very beautiful example of the Myelin figures reported by Sakurai *et al.* [11]. In this case, a very small amount of solid lecithin was sandwiched between a slide glass and a cover glass. A drop of water is contacted with the edge of the cover glass, so

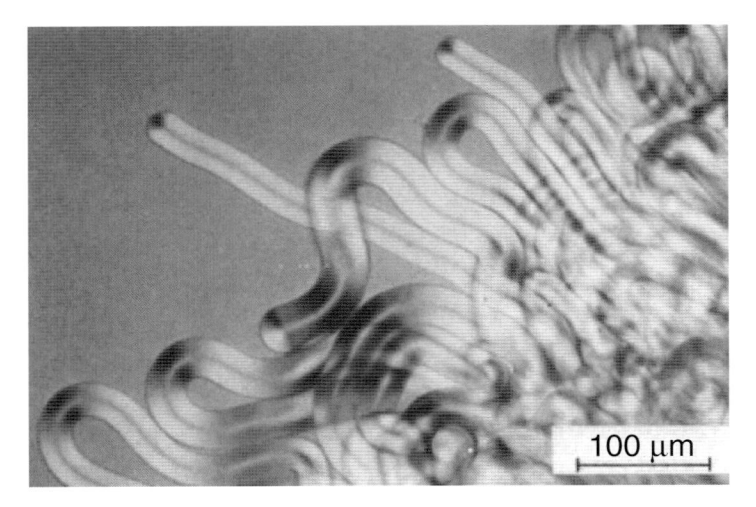

Figure 5.7 Myelin figure of DPPC by polarized microscopy.

that the water penetrates into the gap of the slide and cover glass. At the first stage, water penetrates into the interlayers of the crystallites of lecithin, then forms a hydrated crystal, and finally strips layers from the hydrated crystal. Thus, the Myelin figure grows immediately, which consists of multiple bilayer membranes (about 250 sheets of bilayers) and has a diameter of 20–40 mm. After the first event (growth of Myelin figure), morphological transformation arises, and a rosary-like structure comes into existence (second event). Many researchers consider that the rosary-like structure should be the precursor of the bilayer vesicle. Mechanical action, for example, a vigorous stirring or sonication, should segment the rosary-like structure into bilayer vesicles. However, the crucial point, that is to say scission of the rosary, has not yet been observed.

The bilayer vesicle, thus formed, is not always spherical, because a vesicle dynamically changes its shape, and transformed into various shapes, for example, a flexible bilayer tube as shown in Figure 5.8. Importance is laid on the topology of these shapes, because a spherical vesicle and a bilayer tube are topologically the same structures, whereas the flexible tube and a hollow tube described in Section 5.2 are topologically different structures. The dynamic morphological transformation is observable only when the temperature is above T_c. To observe a small vesicle or a fine filament, an optical microscope equipped with a dark-field condenser should be most suitable. Therefore, Hotani used the dark-field light microscope, and demonstrated time course for the morphological transformation of the Lecithin vesicle [12]. According to this study, a circular biconcave form transforms into a fine flexible filament or small spheres via a variety of regularly shaped vesicles looked like a peanut, a leaf of clover, a brunched tube and so on. The same morphological transformation is observable for all of bilayer membrane system. Figure 5.8 is an example of the morphological transformation of the crystalline complex system given in Figure 5.4, which demonstrates the diversity of aggregate shapes.

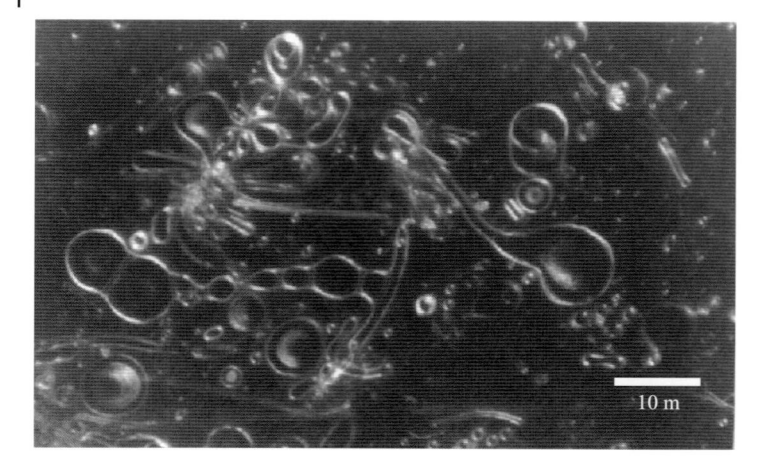

Figure 5.8 Dynamic morphological transformation of crystalline complex of cethyl trimethylammonium chloride and 4-hydroxybiphenyl (structural formulae are given in Figure 5.4).

5.1.2.4 Future of the Bilayer Vesicle

The life of vesicle should be made up of three periods. Growth of Myelin figure is the first stage. The second period includes morphological transformation from the Myelin figure to the bilayer vesicle, and topological changes of vesicular shape. These are the dynamic changes at the temperature above T_c, which are finished within a few hours. Therefore, the reversible morphological transformation of the bilayer vesicle permanently repeats as long as the temperature is maintained above T_c. This situation changes completely, if the temperature decreases below T_c, that is, The dynamic transformation disappears and small brilliant objects is observed in abundance. However, a new type of structural transformation appears despite the low temperature ($<T_c$). Interest is laid on the transformed products. The typical product is a helical aggregate and/or a hollow tube that will be discussed in Section 5.2. Thus, the third period includes the morphological transformation at temperatures below T_c. It is noteworthy that the helical aggregate and the hollow tube are topologically different structure from the bilayer vesicle or its prolonged shape (tubular vesicle).

5.1.3
Cast Films Containing a Bilayer Structure

The amphiphiles that form a synthetic bilayer membrane are chemically and physically stable, and are easily prepared from simple precursors at low cost. Because of these advantages, the synthetic bilayer membrane has been used as a model membrane to elucidate the function of the biomembrane, especially the lipid bilayer membrane. More specifically, permeation of ions or substances, substrate entrapment, fusion and phase-separation phenomena, enzyme model, and the other biomimetic chemistry have been investigated. Some of them are developed into pharmaceutical uses such as a drug-delivery system (DDS). On the other hand, orderly molecular arrangement and nanoscale membrane thickness are other advantages of the bilayer membrane, the former should be applied to exploit a molecular device system, and the latter should be applicable to nanocoating or nanopatterning. For this purpose, the immobilized bilayer membrane was required as an ultrathin film, in which molecules orderly aligned. In this case, water is an obstacle to immobilize the orderly molecular arrangement, although water is necessary to prepare the orderly molecular arrangement.

The cast film is an air-dried film of the solution containing a bilayer membrane, which is spread onto the appropriate substrate. The drying time and temperature affect the molecular array in the cast film, and hence, the surface morphology, the aggregation structure and the physicochemical properties will markedly change. For example, the original aggregate structure is kept intact, when the solvent is quickly removed *in vacuo*, whereas changes into the other structures, when the solvent is slowly evaporated. Talmon demonstrated the bilayer structure is observable in TEM images of the slowly dried specimen prepared from aqueous micelle, and raised an objection against Kunitake's TEM image including a bilayer vesicle (Figure 5.2). Talmon suggested that the TEM images depicted by Kunitake could be an artifact [13]. This is of course misunderstanding, because Kunitake dried the specimen very quickly *in vacuo*, which will produce no artifact. Therefore, a dried specimen for TEM observation has to be dried as fast as possible. If the specimen is slowly dried, almost all aggregate in a diluted solution transformed into other structures that are different from the original structure. It is noteworthy that the slow drying produces a similar structure without regard to whether the original aqueous aggregate is a micelle or a bilayer membrane. Such the structure is a multilamellae structure in most cases, which is consistent with phase diagrams of a variety of amphiphiles.

Being different from TEM observation, a multilamellae structure is more desirable for the cast film rather than the compiled structure of small aggregates. Therefore, slow drying is preferred for the preparation of cast films rather than fast drying. Figure 5.9 is the TEM image of the cast film thus obtained. The image depicted a cross-sectional view of the cast film of a **N^+C_6-Ph-L-Glu(OC$_{12}$)$_2$** bilayer membrane [14]. Orderly lamellae mean that the bilayer structure is maintained if water is removed.

On the other hand, not only a bilayer structure but also physicochemical properties are maintained in the cast film. For example, the cast film exhibits a thermal anomaly, and the gel to liquid crystal phase-transition temperature (T_c) of the cast film is close to that of the aqueous solution containing the original aggregates. Thus, the cast film is treated as a water-free bilayer membrane, and many researchers supposed that chemical or physical processes in the biomembrane should be reproduced without use of water. Although a self-supporting film is obtained if the cast film can be peeled off the substrate, the self-supporting film is extremely brittle. The fragility is a significant disadvantage of the cast film bilayer. In order to overcome the disadvantage, many attempts were made. Polymerization of

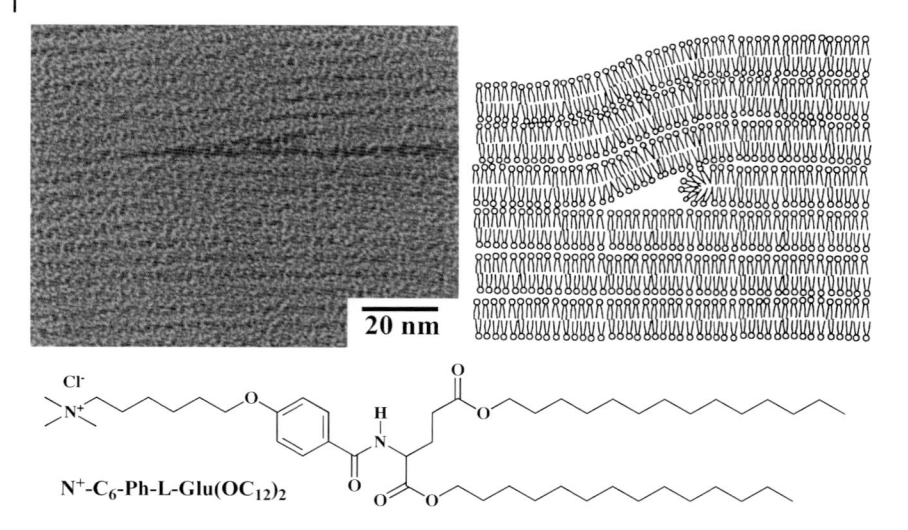

Figure 5.9 A cross-sectional TEM image of an air-dried cast film of **N⁺-C₆-Ph-L-Glu(OC₁₂)₂** and its schemeatic illustration for molecular alignment of components.

bilayer components should be most extensively investigated. O'Brien has reported an excellent review on these study [15]. However, polymerization destroys the molecular arrangement of the bilayer membrane in general. On the contrary, it is difficult to polymerize tightly packed amphiphiles in the bilayer assemblage. These experimental results indicate that the covalent linkage between adjacent molecules is incompatible with the molecular orientation in the bilayer membrane. This is the intrinsic problem, and hence, restricted amphiphiles produce a polymer vesicle successfully.

In contrast, a polyion-complex formed a mechanically strong cast film, which is made of an ionic amphiphile and a polyelectrolyte with an opposite charge by electrostatic attraction [16]. For example, an ammonium amphiphile and potassium poly-styrenesulfonate (PSS⁻K⁺) should be the most frequently used combination. Following is the preparative method of the polyion-complex; that is, into the aqueous solution of the ammonium amphiphile, the aqueous solution of PSS⁻K⁺ possessing equivalent charges is added at once. The solution turns milky by yielding fine precipitates. The precipitates are collected by use of a centrifuge, washed by water, and dried *in vacuo*. The polyion-complex thus obtained is insoluble in water, but soluble in organic solvent to produce a transparent solution. Because electrostatic attraction between the surface of the aggregate and polyelectrolyte do not affect the molecular orientation, a bilayer structure and its inherent properties are maintained intact. Bilayer stability is reinforced by the polyelectrolyte. Using the polyion-complex, Okahata *et al.* exploited the fragrance sensor, which could be the most successful example among applicable usage of the cast film bilayers [17].

5.2
Self-Assembled Fibers, Tubes, and Ribbons

5.2.1
Introduction

Molecules that form a self-assemblage, as well as a bilayer vesicle, produce a micellar fiber, a hollow tube, a ribbon- or tape-like aggregate and so on. The last ribbon (or tape) was sometimes twisted into a helical aggregate. Despite the morphological variation, almost aqueous aggregates were derived from only two fundamental aggregates, namely, a micellar fiber and a bilayer vesicle described in Section 5.1. For example, a micellar fiber and a bilayer vesicle sometimes metamorphoses into a helical aggregate depending upon incubation at temperatures below T_c (Section 5.2.2), and further incubation makes the helical aggregate into a hollow tube, which should be the ultimate morphology. The helical aggregate and the hollow tube are the superstructure, in which a very orderly molecular alignment is involved. Morphological variation is observed not only in water but also in nonaqueous media. A reversed micellar fiber, a helical superstructure, and a hollow tube are also observed as an ordinary aggregate. However, a reversed bilayer structure is the structural basis for these aggregates. Being different from aqueous systems, there are few reports on a reversed bilayer vesicle, and most of aggregates formed in organic solvent were fibrous. The fibrous aggregates are intertwined with each other to form a 3D-network structure, which hardens the medium. This is the organogel. Therefore, Sections 5.2.2 and 5.2.3 chiefly discuss a helical superstructure and the organogel. The former has exploited a new field of chemistry on highly organized molecular assembly, and the latter has contributed to establishment of a methodology on molecular assembly in organic media. Thus, we can now obtain the molecular assemblage that possesses a highly organized molecular alignment by use of a variety of molecules such as a supramolecular polymer. However, in order to array the component molecules, molecular interactions between the components has to be controlled. Section 5.2.4 discusses a way to control the molecular arrangement.

5.2.2
Finding a Helical Superstructure

In 1965, Tachibana and Kambara reported that the aggregates of chiral 12-hydroxyoctadecanoic acid and its soaps with metal ions produced a twisted fiber [18]. This is the first finding of the helical superstructure. Tachibana and coworkers had been diligently studying the fantastic aggregate for about 10 years, but no one took an interest in the helical aggregate at that time. About 20 years later from the first finding by Tachibana, the similar helical aggregates were found within the aged solution of a bilayer vesicle in 1984 [19]. Furthermore, a tubular superstructure was also found in the same year [20]. Since then, the helical superstructure has attracted many research workers' attention, because of the extraordinary regularity of the molecular alignment within the structure, and the structural reversibility between the helix and the vesicle. The helical superstructure is derived from a bilayer membrane by incubation at a temperature below T_c, but returns to the original bilayer membrane with an increase in temperature above T_c. For example, the chiral amphiphile of **N^+-C_{11}-L-Glu(OC_{12})$_2$** (structure is given in Figure 5.11) had been known to form a typical bilayer vesicle in water. However, Nakashima and coworkers have demonstrated that the vesicle transformed into the helical superstructure by the incubation [21]. The vesicles of **N^+-C_{11}-L-Glu(OC_{12})$_2$**, whose T_c is 34 °C, metamorphosed into the helices with micrometer scale, when the aqueous solution was incubated in the temperature range 10–25 °C for 1 day. Further incubation at the same temperature for 1 month yielded hollow tubes. The helices and the hollow tubes are stable only at temperatures below T_c (34 °C in this case) of the bilayer membrane. On the other hand, the helices and the tubes immediately returned to original vesicles when the solution is warmed to 40–50 °C. The morphological transformation is thus reversible, which means that the helical superstructure consists of a bilayer membrane. Figure 5.10 represents pictures of the helix formation of **N^+-C_{11}-L-Glu(OC_{13})$_2$**. Although the

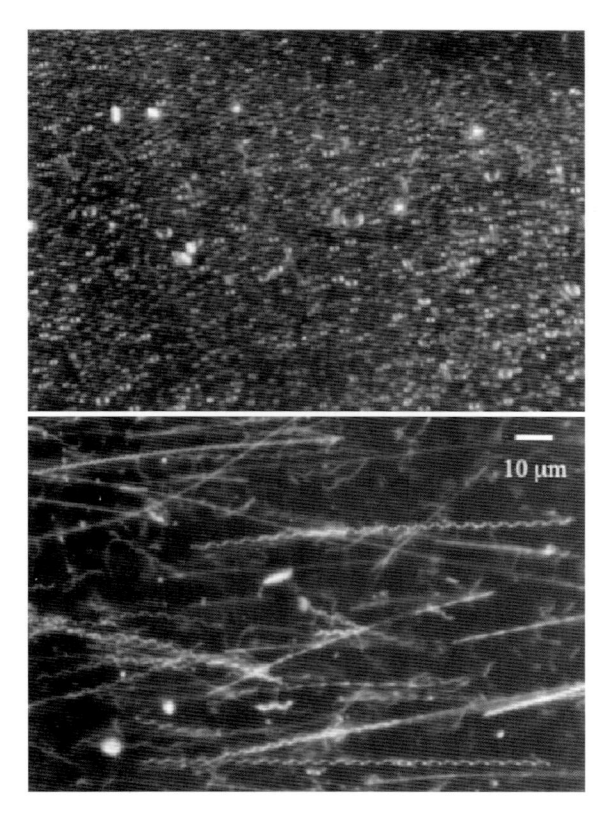

Figure 5.10 Morphological transformation of aqueous aggregate of **N^+C_{11}-L-Glu(OC_{13})$_2$** from vesicles (upper) to helical superstructures (lower).

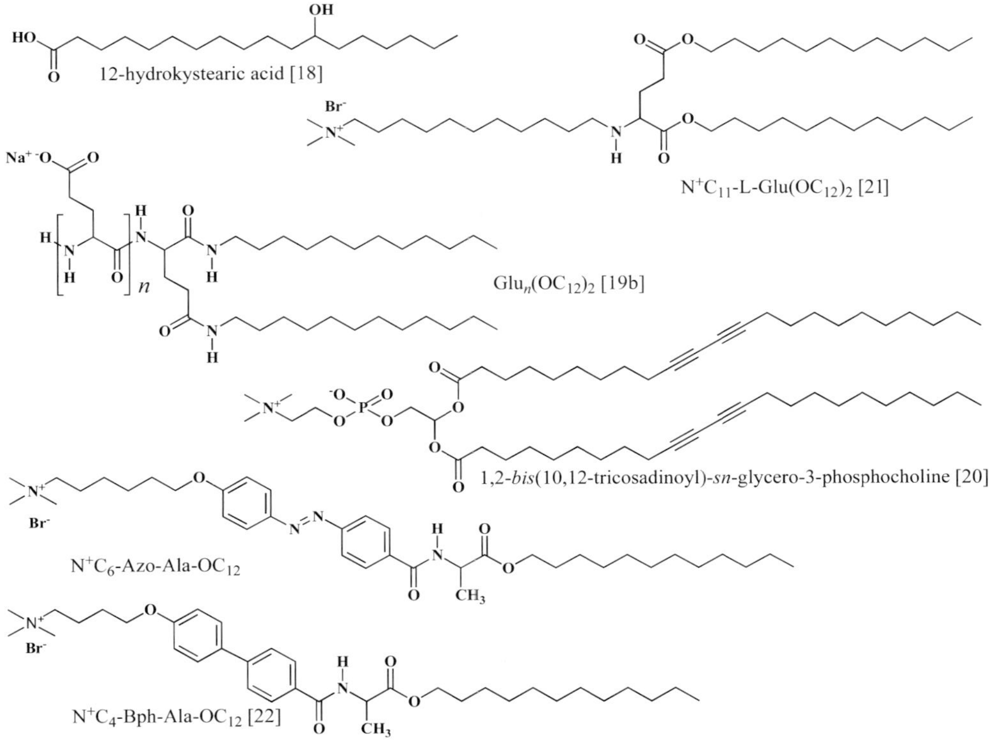

Figure 5.11 Helix-forming molecules that were reported in the early stage of the study.

amphiphile in Figure 5.10 is different from Nakashima's amphiphile in chain length, the same morphological transformation reported by Nakashima et al. [21] was observed.

The helical superstructure is sometimes called a helical microstructure, a helical ribbon, a twisted tape and scrolled tape, which are of the same structures. Figure 5.11 picked up some helix-forming molecules that were reported in the early stage of the study, and TEM images of the typical helices were given in Figure 5.12.

The helical superstructure is a quite ordinary aggregate at the present time, because it has been known that most chiral, rod-like shaped molecules produce the helical aggregate. The structural characteristics of these molecules are summarized as follows.

1) The helical superstructure is peculiar to chiral amphiphiles. Each of the enantiomers forms helices with opposite helical senses. The racemic amphiphile forms a tape-like aggregate, or produces a mixture of helices with opposite senses [23].

2) The helical superstructure exhibits a constant pitch and width, thereby its components arrange very regularly.

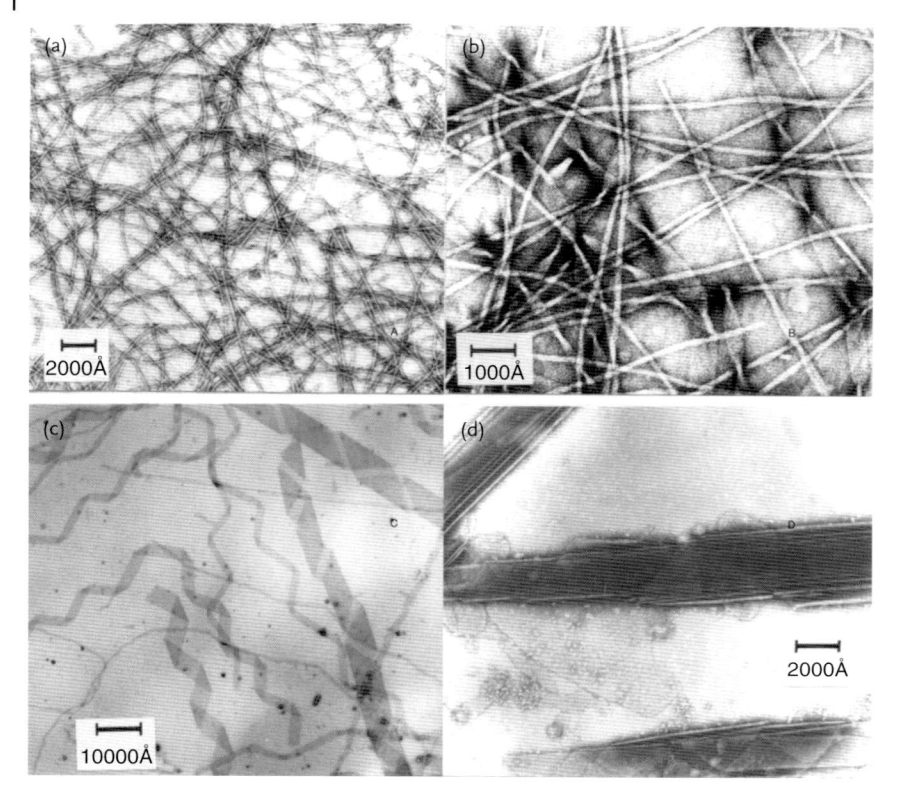

Figure 5.12 Example of TEM images of the typical helical superstructures in water. Morphological transformation of **N$^+$C$_{10}$-Azo-Ala-OC$_{12}$** aggregate; from fibers (a) to helices (b). Narrow and wide helices of **N$^+$C$_4$-Bph-Ala-OC$_{12}$** with the same pitch (c). Rolled helices of **N$^+$C$_6$-Azo-Ala-OC$_{12}$**, which are considered to be the precursor of a hollow tube (d). Structural formulae are given in Figure 5.11.

3) The helical superstructure slowly widens its width, and the edges are fused to form a hollow tube.

4) The helical superstructure and the hollow tube are not flexible, and hence, they should be closely related to the crystallite.

5) The helical superstructure is also found in nonaqueous media, and the helix formation causes gelation of the solution in many cases (see next term).

5.2.3
Organogel

On the analogy with a reversed micelle, many researchers who were studying molecular organization were interested in the formation of a reversed bilayer structure in nonaqueous media. The research report concerning the reversed micellar fiber and the reversed bilayer membrane appeared in the late 1980s [24,

25] and in the early 1990s [26]. Many reports on a reversed bilayer structure appeared one after another, these reports have a common context, that is, the solutions turned into gels, in which a fibrous, helical and/or tubular superstructure was observed abundantly. On the other hand, researchers who were studying organogels, which is a different research field from the former molecular organization chemistry, knew the story already, though many of them didn't know the aggregate contained a reversed bilayer structure, or didn't notice a local structure in the aggregate in those days.

Organogels are viscoelastic materials comprised of an organic solvent and a small amount of low molecular mass molecules called gelators. Because researchers who were studying organogels aimed to elucidate the structural requirements for a molecule to gel an organic solvent, a great many gelators have been synthesized and the ability of gelation examined [27]. Of course, TEM observation is crucial for this purpose, and hence, a fibrous, helical and/or tubular superstructure formed a 3D-network structure within organic solvents. Therefore, the structural framework of the molecules that form a reversed bilayer structure is the same as that of gelators. It is interesting that the same result has been obtained from the different research fields of chemistry. Figure 5.13 shows the molecules forming a superstructure in organic media.

The aggregation in organic solvents can be explained by miscibility and intermolecular interaction. Miscibility means mutual solubility between an arbitrary part of the molecule and the bulk media in this case. For example, a hydrocarbon chain is miscible with a hydrocarbon medium, while a fluorocarbon chain is miscible with a fluorocarbon medium. Because of the miscibility, the compound possessing two hydrocarbon chains and one fluorocarbon chain formed a bilayer vesicle in fluorocarbon media, whereas the compound possessing two fluorocarbon chains and one hydrocarbon chain formed a bilayer structure in hydrocarbon media [28]. When a part of the molecule is not miscible with bulk solvent, it is kept away from the bulk because of the repulsion between them. Molecular assembling based on the miscibility is possible for the molecule composed of structurally different parts, namely, hydrocarbon–fluorocarbon, dipolar–nonpolar, and solvophilic–solvophobic. Such structural asymmetry only assembles the molecules. In order to improve the mechanical strength and/or stability of the molecular assemblage, another interaction that fixes adjacent molecules to each other is needed. Intermolecular H-bonding, van der Waals forces, and metal coordination bonding are considered to be effective interactions. The next section discuss the strategy of strong and stable membrane formation.

5.2.4
Control of Aggregate Morphology

A vesicle, a helical superstructure, and a hollow tube were reversibly available in water by changing the incubation temperature and/or incubation time (Section 5.2.2). A helical superstructure and a hollow tube were also observed in organogels, although their local structure was a reversed bilayer structure (Section 5.2.3). Thus, whether the local structure is a bilayer structure or a reversed bilayer structure

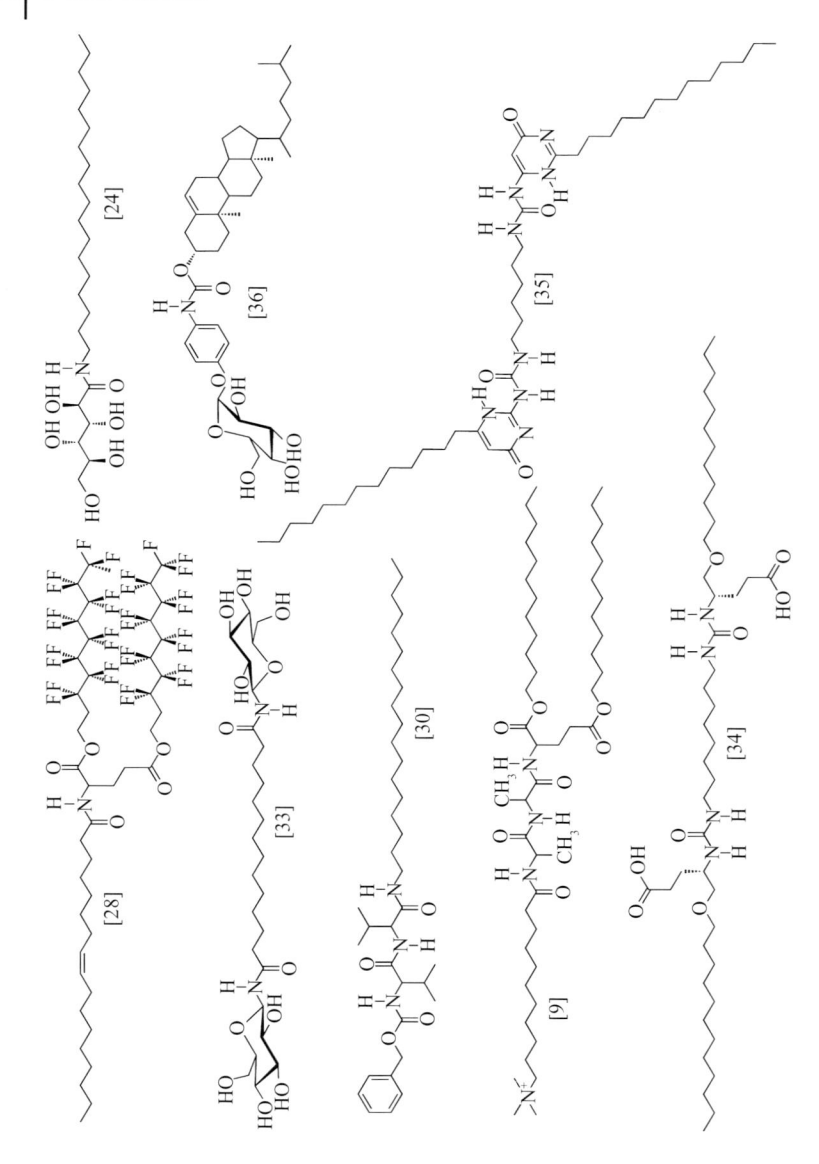

Figure 5.13 The molecules forming a superstructure in organic media.

changes depending upon the kinds of media (aqueous or nonaqueous). Additives sometimes changed the local structures. For example, an aqueous phosphate bilayer membrane yielded a water-insoluble complex by addition of $CaCl_2$, but the complex was dissolved in chloroform and toluene to form a fibrous aggregate, which should be based on a reversed bilayer structure [24]. In this way, aggregation is controllable by temperature, kinds of media, additives, and so on. We can now obtain a wide variety of molecules that form aggregates in water and/or organic solvents. Then, can we prepare the molecule, which forms an imaged aggregate shape in imaged medium, as we require? The answer is "it is not easy, but it is possible". In order to prepare the desired molecules, logical molecular design is required instead of serendipity. Thus, we have to know what kinds of structural elements are needed for the building blocks of the molecule, although the structural aspects are still poorly understood. Therefore, this section refers to structural requirements on aggregation about a rod-like molecule that is the most widespread molecule to posses aggregate-forming ability.

5.2.4.1 Composite Structure with Two or More Different Parts

Tanford has thermodynamically explained the aggregation mechanism of amphiphiles in water as the hydrophobic effect [29]. According to this theory, it seems that the attractive force between a polar head group and water, and repulsive force between a hydrocarbon chain and water produce a micelle or a bilayer membrane. Similar consideration is possible for the aggregation in organic solvents. Self-organization of molecules thus results from synchronous action of attractive and repulsive forces between a molecule and a bulk medium, irrespective of the kinds. A solvophilic and a solvophobic part in a rod-like molecule provides the attractive and the repulsive forces, respectively, because the former part is miscible with the medium, while the latter part is kept away from the bulk. The solvophobic parts form a domain by themselves, which is covered with the solvophilic parts. Solvophilic and solvophobic parts are completely opposite in property, and their structural frameworks are entirely different from each other. For example, an ammonium group and a long hydrocarbon chain are different from each other in polarity.

Therefore, the molecule that forms a molecular assemblage should be synthesized by alternately combination of structurally different functional groups. This is the composite structure. All of the aggregate-forming molecules so far reported involve the composite structure. Figure 5.14 illustrates variation of the composite structure, namely, a simple surfactant type, a double-chain amphiphile type, triple- or more-chain amphiphile type, Gemini amphiphiles (a double-chain and double-head amphiphile type), symmetric and asymmetric bora-amphiphile, and symmetric barbell-type molecules.

5.2.4.2 Hydrogen Bonding to Immobilize Orderly Molecular Arrangement

The aforementioned composite structure is the necessary framework of molecules that form by self-assembly. In order to apply self-assembly to an organic thin film, tighter bonds in the direction of the short axis of the molecule are essentially

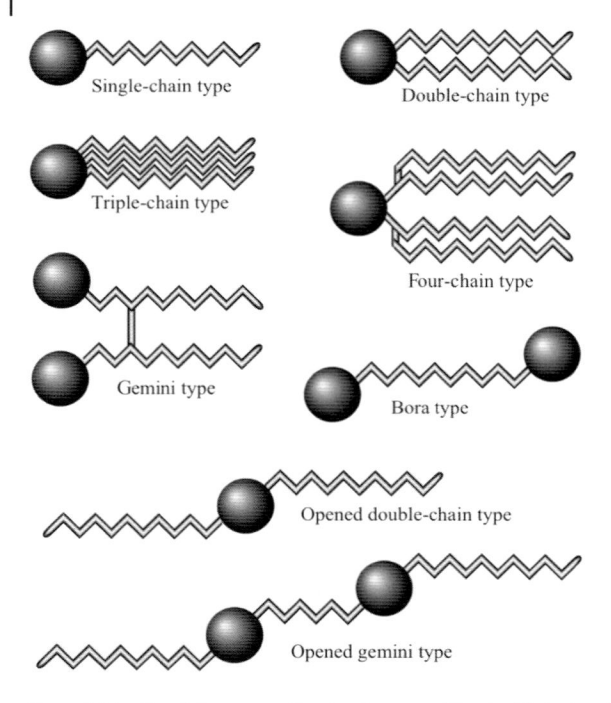

Figure 5.14 Possible composite structure combined with hydrophilic and hydrophobic parts.

needed to reinforce the aggregate structure. The intermolecular H-bonding, especially multiple H-bonding, is the most effective for the purpose, because it brings not only the stability but also a more orderly arrangement of molecules. A peptide or a carbohydrate group is used for the structural unit of multiple H-bonding as it is. For example, N-Protected dipeptide alkyl esters [30] and oligopeptide-containing amphiphiles [9] can form superstructures despite their simple structures. An oligopeptide itself also forms a superstructure, if the oligopeptide is composed of a hydrophilic peptide region and a hydrophobic peptide region [31]. Such oligopeptides are called amphiphilic peptides. Thereby, a peptide group should be the most frequently used building block that can form H-bonding. Moreover, peptide moieties form a β-sheet structure each other when the peptide consists of more than three amino acid residues [9]. The same β-sheet structure has been observed within the β-amyloid fiber, which cause amyloidoses such as mad cow disease (BSE), Alzheimer disease, scrapie, and Creutzfeldt–Jakob disease (CJD) [32]. Therefore, the superstructures formed by a peptide-containing molecule could be applicable to study amyloidses as an amyloid model.

A carbohydrate, particularly sugar, is used as a hydrophilic head group, because the carbohydrate contains many hydroxy groups [26, 33, 34]. The hydroxy groups form intermolecular H-bonding as well. Thus, the carbohydrate derivatives containing hydrophobic parts, for example, an alkyl chain, can form a well-developed molecular assemblage without further interactive groups. N-Alkylaldonamide [26] and bola-form sugars such as 1-glucosamide bola-amphiphile [33], are well-known building blocks of the superstructure.

On the other hand, amide, urea, urethane, ureido, and other related linkages containing a set of H-donor and H-acceptor are used as H-bonding resources [35]. In particular, simultaneous use of two or more groups is very practical to fasten adjacent neighboring molecules, which is applied to "supramolecular polymers" advocated by Meijer and coworkers [36]. The supramolecular polymers do not contain a covalent bond between repeating units. Thus, all of the repeating units in the supramolecular polymer are held together by directional and reversible secondary interactions such as H-bonding. Because the bond energy of a H-bond is about five times smaller than that of a covalent bond, the multiple H-bonding is used to gain a larger bond energy corresponding to the covalent bond. When the repeating molecules bind in the direction of the long axis of the molecule, main-chain H-bonded, linear polymers are obtainable. In contrast, when the repeating molecules bind in the direction of the short axis of the molecule, two-dimensional molecular alignment should be obtained as a monolayer film.

5.2.4.3 Hierarchic Interaction and Further Interaction

Increase of lateral interactions within the layer improves stability or mechanical strength of organic thin films based on either a bilayer or a monolayer membrane. Because intermolecular H-bonding is the most familiar lateral interaction, and the more H-bonding bring the more stability, multiple H-bonding is preferably used. However, one-dimensional H-bonding is not so effective to improve the quality of thin films, even if multiple H-bonds are formed. In order to make fine-quality films, the component molecules must be held together within a two-dimensional layer.

There are a few means for two-dimensional lateral stabilizing. Use of diglycine derivatives containing three amide linkages is one of the effective methods, because the three amide linkages form a crosslinked structure within the layer [37]. This crosslinked structure results from the unique structure, in which each amide is rotated about the long axis of the molecule by 120° with respect to neighboring amides. The ternary H-bonding is called a 3_1 helical structure, and it was first applied to stabilizing SAM by Hutchison and coworkers [37], then to form a vesicular tube by Shimizu et al. [38].

Participation of secondary interaction should be capable of improving the layer's property. In this case, the direction of the secondary interaction must be at right angles to that of the primary interaction, and the primary and the secondary interactions act in that order. Thus, all molecules are fixed in the direction of the X-axis according to the primary interaction, then the secondary interaction act in the vertical direction to the X-axis. A conclusive example is now drawn by use of a series of peptide-containing ammonium amphiphiles. Ammonium amphiphiles without any amino acid residues form a bilayer vesicle in water (Figure 5.2). The same vesicle is formed from the amphiphiles that contain one amide linkage and a dipeptide group (Figure 5.15a). The peptide-containing amphiphiles orderly align within the vesicle because of the intermolecular H-bonding. Therefore, the peptide-containing amphiphiles produce a very regular lamella structure in the

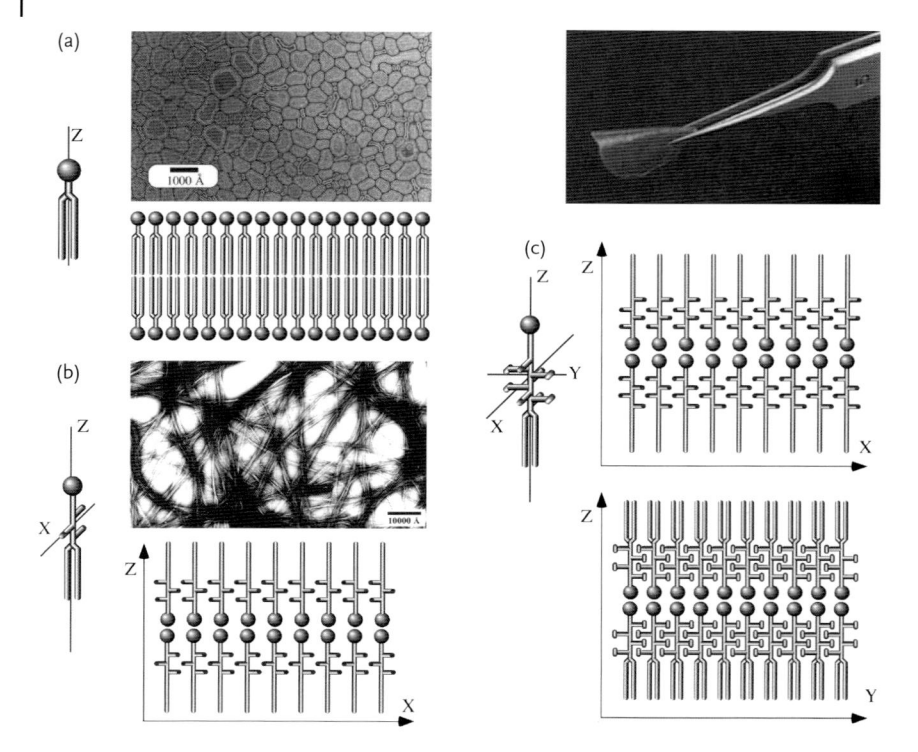

Figure 5.15 Evolution of an aggregate-forming amphiphile. A composite structure (a polar head and alkyl chains) itself provides a bilayer vesicle (a). Addition of H-bonding units derived a micellar fiber, in which linear immobilization of components was achieved in the direction of a long axis of the fiber (b) [9]. Further molecular interaction, molecular zipper in this case, also immobilized components in the vertical direction of the b-sheet plane, which promoted mechanical strength of the air-dried cast film (c) [39].

cross section of the cast film (Figure 5.9), and form a helical superstructure that involves highly orderly molecular alignment, when the aqueous solution was aged at temperatures below T_c. A marked change arises, when amphiphiles contain a tripeptide group. The tripeptide-containing amphiphile scarcely forms a vesicle, but forms a fibrous aggregate not only in water but also in nonpolar organic solvents. The fibrous aggregate in water is a micellar fiber and that in organic solvents is a reversed micellar fiber (Figure 5.15b).

Interestingly, these fibers do not yield a helical superstructure, although many other fibrous aggregates grow into helical aggregates. This could arise from strong multiple H-bonding. The tripeptide-containing amphiphiles form a parallel chain pleated β-sheet structure, which is very easily diagnosed by means of FT-IR spectroscopy. The β-sheet is a two-dimensional planar structure, and respective molecules oriented with parallel. Thus, the parallel chain β-sheet structure is formed in the direction of the long axis of the fiber, hence, the bilayer structure should be found in the longitudinal section, while the micelle structure in the cross section.

On the other hand, the side chains of the adjacent amino acid residues locate in opposite directions across the β-sheet plane. Therefore, the side chains could interlock with counterparts of the other β-sheet planes. This specific interdigitated structure should fasten the β-sheet planes when the peptide part contains at least three consecutive leucine residues, as illustrated in Figure 5.15c. This is a new type of molecular zipper called leucine fastener [39]. The leucine fastener will not be formed if a β-sheet structure is absent. Emphasis is placed on this hierarchy. Because the secondary interaction never acts unless the first structure due to the primary interaction is formed, these interactions are strongly correlated and hierarchic.

5.3
Polymer Brush Layer

5.3.1
Definition of Polymer Brushes

Modification of material surfaces with synthetic polymers attracts much attention for fundamental science and practical applications in various technological fields. Organized ultrathin films with nm–μm thickness can be prepared by synthetic polymers end-grafted chemically or physically at a certain graft density on a solid substrate. The formation of polymer-grafted layers is significant to control the physicochemical properties of material surfaces such as adhesion, separation, transportation, lubrication, abrasion, and so forth. In particular, the covalent binding of a graft polymer terminus enable the polymer films to be stable against chemical and mechanical stimuli, and therefore give various unique properties to the polymer films.

The conformation of graft-polymer chains in a good solvent is strongly dependent on the 2-dimensional graft density of polymer chains (Figure 5.16) [8, 40, 41]. At relatively low graft density, polymer chains do not overwrap with each other at a substrate surface, and form a "mushroom" conformation similar to free polymer chains with a random-coil conformation in solutions of a good solvent. The graft

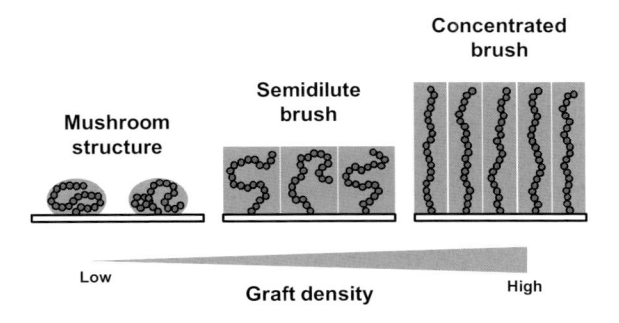

Figure 5.16 Density-dependent structures of graft polymers.

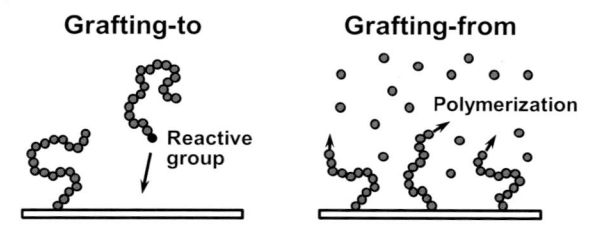

Figure 5.17 Methods to prepare polymer brush layers.

density for mushroom conformations may be less than ~0.01 chains/nm^2. At greater graft density, which may be more than ~0.1 chains/nm^2, polymer chains overwrap with each other, and are constrained to stretch away from a substrate surface due to the osmotic pressure of solvents penetrated into the graft polymer layers, thus forming organized ultrathin films. The resulting films composed of end-grafted polymer chains with stretched conformations are defined as a "polymer brush". Depending on graft density, two types of polymer brushes are proposed; one is "semidilute" polymer brushes and the other is "concentrated" (or high-density) polymer brushes.

5.3.2
Preparation of Polymer Brushes

Polymer brush layers can be prepared either by "grafting-to" or "grafting-from" methods (Figure 5.17). The grafting-to method is performed by chemical or physical reactions of end-functionalized polymer chains onto a substrate surface. It is difficult to prepare concentrated polymer brushes by this method, because the reactivity of end-functionalized polymer chains decreases with an increase in graft density due to kinetic hindrance for reacting new polymer chains. In fact, the kinetic effect is promoted by increasing the molecular weight of end-functionalized polymer chains. Therefore, polymer brushes with mushroom or semidilute structure are commonly prepared by the grafting-to method.

On the other hand, the grafting-from method is achieved by polymerization of certain monomers from initiators immobilized onto a substrate surface. Since each graft polymer chain grows independently on a substrate surface in a good solvent due to low kinetic hindrance of small-size monomers and catalysts, the surface-initiated polymerization is more appropriate to prepare polymer brushes with high graft density and high molecular weights of polymer chains. Conventionally, the grafting-from method is based on free-radical polymerization. Although thick polymer brushes can be prepared by surface-initiated free-radical polymerization, the resulting polymer brushes normally have the semidilute structure. It is therefore difficult to precisely control the molecular weight and its distribution of graft polymer chains by the free-radical polymerization. The efficiency of initiation from a substrate surface is also low. When polymer chain lengths have large distributions, the polymer brush layers have a heterogeneous height distribution.

The grafting-from method is rapidly advanced by applying living polymerizations such as living radical, anionic, cationic, and ring-opening polymerizations [42]. Surface-initiated living-radical polymerizations drastically increase graft density, thereby preparing concentrated polymer brushes. Since living polymerizations precisely control molecular weights and their distributions of graft polymers, the resulting polymer brushes ideally have uniform bulk and surface structures. Among the aforementioned living polymerizations, living-radical polymerization is widely utilized to prepare polymer brushes due to its applicability to various monomers and its easiness of polymerization.

Living-radical polymerizations are essentially based on a reversible activation/deactivation process of the growing terminus of polymer chains (Scheme 5.1). The end-capped and unreactive dormant chain P-X is normally predominant under a pseudoequilibrium state, and is activated to the active polymer chain P˙ with a radical terminus, followed by stepwise growth of the polymer chain (propagation of monomer, M). The activation is performed by various chemical or physical stimuli, depending on polymerization methods. When each polymer chain similarly repeats the activation/deactivation process during the polymerization, the polymer chains with similar molecular weights are obtained principally. As capping groups, stable nitroxide, halogens with transition-metal catalysis, and dithioester compounds are utilized for nitroxide-mediated polymerization (NMP), atom-transfer radical polymerization (ATRP), and reversible addition-fragmentation chain transfer (RAFT), respectively [43–46]. Various types of monomers can be applied to living-radical polymerizations.

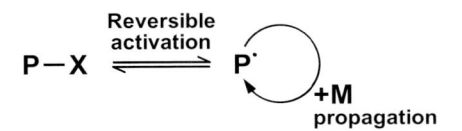

Scheme 5.1 Representation of living-radical polymerizations.

These living-radical polymerizations are successfully applied to prepare polymer brushes with the well-defined regular structures after the immobilization of initiators onto a substrate surface. The representative initiators that can be immobilized onto the metal oxide (with silanol groups) or gold (with thiol groups) surfaces are shown in Figure 5.18. Various shapes of substrates such as films, particles, fibers, and so forth can be applied to prepare polymer brushes based on surface-initiated graft polymerization. Not only simple and linear polymers but also other polymers such as polymacromonomers, hyperbranched polymers, and crosslinked polymers are prepared by surface-initiated graft polymerization. The mixed immobilization of initiators with noninitiative compounds prepares the substrate surface with different initiator density, followed by changes in polymer graft density.

ATRP is most widely utilized to prepare well-defined polymer brushes. In order to control the growth of graft polymers on the dormant-immobilized substrate surface, the addition of reversibly deactivating reagents such as $Cu^{II}Br_2$ and small-size dormant species into the solution is known to be effective. The resulting

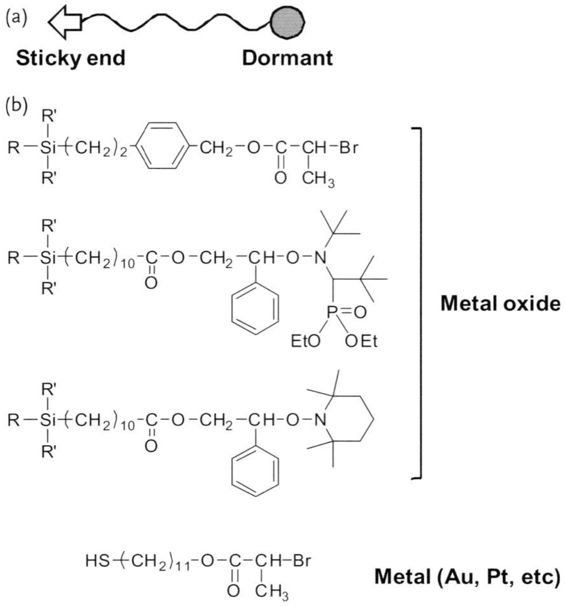

Figure 5.18 (a) An essential structure of initiators for surface-initiated living-radical polymerizations and (b) representative initiators.

molecular weight of graft polymers is confirmed to be similar to that of free polymer chains simultaneously prepared in bulk solutions in certain cases. Therefore, the analysis of free polymer chains could be a method to potentially reveal the properties of graft polymers. When methyl methacrylate (MMA) is grafted by the surface-initiated ATRP under optimized conditions, the number-average molecular weight with narrow distribution is increased with an increase in reaction time. The total amount of graft polymers is proportional to the molecular weight of the graft polymer. These observations suggest that living-radical polymerization successfully proceeds under the constant graft density. The graft density is reached to 0.6 chains/nm^2, thus indicating the preparation of concentrated polymer brushes [47].

5.3.3
Properties and Applications of Concentrated Polymer Brushes

In the case of "semidilute" polymer brushes, the brush structure is predicted theoretically by the following equation:

$$L_e \propto L_c \sigma^{\frac{1}{3}}$$

where L_e is the equilibrium thickness of the semidilute polymer brush in a good solvent, L_c is the length of fully extended polymer chains (correlated to the degree of polymerization or the ideal thickness produced by fully extended polymer chains), and σ is the surface density of the graft polymer. This equation suggests that L_e is linearly dependent on L_c with a change in σ. Many studies of the semidilute polymer brushes experimentally supported the aforementioned theoretical prediction. On the other hand, in the case of concentrated polymer brushes, polymer chains interact more with each other. The brush structure is theoretically predicted by the following equation:

$$L_e \propto L_c \sigma^{\frac{1}{2}}$$

This equation implies that L_e is more increased with an increase in σ. Recent studies of the concentrated polymer brush supported this relationship [40, 41]. In fact, it is known that L_e of poly(methyl methacrylate) (PMMA) brushes reaches 80–90% of L_c in a good solvent, e.g., toluene, implying that the chain length of PMMA in well-defined brush layers is seven times greater than that in the dry state [47]. Note that the formation of polymer brushes is limited in a good solvent, but this is not the case in the dry state.

Concentrated polymer brushes composed of highly extended polymer chains show a superb tribological property. For instance, concentrated PMMA brushes on the silica particle show an extremely small friction coefficient against those on the planar substrate in a good solvent toluene. Meanwhile, a greater coefficient is observed for semidilute PMMA brushes with an increase in load. It is considered that graft polymer chains of concentrated polymer brushes do not interdigitate each other due to the high osmotic pressure [43]. On the other hand, the formation of hydrophilic polyelectrolyte brushes also decreases the friction coefficients in aqueous phases due to the ionic osmotic pressure and strong hydration. The reduced friction of polymer brushes is significant in designing the surface of biomedical materials. In fact, the twitter-ionic polymer brushes of well-known biocompatible poly(2-methacryloyloxyethyl phosphorylcholine) drastically improved durability of an artificial joint composed of ultrahigh molecular weight polyethylene [48].

Concentrated polymer brushes show a unique size-exclusion property against solute molecules (Figure 5.19). When the size of solute molecules is greater than the distance between graft polymer chains, either the incorporation of the solute molecules into the polymer brush (and/or direct adsorption of the solute molecules onto the substrate surface) or the adsorption of the solute molecules onto the brush surface is prevented. The size-exclusion limit of concentrated polymer brushes is smaller than that of conventional semidilute polymer brushes, and is essentially decreased with an increase in graft density. A solute molecule, which is difficult to separate based on conventional chromatography, may be resolved by concentrated polymer brushes prepared on the stationary phase [43]. The size-exclusion limit of concentrated brushes prepared from hydrophilic polymers is also applicable to physical adsorptions of proteins, suggesting that concentrated polymer brushes have potential for antifouling coatings [49].

Fine tuning of the surface properties is significant in developing novel functional surfaces. Patterned polymer brushes with the different properties are prepared by stepwise graft polymerization from the patterned surface, onto which

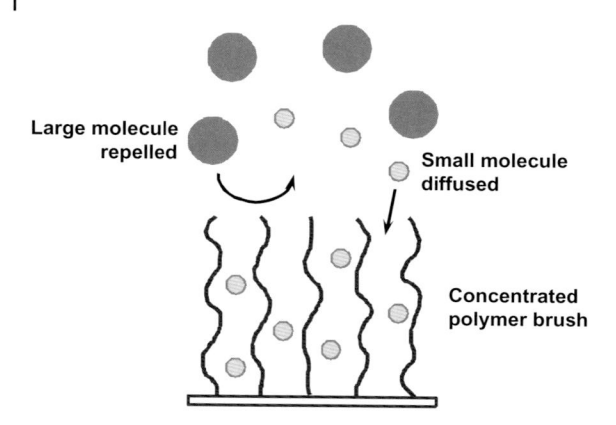

Figure 5.19 Size exclusion properties of polymer brush layers.

different activation types of initiators (e.g., a combination ATRP and NMP) are immobilized onto the substrate surface based on photolithography. Diblock-copolymers are also synthesized by surface-initiated graft polymerization. When polymer brushes composed of diblock-copolymers with hydrophobic and hydrophilic segments are synthesized, each segment swells favorable solvents, resulting in selective extension of the segment. Polymer brushes composed of randomly mixed different polymers are prepared by randomly immobilizing different activation types of initiators. When PMMA and polystyrene, which are not miscible with each other, are randomly grafted in concentrated polymer brushes, the surface shows unique morphologies after solvent treatment, which are dependent on the composition between PMMA and polystyrene as well as the solvents [43].

Since concentrated polymer brushes have a unique structure with extended polymer chains, the brush films in the "dry" state also show different properties from conventionally spin-cast or solution-cast films. It is known that the glass-transition temperature of the brush films with sufficiently high molecular weights is higher than the conventional films, possibly due to the chemical binding of the chain ends on the substrate surface. It is also known that the brush films in their molten state are more resistant against plate compression than that of the conventional films. This is possibly due to a strain-hardening effect of the highly stretched and entangled polymer chains. On the other hand, the polymer brushes are hardly miscible with the conventional films even after contacting the films above the glass-transition temperature, although semidilute polymer brushes are not the case. These properties are significantly different from those of spin-cast or solution-cast films.

5.4
Organic–Inorganic Hybrids

Not limited to organic components, organic–inorganic hybrids are also useful for the components of ultrathin films. Addition of inorganic components to organic materials has attracted much attention in industrial fields because they can add higher material properties such as increase of mechanical strength and flame resistance to the resulting hybrid materials. In addition, unique functions such as optical or magnetic properties can also be given by certain inorganic components. A wide variety of materials can be the components of hybrid ultrathin films. For example, various kinds of synthetic and natural polymers including polyelectrolytes, amphiphilic molecules (lipids, surfactants), other small molecules, carbon nanomaterials such as carbon nanotubes and graphene are used for the organic components, whereas metal oxides, clay minerals, layered double hydroxides (LDHs), semiconductor nanoparticles are used for the inorganic components. Metal nanoparticles are also used as other components instead of the inorganic components.

Briefly, preparation methods for organic–inorganic nanohybirds are classified as follows: (i) mixing preformed organic- and inorganic (nano)components, (ii) synthesis of the organic components in the presence of the inorganic components, (iii) synthesis of the inorganic components in the presence of the organic components, and (iv) simultaneous synthesis of both components. Preparation of organic–inorganic (nano)hybrids itself is well reviewed in several papers and books [50–52]. In this section, such hybrids having (ultra)thin-film shapes are described. In all cases, various fabrication techniques for organic ultrathin films, which are already described in the previous chapters in this book, are utilized to form the resulting hybrids into ultrathin films. Typical distributions of the organic and the inorganic components in the hybrid ultrathin films thus obtained are schematically illustrated in Figure 5.20. Various interactions are possible between the interface of the organic and the inorganic components: noncovalent interactions such as electrostatic interactions, hydrogen bonding, hydrophobic interactions, coordination, and covalent bonding. The surfaces of the inorganic components are sometimes functionalized to increase dispersibility or to realize effective interactions to the organic components before hybridization.

In method (i), direct mixing of the organic and the inorganic components with or without solvents results in formation of bulk hybrid solids that can be formed into various shapes, including ultrathin films using conventional processes such as hot pressing and molding, and solution-based techniques including casting,

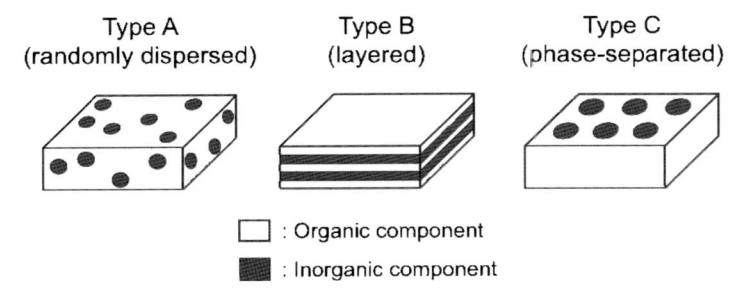

Figure 5.20 Illustrations of typical distribution of the organic and the inorganic components in hybrid ultrathin films. Sizes of each component and area-to-thickness ratios are not to scale, and vary depending on systems.

spin coating, and dip coating. For example, blending of polymers and silica particles resulted in formation of hybrid thin films like type A (Figure 5.20). Surface modification of the silica particles is effective to prevent their agglomeration in the resulting hybrid films. On the other hand, casting of a mixed dispersion of bilayer-forming amphiphiles and clay minerals tends to gives hybrid thin films like type B (Figure 5.20) due to the combination of the self-assembling nature of the amphiphiles and electrostatic interaction between the polar head group of the amphiphiles and charged clay mineral surfaces. Plate-like structures of clay minerals also assist such layer-structure formation. In such hybrids, the layer thickness of each organic components can be minimized to their molecular sizes, whereas that of the inorganic components is much thicker (at least a few nanometers).

Another approach is that stepwise lamination of the organic and the inorganic components on solid substrates, which resulted in formation of hybrid thin films like type B in Figure 5.20. For example, Langmuir–Blodgett (LB) techniques (see Chapter 3) were used to obtain various multilayer thin films having lamellar structures consisted of lipids and plate-like shapes of clay minerals and LDHs. Spreading of preformed surfactant/inorganic nanocomponent complexes at the air/water interfaces are also available. Layer-by-layer (LbL) assembly (see Chapter 4) is much widely used for preparation of hybrid thin films consisted of various organic and inorganic components such as metal oxide particles, clay minerals, and LDHs. Poly(diallyldimethylammonium chloride)/montmorillonite multilayered films exhibit remarkable mechanical properties such as high strength, flexibility, and resistance to crack propagation. Organic–inorganic hybrid hollow capsules prepared using LbL assembly followed by dissolution of the colloidal templates are also successfully obtained. Electrostatic interactions are mainly employed for hybridization in these systems.

One example of method (ii) is that radical polymerization of methacrylate derivatives in the presence of nanosized clay (hectorite) [53]. The resulting hybrid gels can be formed into film morphology. Although it was not mentioned, much thinner films should be prepared. When both components are well mixed homogeneously at the nanoscale level (i.e., the ideal case of type A in Figure 5.20), the resulting hybrids gain outstanding mechanical properties as soft materials.

In method (iii), various solution-based techniques such as a sol-gel process, precipitation from aqueous solution, biomimetic mineralization, are applicable to form the inorganic components in the presence of the organic components. One simple example is synthesis of the inorganic components in solutions containing the organic components. The resulting hybrids are subsequently formed into thin films. Immersion of polymer thin films in reaction solutions for the inorganic components is another simple procedure to obtain hybrid thin films. On the other hand, synthetic procedures that can control hybrids' microstructures have also been investigated. In such cases, the organic parts are preformed into ultrathin films using various techniques as already described in previous chapters. The organic parts act as the templates for precisely designed formation of the inorganic components. For example, multilayer films of cadmium arachidate prepared using

5.4 Organic–Inorganic Hybrids

the LB technique are employed to form CdS nanoparticles in the lattice of the organic layers, which resulted in formation of hybrid thin films. Multibilayer cast films of bilayer-forming lipid molecules can also act as the scaffold for preparation of various states of inorganic nanocomponents such as CdS, iron oxides, and cyano-bridged Cu–Ni bimetallic aggregates. It should be noted that the last example gives inorganic two-dimensional network structures with nanometer thicknesses [54]. In these studies, the resulting structures of the hybrid films are like type B (Figure 5.20). Coating of inorganic shells with nanometer thicknesses on lipid bilayer vesicles using solution-based process were also reported, although maintenance of the organic vesicular structure after the reaction was not clearly proved.

Biomimetic mineralization of calcium carbonates or calcium phosphates on organic film surfaces resulted in formation of organic/biomineral hybrid films [52, 55]. This is based on heterogeneous nucleation of these biominerals that occur at interfaces between the organic film surfaces and the mineralization solutions that are supersaturated with respect to corresponding biominerals. Displaying ionic functional groups such as carboxylic acid groups on the organic film surfaces are effective to induce heterogeneous nucleation because they can bind calcium ions, one component of these biominerals. Although studies have demonstrated that site-selective deposition or precise morphological control of the biominerals was achieved in the resulting hybrid films [56], size control and patterning of the biominerals at the nanoscopic level are still challenging issues.

Phase separation or domain structures in polymer thin films prepared by block-copolymers such as poly(isoprene-*b*-ethylene oxide) (PI-*b*-PEO) can be templates for hybrid thin films [57]. In these cases, the inorganic precursors such as (3-glycidyloxypropyl)trimethoxysilane with aluminum *s*-butoxide are selectively accumulated into one of the organic components (PEO). If the phase-separation structures are maintained throughout the reaction, hybrid thin films like type B or C (Figure 5.20) will be obtained. Actually, these reported examples employed coassembly of block-copolymers and inorganic precursors, which resulted in formation of hybrid thin films having precise periodic structures.

Mesoporous silica structures, prepared from sol-gel reaction of silicon alkoxides in the presence of tubular surfactant micelles templates, are used as the component of hybrid thin films by incorporating organic molecules inside the pores. Thin layers of mesoporous silica themselves can be a component of hybrid thin films. When they are prepared on the organic thin-film surfaces, orientation of mesoporous silica layers can be controlled by molecular arrangement of the outmost surface of the organic layers [58]. Shinkai and coworkers extensively investigated preparation of organic/silica nanohybrids using supramolecular organic templates, some of which resulted in formation of hybrids having interesting ultrathin film-like structures such as helical tubes [50].

Typical examples for method (iv) is that use of solutions containing an organic monomer and metal alkoxides. Before/after film formation using spin coating or other techniques, polymerization and sol-gel processes are conducted to obtain hybrid thin films. Such procedures generally obtained type A (Figure 5.20) structures.

Table 5.1 Examples of organoalkoxysilanes and structures of the resulting hybrids.

Organoalkoxysilanes	Structure in the hybrid
(n = 1–7) [structure]	Lamellar [59]
(n = 1–13) [structure]	Hexagonal (n = 1–5) [59] Lamellar (n = 9–13) [59]
[structure]	Hexagonal [60] Mesoporous (after hydrolysis) [60]
[structure]	(prepared with surfactant template) Mesoporous [61]
[structure]	Helical [62]
(n = 1, 3) [structure]	Bilayer vesicle (n = 1) [63] Monolayer, LB film (n = 3) [64]
[structure] (R = H, Me)	Bilayer vesicle [63] Bicelle [65]

The use of organic–inorganic hybrid molecules like organoalkoxysilanes is another approach to obtain hybrid ultrathin films having precise periodic structures. Typical structures of organoalkoxysilanes used for such purposes and the resulting structures are summarized in Table 5.1. Self-assembly of monoalkyl-type organoalkoxysilanes, including dual-head type, resulted in formation of lamellar or hexagonal hybrids having precise periodicity at the molecular level depending on their molecular structures and preparative conditions [59–65]. Ariga and cow-

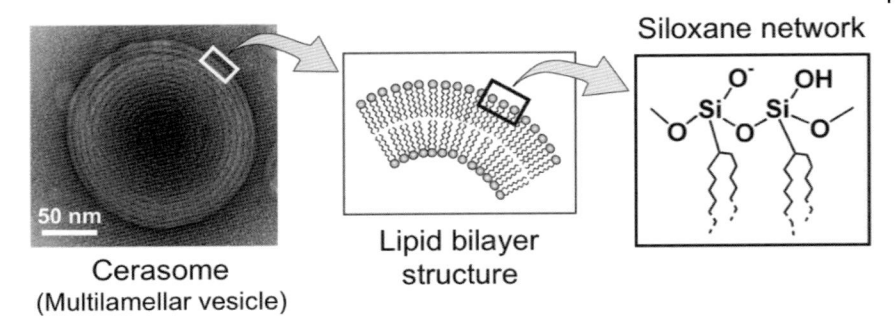

Figure 5.21 Organic–inorganic nanohybrid vesicle "cerasome".

orkers prepared mesoporous hybrids having organically functionalized pores by hydrolyzing the alkyl chains in the hexagonal hybrids [60].

In addition to lipid monolayers at the air/water interface [64], dialkyl-type alkoxysilanes can from lipid bilayer structure in water after proper hydrolysis of their ethoxysilyl groups that makes the molecule amphiphilic. Kikuchi and coworkers have investigated preparation of such vesicular-type hybrid ultrathin films named cerasome (Figure 5.21) [63]. In the bilayer assembly of the lipid-like organoalkoxysilanes, their neighboring alkoxysilane groups can form siloxane bonds as in sol-gel processes. The cerasomes thus obtained have remarkably high morphological stability compared to conventional lipid bilayer vesicles due to formation of surface siloxane network structures, while maintaining their alkyl chains' phase-transition behaviors. The surface siloxane layers act as scaffolds for further surface functionalization of cerasomes. 3-aminopropyltriethoxysilane (APS)-coated and titania-coated cerasomes are obtained using sol-gel processes. Ultrathin metallic and metal alloy layers can also be formed on the cerasome surface using electroless plating techniques. Conventional lipid vesicles could not form such surface-modified hybrid vesicles because they lost their vesicular shape during the preparation processes. High morphological stability of cerasomes bring them other advantages: they act as efficient carriers for drug-delivery systems and higher hierarchical assembling thin films of cerasomes are successfully formed using LbL assembly. Organic–inorganic hybrid bicells (lipid bilayer nanodiscs) are also obtained using mixtures of such organoalkoxysilanes and phospholipids having short alkyl chains [65].

Formation of polymer/metal oxide multilayer thin films (type A in Figure 5.20) using surface sol-gel processes [66] is regarded as a combination of process (iii) and (iv). Polyvinyl alcohol (PVA) and poly(acrylic acid) are useful for this purpose because they display hydroxyl or carboxyl groups on the film surface densely and homogeneously, which is effective for the following surface sol-gel process of metal oxides. The thickness of the multilayered hybrid films can be controlled at molecular precision. Cationic polymers are also effective for alternating surface sol-gel processes because of electrostatic interactions between partially hydrolyzed alkoxides and cationic polymer surfaces.

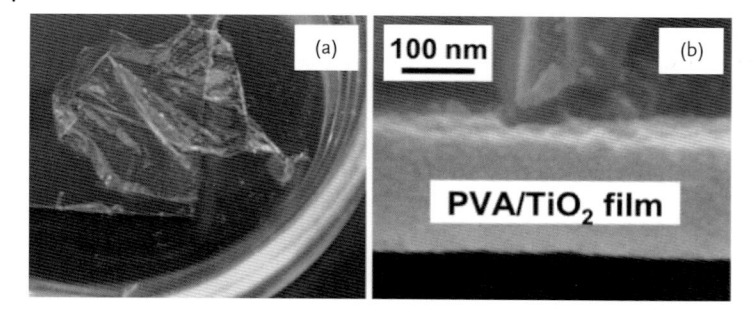

Figure 5.22 Macroscopic image (a) and scanning electron microscopic cross-sectional image (b) of free-standing ultrathin PVA/TiO$_2$ film.

Free-standing (or self-supporting) ultrathin films (see Section 4.2.4) of organic–inorganic nanohybrids are also readily formed by detaching hybrid ultrathin films from solid substrates by dissolving sacrificial layers. For example, the LbL technique gives various kinds of free-standing hybrid films such as those of polyelectrolyte and magnetite nanoparticles having layered structures like type B (Figure 5.20). Hybrid ultrathin films consisting of only one layer of metal nanoparticles are also successfully obtained [67]. Polyelectrolyte/clay hybrid thin films obtained using LBL techniques exhibit remarkably high mechanical properties. Surface sol-gel processes are also applicable to obtain self-supporting polymer/titania multilayered ultrathin films and tubes. Spin coating is a powerful technique to obtain free-standing films conveniently. Kunitake and coworkers prepared various free-standing hybrid films having thicknesses from less than 20 nm to ca. 200 nm depending on the preparative conditions [68]. First, they prepared bilayer type (one bilayer of type B in Figure 5.20) ultrathin hybrid firms of PVA/metal oxides (Figure 5.22). Such films exhibited potential as permselective membranes when the metal-oxide layers were prepared in the presence of target molecules that were later removed after film formation (i.e., molecular imprinting). However, the mechanical strengths of these films were not sufficient for practical use. They soon prepared free-standing hybrid ultrathin films consisted of methacrylate polymers and metal oxides using spin-coating techniques followed by simultaneous photopolymerization and sol-gel reaction. The resulting few-tenths of a nm thicknesses of ultrathin films have interpenetrating network (IPN)-like structures, which give the film remarkable mechanical strength. Epoxy polymers are also useful to form robust free-standing ultrathin hybrid films with the combination of silane coupling agents such as APS. Other approaches are also employed to fabricate free-standing hybrid thin films. Ichinose and coworkers fabricated free-standing hybrid ultrathin films, having high filtration performance, consisting of inorganic nanowires and ferritin by utilizing filtration process and crosscoupling reactions [69]. Close-packed two-dimensional arrays of surface-modified metal nanoparticles can be formed into their free-standing films, although their sizes are a few micrometers square [70].

In summary, various techniques, most of them are described in this book, are readily applicable to fabricate organic–inorganic hybrid ultrathin films. Since these films possess both advantages coming from the organic and the inorganic components, they are expected to act as novel functional nanomaterials. Studies along this line are still expanding.

5.5
Colloidal Layers

Not limited to molecules, organic materials having colloidal dimensions, from a few nanometers to submicrometer, can be the components of organic (ultra)thin films. Some colloidal materials themselves can form thin films consisting of their monolayer or multilayers, which are known as colloidal layers. Those having high structural periodicity are sometimes called "colloidal crystals". Colloidal layers (or colloidal crystals) have attracted much attention because of their potential applications, such as photonic crystals, imaging devices, artificial opals and templates for inverse opals, and masks for lithographic techniques. Developments and recent progresses for fabrication of colloidal layers and their applications are well summarized in the literature [71, 72]. In this section, the basic system, that is, colloidal layers formed by spherical, isotropic colloidal organic particles are mainly described.

A simple approach for preparation of colloidal layers is one that utilizes the self-assembly nature of colloidal particles in their dispersions. When self-assembly of colloidal particles initiates from the surfaces of solid substrates (e.g., the bottom of a reaction vessel or immersed substrates), thin films consisted of mono- or multilayers of colloidal layers are formed. Self-assembling behaviors of colloidal particles are affected by various factors such as particle size, surface property (hydrophobicity or hydrophilicity, charges), density, concentration, dispersion medium, reaction temperature, and reaction time. These factors affect interparticle forces, diffusion properties or the mobility of particles (e.g., Brownian motion), sedimentation time, which will decide the assembly structure of the particles. The balance of interparticle forces including van der Waals interactions, steric repulsion forces, and electrostatic interactions are some of the key factors to decide whether the resulting assembling is simple aggregation or crystal-like assembly having good periodicity. Very strong attractive forces between particles are not favored to achieve disorder-to-order phase transition in the assembling structures. Colloidal layer formation is a dynamic process.

Several approaches are established to obtain monolayer and multilayers of colloidal particles from the particle dispersion, based on the physicochemical properties of the particles. Some of them are schematically illustrated in Figure 5.23 and described below.

One process is sedimentation of colloidal particles in the gravitational fields (Figure 5.23a). This process is employed in biological systems: that is, formation of opals. Particle sizes and concentrations, sedimentation speeds affect quality in orderliness (periodicity) of the resulting colloidal layers formed on the bottom of reaction vessels. Although this process is simple and low cost, it is rather time

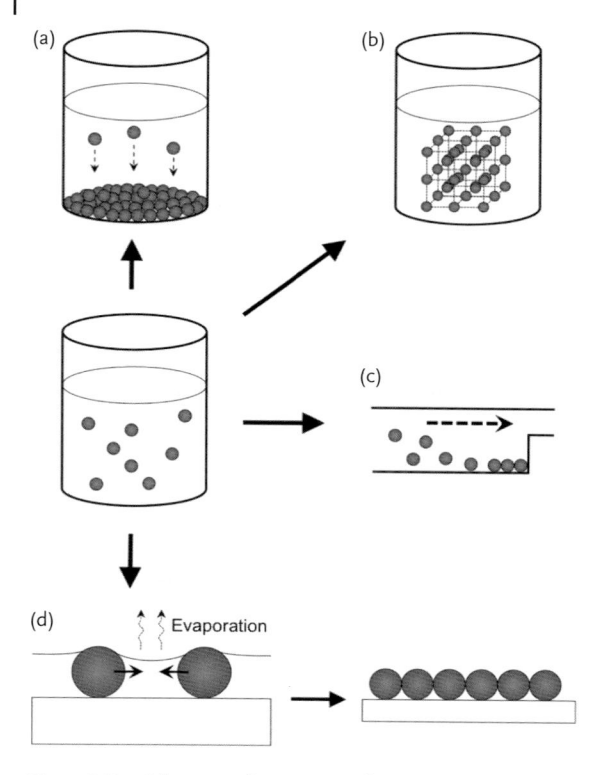

Figure 5.23 Schematic illustrations of representative preparation processes to obtain colloidal layers from particle dispersion. (a) By sedimentation, (b) by crystallization, (c) by laminar flows, and (d) by capillary forces.

consuming. In addition, several problems such as homogeneity of thickness (layer numbers) and crystallinity of the resulting films are also present in many cases.

The second one is crystallization of colloidal particles in dispersion media, which results in formation of three-dimensional (3D) colloidal crystals (Figure 5.23b). This process is mainly controlled by the balance of interparticle forces: long-range attractive van der Waals interactions and short-range steritic repulsive forces. When the particle surfaces are charged, electrostatic interactions (long-range Coulomb repulsion) are also involved. Therefore, addition of electrolytes in the dispersion media largely affects the assembly behavior of the particles, because of their shielding effect in the electric double layer. The electrostatic screening lengths (or the Debye–Hückel lengths) of the particles also affect the interparticle distances (crystal lattice distances) of the resulting colloidal crystals. Crystal structures of the resulting colloidal crystals such as face-centered cubic (fcc), body-centered cubic (bcc), hexagonal close-packed (hcp), are dependent on experimental conditions.

The third one is that using laminar flows (Figure 5.23c), which has been extensively investigated by Xia and coworkers [73]. This process is simple and relatively fast, and the thicknesses (layer numbers) of the colloidal layer films are controllable. However, fabrication of the reaction chamber is necessary.

The fourth one is that utilizing attractive capillary forces (Figure 5.23d). This process was originally developed by Nagayama and coworkers to obtain two-dimensional (2D) arrays of colloidal particles on various substrates [74]. This process has wide applicability and various modified processes have been proposed. For example, continuous, large areas of colloidal layers are successfully prepared by moving or lifting the solid substrates while dispersion is continuously provided at the same place. Monolayers and multilayers of colloidal assemblies are successfully prepared depending on experimental conditions such as particle concentrations and evaporation speeds. Free-standing colloidal layer films are also obtainable from polystyrene particles having photocrosslinking ability.

Other approaches such as spin coating of particle suspensions and electric-field-induced layer formation, and technical improvements of the above-mentioned approaches have also been widely investigated to obtain colloidal layers on solid substrates. The use of substrates having topological patterns results in formation of patterned colloidal layers.

Some kinds of colloidal particles behave like surfactants: that is, they assemble at the oil/water interfaces to reduce the total surface energy. For example, hollow capsules consisted of a monolayer of colloidal layer shell are successfully formed at the oil/water interfaces, so-called "Pickering emulsion" [75] or "colloidosome" [76]. Hydrophobic particles (water contact angle $(\theta) > 90°$) tend to form water-in-oil emulsions, whereas hydrophilic particles ($\theta < 90°$) tend to form oil-in-water emulsions (Figure 5.24a and b). Similar to surfactant monolayers, hydrophobic colloidal particles such as polystyrene particles can form Langmuir monolayers (see Chapter 3) at the air/water interfaces (Figure 5.24c).

Adsorption of colloidal particles to solid phase is another practical approach to form their colloidal layers. When electrostatic interactions are employed, the assembly process on solid substrates form particle dispersion is similar to that of Layer-by-layer (LbL) assembly (see Chapter 4). Generally, the resulting colloidal layers are not closely packed in the planar direction because of charge repulsion

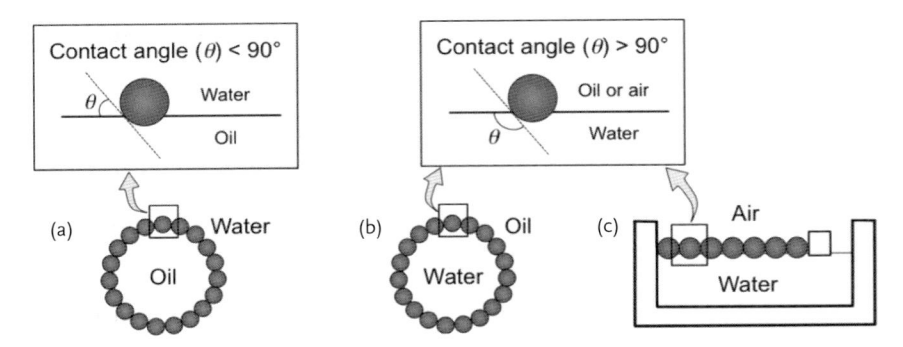

Figure 5.24 Schematic representations of water contact angle (θ) and examples of colloidal layer formation at interfaces. (a) Water-in-oil emulsion-type hollow capsules by hydrophilic colloidal particles ($\theta < 90°$), (b and c) oil-in-water emulsion type hollow capsules (b) and monolayer formation at the air/water interface (c) by hydrophobic colloidal particles ($\theta > 90°$).

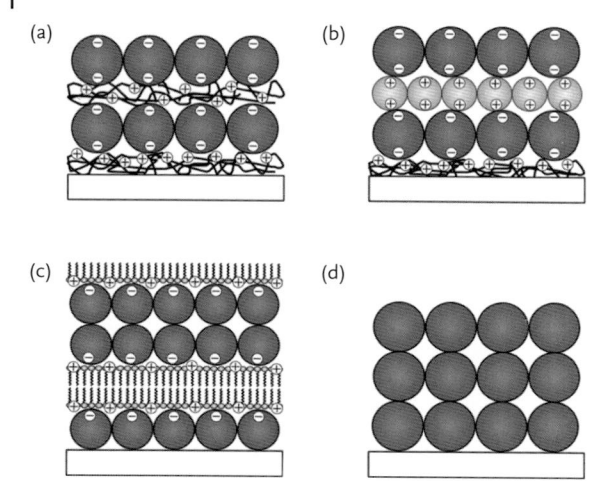

Figure 5.25 Multilayer formation of colloidal layers. (a) LbL assembly of colloidal particles and oppositely charged polymers, (b) LbL assembly of anionic colloidal particles and cationic colloidal particles, (c) LB transfer of lipid/colloidal particle monolayers (Y-type transfer), and (d) LB transfer of colloidal particle monolayers. The sizes of each component are not to scale.

among the particles. To enhance interparticle packing, addition of electrolytes is necessary to decrease interparticle electrostatic repulsion, as in the case of formation of colloidal crystals. These colloidal layers can be laminated to from their multilayers using LbL assembly. In addition to alternate assembly of charged colloidal layers and oppositely charged polymers (Figure 5.25a), alternate assembly of the colloidal layer of cationic particles and that of anionic particles is also applicable (Figure 5.25b). Hydrophilic, charged particles can also adsorb to lipid monolayers having oppositely charged headgroups at the air/water interfaces to form their colloidal layers, which can be laminated using conventional Langmuir–Blodgett (LB) transfer processes (see Chapter 3) (Figure 5.25c). However, such systems have not been widely investigated. When colloidal particles adsorb to the monolayers, the resulting lipid–particle complexes may move to the air/water interface because of their increased hydrophobicity. Therefore, creation of lipid/colloidal particle layered structures like Figure 5.25c seems difficult in such systems. If colloidal particles themselves can form monolayers at the air/water interfaces, their multilayers can be laminated using LB techniques (Figure 5.25d). LbL assembly and LB techniques can laminate colloidal layers at monolayer precision. In these systems, other driving forces such as chelating or molecular recognitions are sometime used instead of electrostatic interactions, where the surfaces of colloidal particles are modified with proper functional molecules.

In addition to isotropic, spherical colloidal particles, other kinds of particles such as anisotropic particles and nonspherical particles, and their mixtures can also be the components of colloidal layers. Similar to organic particles, inorganic colloidal particles can form colloidal layers [71], although they are not highlighted in this book. Colloidal layers of metal or semiconductor nanoparticles have been given much attention because of their environment-sensitive optical or magnetic properties. For example, free-standing ultrathin films consisting of monolayer of metal nanoparticle 2D arrays have been prepared recently. Including these studies, most colloidal layers consisting of metal or semiconductor nanoparticles are actually organic-metallic (or inorganic) hybrid layers, because generally most of these nanoparticles' surfaces are capped with organic molecules such as thiol compounds or ultrathin polymer layers to prevent agglomeration.

Biological substances having colloidal dimensions have also been used as the components of colloidal layers. For example, 2D crystals of proteins, which can be regarded as colloidal layers, originally have been given attention to be used in crystallographic studies by transmission electron microscopy. Fromherz firstly reported fabrication of 2D crystals of proteins with the aid of lipid monolayers at the air/water interfaces [77]. Later Ringsdorf and coworkers employed specific avidin–biotin interactions to create 2D crystals of avidin on the hydrophilic part of lipid monolayers containing biotinylated lipids [78]. Similarly, proteins containing His-tag sequences can form their 2D crystals on the hydrophilic part of lipid monolayers containing Ni^{2+}-chelated NTA(nitrilotriacetic acid)-lipids. Polymer thin films spread at the air/water interfaces can also act as the scaffold for 2D protein crystals formation. Instead of water, the use of mercury as the subphase was also investigated, which revealed that such systems gave well-ordered 2D protein crystals [79]. In these systems, the crystal-forming nature of protein is important to obtain 2D protein layers having higher crystallinity, in addition to the driving forces of proteins to adsorb substrate surfaces. Alternately, LbL assembly is readily applicable to form protein layers. Because of their sizes and lower structural rigidity, the adsorption and assembly behaviors of proteins at air/water interfaces and on solid substrates are somewhat different from the cases of spherical colloidal particles. Recent studies also demonstrated successful formation of colloidal layers of other biological colloidal substances such as viruses, phages, and yeast cells.

It should be noted that theoretical studies based on colloid and surface chemistry [71] to understand the spatiotemporal, physicochemical behavior of the target particles in the reaction media, has supported successful fabrication of various colloidal layer structures. These theoretical and experimental developments relating to fabrication and application of colloidal layers as organic ultrathin films are still continuing.

5.6
Newly Appearing Techniques

5.6.1
Material-Binding Peptide

In 1985, Smith reported that peptides could be genetically displayed on the coat proteins of filamentous phages, based on the insertion of the DNA fragment into

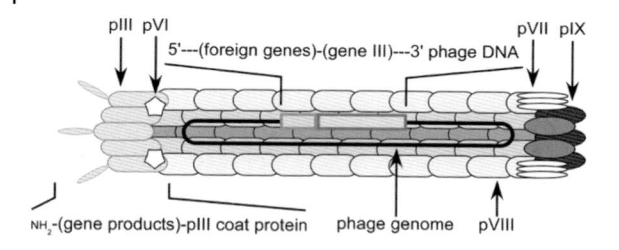

Figure 5.26 Illustration of a M13 filamentous phage.

the phage genome [80]. The resulting genetically engineered phages infect a host cell such as *Escherichia coli* (*E. coli*) similarly to wild-type phages, followed by phage proliferation. This method to display desired peptides on the phage surfaces as fused proteins is called the phage-display (PD) method. The most commonly used bacteriophage is a M13 filamentous phage, which has five coat proteins such as pIII, pVI, pII, pVIII, and pIX on its surface (Figure 5.26). When peptides with different sequences are displayed on the surface of each different phage, a peptide library displayed on the phage surface is prepared. Based on the affinity selection of certain phages against a target molecule and the subsequent DNA sequencing, peptides that specifically bind to the target molecule are identified. The genetic engineering of *Escherichia coli* (*E. coli*) cells also enables development of cell-surface display (CSD) methods, in which peptides are fused into proteins on the membrane surface [81].

The methods to prepare phage-displayed peptide libraries have already been established; therefore, random peptides can be fused at the desired positions of phage coat proteins. The process of enriching phage libraries to certain phage pools is called "biopanning" (Figure 5.27). The biopanning experiment is composed of four steps as follows. In Step 1 (affinity), an aqueous solution of the phage libraries is mounted onto the target surface (film, plate, particle, or other solid) at ambient temperature for an appropriate time. In Step 2 (wash), to remove unbound or weakly bound phages, the target surfaces are washed with a buffer solution several times. The washing buffer sometimes contains surfactants to more effectively wash out the phages. In Step 3 (elution), the bound phages are eluted from the target surfaces by an elution buffer. In Step 4 (amplification), to amplify the phages, the eluted phages are proliferated within *E. coli*. These steps are repeated until the phage libraries are concentrated into the desired phage clones. Finally, the DNA sequences of each phage clone, which correspond to each amino acid sequence, are obtained.

The enrichment of the phage clones is evaluated by the yield – the amount of eluted phages divided by the amount of mounted phages. The enrichment of certain clones is also evaluated by a determination of the amino acid sequences of an adequate number of clones. A single clone should be isolated after an adequate number of biopanning cycles. However, at least several clones are usually isolated. Therefore, all of the parameters for the affinity, wash, elution, and biopanning cycles must be optimized for each target. The binding of the selected phage clones on the surface of the target or reference polymers is analyzed by a semi-

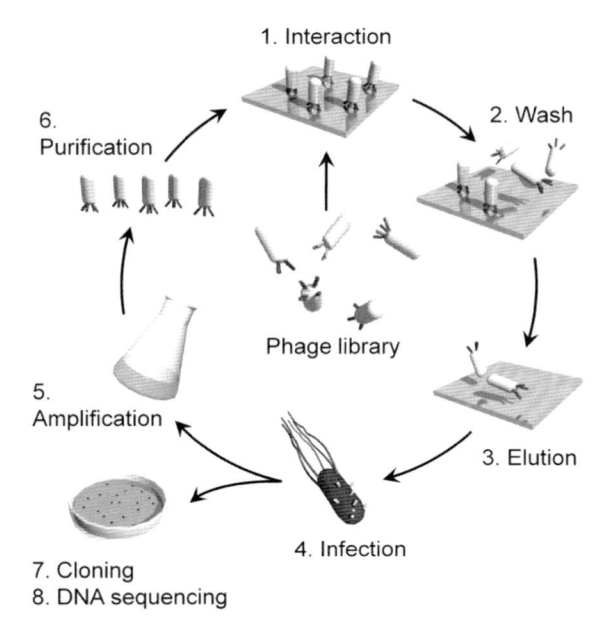

Figure 5.27 Scheme of the PD method (biopanning).

quantitative titration analysis of the bound phages, enzyme-linked immunosorbent assay, and so on. After the possible clones are revealed, the corresponding peptides are synthesized by a standard Fmoc solid-phase method. The amounts of synthetic peptides bound on the surface of the target materials can be evaluated by a quantification of the unbound peptides in the supernatants or by a quantification of the bound phages.

Over the past decade, the PD and CSD methods have been applied to artificial materials, and unique peptides that bind specifically to material surface were developed successfully [82–84]. The inorganic surfaces of metal oxides, semiconductors, metals, and magnetics can be peptide targets. Furthermore, organic molecules such as carbon nanotubes, carbon nanohorns, fullerenes, molecular assemblies, and synthetic polymers are also applied to these methods. The resulting peptides have the potential for noncovalently constructing a stable interface between these materials and biomolecules. Therefore, peptides can be used as novel organic nanomaterials such as catalysts for the preparation and dispersion stabilization of inorganic nanoparticles, nanocrystal assembly, adsorbents for patterning, surface modifiers, modifiers of phages/proteins, and so on.

5.6.2
Block-Copolymer Films

Surface nanopatterning of solid substrates is achieved either by top-down or bottom-up technologies. The former technology is performed by conventional

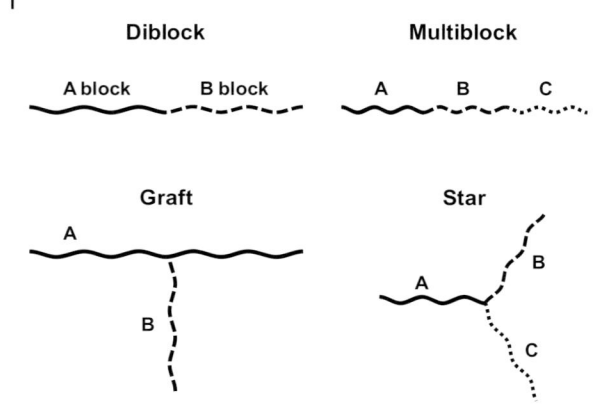

Figure 5.28 Illustration of block-copolymers.

lithography, and has been used for the development of electronic devices such as memories, integrated circuits, microprocessors, and so on. The latter technology is based predominantly on self-assembly of adequate molecules on solid substrates, and gathers great attention due to the potential of low-cost, straightforward, and mass productivities. Among recently developed bottom-up and self-assembly technologies for surface nanopatterning [85], the thermodynamically controlled microphase separation of block-copolymer films for the preparation of patterned morphologies is mentioned here.

Block-, graft-, and star-shaped copolymers are composed of two or more polymer chains chemically linked together in a single molecule (Figure 5.28). When component chains in the copolymers have a mutual repulsion and do not mix with each other in a solid state, the ordered structures such as lamellae, spheres, and cylinders with domain sizes of 5–50 nm are thermodynamically formed by microphase separation. The domain size and shape are strongly dependent on the molecular weight and composition of the copolymers, thereby realizing straightforward control of the morphologies [86]. The domain size essentially increases on increasing the molecular weight of the corresponding chain. The representative phase diagram for AB diblock-copolymers is shown in Figure 5.29. A-rich sphere phases are formed at relatively small composition of A chains. A-rich cylinder phases are formed with increasing the composition. A/B lamella phases are formed at the similar composition of A and B chains. At the larger composition, the phases are reversed, followed by the appearance of B-rich cylinder and sphere phases. In addition, an ordered bicontinuous double-diamond phase, so-called gyroid, appears between cylinder and sphere phases. Interestingly, ABC star-shaped terpolymers show the complicated but regular structures such as the Archimedean tiling patterns [87].

Block-copolymer films with patterned nanostructures have the potential for utilization as large-scale magnetic recording media, photonic crystals, nanoporous filters, and so on. For sophisticated applications of block-copolymer films, the precise control of the structural orientation against the substrate surface is an

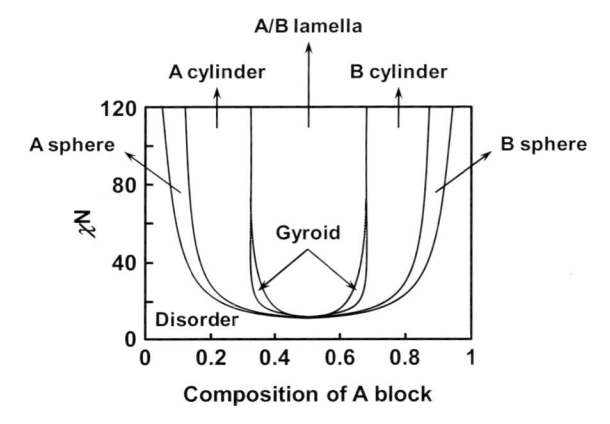

Figure 5.29 Representative phase diagram of microphase separation of AB diblock-copolymer. N and c indicate the degree of polymerization and interaction parameter representing repulsive forces between A and B blocks, respectively.

essential requirement. In particular, it is interesting to prepare the lamella and cylinder structures either parallel or perpendicular to the substrate surface. Since certain polymer chains are preferentially accumulated at the substrate or outermost film surface due to minimization of the interfacial free energy, the lamella and cylinder structures parallel to the substrate surface tend to be predominant. In addition, the lamellae and cylinders have many domains that have different inplane directions. Therefore, the structural regulation out-of-plane and inplane is significant. Representative useful methods to control the structural orientation are mentioned below.

The modification of substrate surfaces changes the structural orientation of lamellae and cylinders, based on the change in interactions between block-copolymers and the substrate surface [88]. In the case of diblock-copolymer films composed of poly(styrene-*b*-methyl methacrylate) (PS-*b*-PMMA), the modification of substrate surfaces with graft copolymers of styrene and methyl methacrylate results in lamella and cylinder structures perpendicular to the substrate surface, when the graft surfaces show similar wettabilities against two polymer chains. The structures parallel to the substrate surface are obtained by changing the composition of graft copolymers.

The regulation of lamella and cylinder structures can also be achieved by microfabrication of substrate surfaces based on conventional photolithography. When the rough and nonflat substrate surfaces are used for PS-*b*-PMMA films, the lamellae perpendicular to the substrate surface are formed due to instability of the parallel structure [89]. On the other hand, the patterning of substrate surfaces with hydrophobic and hydrophilic regions results in the perpendicular structure of PS-*b*-PMMA films due to preferential accumulation of PS and PMMA chains, respectively [90]. The most important advantage is to minimize the structural defect derived from lithography by replicating the structure to block-copolymers. The aforementioned methods that combine conventional top-down lithography with

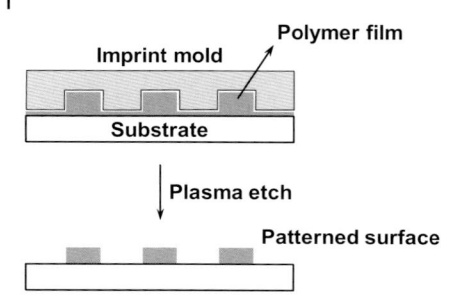

Figure 5.30 Scheme of nanoimprint lithography.

bottom-up self-assembly are called graphoepitaxy. Other methods such as utilizations of solvent annealing [91], external stimuli [92], and liquid crystallinity [93] have been developed for precise control of the structural orientation.

5.6.3
Nanoimprint Lithography

High-throughput, reproducible, and low-cost nanopatterning of polymer surfaces is strongly required for device fabrication in electronics, photonics, biotechnologies, and so on. Nanoimprint lithography is an alternative nanopatterning method to conventional lithography, and precisely and simply fabricates nanopatterned polymer surfaces [94]. The fundamental protocol of nanoimprint lithography is schematically shown in Figure 5.30. A mold with a surface-relief nanostructure, which is composed of hard materials such as silicon, silicon dioxide, nickel, and so on, is prepared by conventional lithography. The mold is mechanically pressed onto the surfaces of polymer films prepared on planar solid substrates at adequate temperatures, followed by imprint of the relief to the polymer surface. In order to successfully remove the mold from the imprinted polymer films, control of the interaction between the surfaces of the mold and polymer films is significant. Thermoplastic or light-curable resins are utilized as polymer components. Nanoimprint lithography is really high throughput, and is performed by micrometer to sub-hundred-nanometer resolution. Defect-free fabrication and production-level throughput have recently been the focus. Various applications including biological applications such as protein array and cell culture on patterned surfaces are anticipated. Not only 2-dimensional patterned surfaces but also 3-dimensional structures can also be prepared by nanoimprint lithography [95].

References

1 Singer, S.J., and Nicolson, G.L. (1972) Science, **175**, 720–731.
2 Bangham, D., and Horne, R.W. (1964) J. Mol. Biol., **8**, 660.
3 (a) Kunitake, T., and Okahata, Y. (1977), Chem. Lett., 1337–1340; (b) Kunitake, T., and Okahata, Y. (1977) J. Am. Chem. Soc., **99**, 3860.

4 Kunitake, T. (1992) *Angew. Chem. Int. Ed. Engl.*, **31**, 709.
5 Kaler, E.W., Murthy, A.K., Rodriguez, B.E., and Zasadrinski, T.A.N. (1989) *Science*, **245**, 1371.
6 Yamada, N., Iijima, M., Vongbupnimit, K., Noguchi, K., and Okuyama, K. (1999) *Angew. Chem.*, **111**, 969; (1999) *Angew. Chem. Int. Ed. Engl.*, **38**, 916.
7 Kimizuka, N., Kawasaki, T., Hirata, K., and Kunitake, T. (1998) *J. Am. Chem. Soc.*, **120**, 4049–4104.
8 Israelachvili, J.N. (1992) *Intermolecular and Surface Forces*, 2nd edn, Academic Press, London.
9 Yamada, N., Ariga, K., Naito, M., Matsubara, K., and Koyama, E. (1998) *J. Am. Chem. Soc.*, **120**, 12192.
10 Fendler, J.H. (1982) *Membrane Mimetic Chemistry*, John Wiley & Sons, Inc., New York.
11 Sakurai, I., Kawamura, Y., Sakurai, T., Ikegami, A., and Seto, T. (1985) *Mol. Cryst. Liq. Cryst.*, **130**, 203.
12 Hotani, H. (1984) *J. Mol. Biol.*, **178**, 113.
13 Talmon, Y. (1983) *J. Colloid Interface Sci.*, **93**, 366; Kilpatrick, P.K., Miller, W.G., Talmon, Y. (1985) *J. Colloid Interface Sci.*, **107**, 146.
14 Wakayama, Y., and Kunitake, T. (1993) *Chem. Lett.*, 1425.
15 O'Brien, D.F. (2002) *Chem. Rev.*, **102**, 727.
16 Kunitake, T., Tsuge, A., and Nakashima, N. (1984) *Chem. Lett.*, 1783–1786.
17 Okahata, Y., En-na, G., and Ebato, H. (1990) *Anal. Chem.*, **62**, 1431.
18 Tachibana, T., and Kambara, H. (1965) *J. Am. Chem. Soc.*, **87**, 3015.
19 (a) Nakashima, N., Asakuma, S., Kim, J.-M., Kunitake, T. (1984), *Chem. Lett.*, 1709; (b) Yamada, K., Ihara, H., Ide, T., Fukumoto, T., and Hirayama, C. (1984) *Chem. Lett.*, 1713.
20 Yager, P., and Schoen, P.E. (1984) *Mol. Cryst. Liq. Cryst.*, **106**, 371.
21 Nakashima, N., Asakuma, S., and Kunitake, T. (1985) *J. Am. Chem. Soc.*, **107**, 509.
22 Kunitake, T., and Yamada, N. (1986) *J. Chem. Soc. Chem. Commun.*, 655–656.
23 Yamada, N., Sasaki, T., Murata, H., and Kunitake, T. (1989) *Chem. Lett.*, 205–208.
24 Fuhrhop, J.-H., Schnieder, P., Boekema, E., and Helfrich, W. (1988) *J. Am. Chem. Soc.*, **110**, 2861–2867.
25 Kim, J.-M., and Kunitake, T. (1989) *Chem. Lett.*, 959.
26 Kunieda, H., Nakamura, K., and Evans, D.F. (1991) *J. Am. Chem. Soc.*, **113**, 1051.
27 Terech, P., and Weiss, R.G. (1997) *Chem. Rev.*, **97**, 3133; Terech, P., and Weiss, R.G. (eds) (2006) *Molecular Gels*, Springer, Inc., Netherlands.
28 Kuwahara, H., Hamada, M., Ishikawa, Y., and Kunitake, T. (1993) *J. Am. Chem. Soc.*, **115**, 3002–3003.
29 Tanford, C. (1973) *The Hydrophobic Effect*, John Wiley & Sons, Inc., New York.
30 Hanabusa, K., Tange, J., Taguchi, Y., Koyama, T., and Shirai, H. (1993) *J. Chem. Soc. Chem. Commun.*, 390.
31 Aggeli, A., Bell, M., Boden, N., Keen, J.N., Knowles, P.F., McLeish, T.C.B., Pitkeathly, M., and Radford, S.E. (1997) *Nature*, **386**, 259.
32 Cohen, F.E., Pan, K.-M., Huang, Z., Baldwin, M., Fletterick, R.J., and Prusiner, S.B. (1994) *Science*, **264**, 530–531.
33 Shimizu, T., and Masuda, M. (1997) *J. Am. Chem. Soc.*, **119**, 2812–2818.
34 Kobayashi, H., Amaike, M., Jung, J.-H., Friggeri, A., Shinkai, S., and Reinhoudt, D.N. (2001) *Chem. Commun.*, **2001**, 1038.
35 Estroff, L.A., and Hamilton, A.D. (2000) *Angew. Chem. Int. Ed.*, **39**, 3447.
36 Brunsveld, L., Folmer, B.J.B., Meijer, E.W., and Sijbesma, R. (2001) *Chem. Rev.*, **101**, 4071.
37 Clegg, R.S., Reed, S.M., Smith, R.K., Barron, B.L., Rear, J.A., and Hutchison, J.E. (1999) *Langmuir*, **15**, 8876–8883.
38 Shimizu, T., Kogiso, M., and Masuda, M. (1997) *J. Am. Chem. Soc.*, **119**, 6209–6210.
39 Yamada, N., Komatsu, T., Yoshinaga, H., Yoshizawa, K., Edo, S., and Kunitake, M. (2003) *Angew. Chem. Int. Ed.*, **42**, 5496.
40 Halperin, A., Tirrell, M., and Lodge, T.P. (1992) *Adv. Polym. Sci.*, **100**, 31.
41 Kawaguchi, M., and Takahashi, A. (1992) *Adv. Colloid Interface Sci.*, **37**, 219.
42 Edmondson, S., and Huck, W.T.S. (2004) *J. Mater. Chem.*, **14**, 730.
43 Tsujii, Y., Ohno, K., Yamamoto, S., Goto, A., and Fukuda, T. (2006) *Adv. Polym. Sci.*, **197**, 1.

44 Edmondson, S., Osborne, V.L., and Huck, W.T.S. (2004) *Chem. Soc. Rev.*, **13**, 14.
45 Pyun, J., Kowalewski, T., and Matyjaszewski, K. (2003) *Macromol. Rapid Commun.*, **24**, 1043.
46 Hawker, C.J., Bosman, A.W., and Harth, E. (2001) *Chem. Rev.*, **101**, 3661.
47 Yamamoto, S., Ejaz, M., Tsujii, Y., and Fukuda, T. (2000) *Macromolecules*, **33**, 5608.
48 Moro, T., Takatori, Y., Ishihara, K., Konno, T., and Takigawa, Y. (2004) *Nature Mater.*, **2**, 829.
49 Yoshikawa, C., Goto, A., Tsujii, Y., Fukuda, T., Kimura, T., Yamamoto, K., and Kishida, A. (2006) *Macromolecules*, **39**, 2284.
50 van Bommel, K.J.C., Friggeri, A., and Shinkai, S. (2003) *Angew. Chem. Int. Ed.*, **42**, 980.
51 Sanchez, C. (ed.) (2005) *J. Mater. Chem.*, **35–36**, 3541. (Theme Issue: Functional Hybrid Materials).
52 Ruiz-Hitzky, E., Ariga, K., and Lvov, Y. (eds) (2008) *Bio-Inorganic Hybrid Nanomaterials*, Wiley-VCH Verlag GmbH, Weinheim.
53 Haraguchi, K., Ebato, M., and Takehisa, T. (2006) *Adv. Mater.*, **18**, 2250.
54 Kimizuka, N., Handa, T., Ichinose, I., and Kunitake, T. (1994) *Angew. Chem. Int. Ed. Engl.*, **33**, 2483.
55 Mann, S. (2001) *Biomineralization: Principles and Concepts in Bioinorganic Materials Chemistry*, Oxford University Press, New York.
56 Sugawara, A., Ishii, T., and Kato, T. (2003) *Angew. Chem. Int. Ed.*, **42**, 5299.
57 Templin, M., Franck, A., Du Chesne, A., Leist, H., Zhang, Y., Ulrish, R., Schädler, S., and Wiesner, U. (1997) *Science*, **278**, 1795.
58 Kawashima, Y., Nakagawa, M., Seki, T., and Ichimura, K. (2002) *Chem. Mater.*, **14**, 2842.
59 Shimojima, A., and Kuroda, K. (2006) *Chem. Record*, **6**, 53.
60 Zhang, Q., Ariga, K., Okabe, A., and Aida, T. (2004) *J. Am. Chem. Soc.*, **126**, 988.
61 Inagaki, S., Guan, S., Ohsuna, T., and Terasaki, O. (2002) *Nature*, **416**, 304.
62 Moreau, J.J.E., Vellutini, L., Wong Chi Man, M., and Bied, C. (2001) *J. Am. Chem. Soc.*, **123**, 1509.
63 Katagiri, K., Hashizume, M., Ariga, K., Terashima, T., and Kikuchi, J. (2007) *Chem. – Eur. J.*, **13**, 5272.
64 Ariga, K., and Okahata, Y. (1989) *J. Am. Chem. Soc.*, **111**, 5618.
65 Yasuhara, K., Miki, S., Nakazono, H., Ohta, A., and Kikuchi, J. (2011) *Chem. Commun.*, **47**, 4691.
66 Ichinose, I., Kuroiwa, K., Lvov, Y., and Kunitake, T. (2003) Recent progress in the surface sol-gel process and protein multilayers, in *Multilayer Thin Films* (eds G. Decher and J.B. Schlenoff), Wiley-VCH Verlag GmbH, Weinheim, pp. 155–175.
67 Jiang, C., Markutsya, S., Pikus, Y., and Tsukruk, V. (2004) *Nature Mater.*, **3**, 721.
68 Watanabe, H., Vendamme, R., and Kunitake, T. (2007) *Bull. Chem. Soc. Jpn.*, **80**, 433. (Accounts).
69 Peng, X., Jin, J., Nakamura, Y., Ohno, T., and Ichinose, I. (2009) *Nature Nanotechnol.*, **4**, 353.
70 Mueggenburg, K.E., Lin, X.-M., Goldsmith, R.H., and Jaeger, H.M. (2007) *Nature Mater.*, **6**, 656.
71 Caruso, F. (ed.) (2004) *Colloids and Colloid Assemblies*, Wiley-VCH Verlag GmbH, Weinheim.
72 Zhang, J., Li, Y., Zhang, X., and Yang, B. (2010) *Adv. Mater.*, **22**, 4249. and references therein.
73 Park, S.H., Qin, P.D., and Xia, Y. (1998) *Adv. Mater.*, **10**, 1028.
74 Denkov, N.D., Velev, O.D., Kralchevsky, P.A., Ivanov, I.B., Yoshimura, H., and Nagayama, K. (1992) *Langmuir*, **8**, 3183.
75 Pickering, S.U. (1907) *J. Chem. Soc.*, **91**, 2001.
76 Dinsmore, A.D., Hsu, M.F., Nikolaides, M.G., Marquez, M., Bausch, A.R., and Weitz, D.A. (2002) *Science*, **298**, 1006.
77 Fromherz, P. (1971) *Nature*, **231**, 267.
78 Blankenburg, R., Meller, P., Ringsdorf, H., and Salesse, C. (1989) *Biochemistry*, **28**, 8214.
79 Yoshimura, H., Matsumoto, M., Endo, S., and Nagayama, K. (1990) *Ultramicroscopy*, **32**, 265.
80 Smith, G. (1985) *Science*, **228**, 1315.

81 Georgiou, G., Stathopoulos, C., Daugherty, P.K., Nayak, A.R., Iverson, B.L., and Curtiss, R., III (1997) *Nature Biotechnol.*, **15**, 29.

82 Sarikaya, M., Tamerler, C., Jen, A.K.-Y., Schulten, K., and Baneyx, F. (2003) *Nature Mater.*, **2**, 577.

83 Patwardhan, S.V., Patwardhan, G., and Perry, C.C. (2007) *J. Mater. Chem.*, **17**, 2875.

84 Baneyx, F., and Schwartz, D.T. (2007) *Curr. Opin. Biotechnol.*, **18**, 312.

85 Galatsis, K., Wang, K.L., Ozkan, M., Ozkan, C.S., Huang, Y., Chang, J., Monbouquette, H.G., Chen, Y., Nealey, P., and Botros, Y. (2010) *Adv. Mater.*, **22**, 769.

86 Matsen, M.W., and Bates, F.S. (1996) *Macromolecules*, **29**, 1091.

87 Matsushita, Y. (2007) *Macromolecules*, **40**, 771.

88 Huang, E., Russell, T., Harrison, C., Chaukin, P.M., Register, R.A., Hawker, C.J., and Mays, J. (1998) *Macromolecules*, **31**, 7641.

89 Sivaniah, E., Hayashi, Y., Matsubara, S., Kiyono, S., Hashimoto, T., Fukunaga, K., Kramer, E.J., and Mates, T. (2005) *Macromolecules*, **38**, 1837.

90 Kim, S.O., Solak, H.H., Stoykovich, M., Ferrier, N.J., de Pablo, J.J., and Nealey, P.F. (2003) *Nature*, **424**, 411.

91 Kim, S.H., Misner, M.J., Xu, T., Kimura, M., and Russell, T.P. (2004) *Adv. Mater.*, **16**, 226.

92 Xu, T., Zhu, Y., Gido, S., and Russell, T. (2004) *Macromolecules*, **37**, 2625.

93 Watanabe, S., Fujiwara, R., Hada, M., Okazaki, Y., and Iyoda, T. (2007) *Angew. Chem. Int. Ed.*, **46**, 1120.

94 Guo, J. (2007) *Adv. Mater.*, **19**, 495.

95 Ofir, Y., Moran, I.W., Subramani, C., Carter, K.R., and Rotello, V.M. (2010) *Adv. Mater.*, **22**, 3608.

Index

Page numbers in *italics* refer to entries in figures or tables.

a

activation/deactivation process   181
adsorbed monolayers   43–45, *46*
adsorption process
– electrostatic   4, 107, *108*
– LbL assembly   114, 115, *116*, *119*, *120*, 127
– readsorption   13, 14
– sulfur   10
affinity step   196, *197*
aggregate morphology
– bilayer membrane   *162*, 162, 166
– control of   173, 175–179
air/water interface
– chiral discrimination   77, 78, *79*
– H-bonding and electrostatic interactions   73–77
aldehyde removal   139
alkanethiols
– coverage and orientation   12, *13*
– desorption and readsorption   13, 14
– electron transfer   28, *29*
– ferrocenylalkanethiols   28, 29–32
– gold reaction   7, 8–15
– nanotribology   26, *27*
– photoinduced electron transfer   31, 32, *33*
– surface coating and printing   21–22, *23*, *24*
N-alkyl carbazole   122, *123*
alkyl chain   8
– IR vibrational bands   60, *61*
alkyl-tri-chlorosilanes   15, *16*
alkyl-tri-methoxysilanes   15
alumina membranes   132
Alzheimer disease   176
ammonia removal   139
ammonium amphiphile   168, 177

amphiphiles
– aggregate-forming   178, *178*
– ammonium   168, 177
– bilayer forming   159–162
– bola-amphiphiles   112, *113*
– chiral   171
– groups   *160*
– solubility in water   160, 161
– turnover during deposition   67–69, *71*
– types   175, *176*
amplification   196, *197*
amyloidoses   176
AND logic gates   88
anionic polyelectrolytes   109, *194*
anodic peak   14, 15
anthracene (An)   87, *88*
antibody capture   *152*
antibody recognition   149
anticancer drugs   153
antifouling coatings   183
arachidic acid   *51*, 54, 55, 62
aragonite   94, *96*
area modulus (k^s)   52
area reduction, A–T isobars   56
area relaxation, isothermal   55
A–T isobars, *see* molecular area–temperature (A–T) isobars
atom-transfer radical polymerization   181
atomic force microscopy (AFM)
– horizontal scooping-up prior to   71
– H–Si(111) surface   18, *19*
– LbL films   117, *118*, 136
– molecular recognition   76, *77*
– organosilane SAMs   *16*
– SAMs   23, 26–27
azimuthal angle   *18*
azobenzene derivatives   98

Organized Organic Ultrathin Films: Fundamentals and Applications, First Edition. Edited by Katsuhiko Ariga.
© 2013 Wiley-VCH Verlag GmbH & Co. KGaA. Published 2013 by Wiley-VCH Verlag GmbH & Co. KGaA.

b

BAM, *see* Brewster-angle microscopy (BAM)
barbituric acid 76
behenic acid 55, *61*, 63
benzene 141, 143
benzonitrile 23
β-sheet structure 162, 176, 178
bilayer membrane 157–159
– aggregate morphology 161, *162*, 162, 165–166
– formation, mechanism and preparation 164–165
– properties 162, 163
bilayer structure 157–159
– cast films containing 166–168
– formation 159–166
– reversed 169, 172
– *see also* lipid bilayers
bilayer vesicle 157–159
– formation *160*, 162, 164
– helical superstructure within 169, *170*
– hierarchic interaction *178*
– life of 166
binary systems *160*, 161
binding constant (K) 74
binding energy (ΔG) 74
biomedical applications
– LbL films 143–153, 154
– polymer brushes 183
biomembrane models 89–93, *158*
biomimetry
– LB films 88, 89
– LbL assembly 134
– mineralization 94, 95, 96, 97, 187
biopanning 196, *197*
biopolymers, LbL assembly 110, 131, *133*
biosensors 148, *149*
biospecific interactions/recognition, LbL assembly 125
biotin–avidin 125
bis(ethylene glycol)mono-n-tetradecyl ether (C14E2) 44, 45
block copolymers 187, 197–200
– diblock 184, *198*, 199
bola-amphiphiles 112, *113*
bora amphiphiles 175, *176*
boronic acid 97, *98*, 150
bottom-up approach 1, *2*, 198, 200
Brewster angle 56
Brewster-angle microscopy (BAM) 56, 57
– Gibbs monolayers 44, 45, 46
– Langmuir monolayers 53
– molecular recognition 76, 78, 79
bulk modulus 52

c

cadmium arachidate 59, 60, 67, 71, 186
cadmium stearate 57, *58*, 67, 68
cadmium sulfide (CdS) 33, 186
calcite 94, 96
calcium arachidate 55, 56
calcium carbonate 94, 95, 96, 97, 187
calcium phosphate 187
calyx[n]arenes (CAs) 80, *81*
capillary forces, colloidal particles 192, 193
capsules
– carbon 141, *142*, 143
– LbL assembly 129, 130
– mesoporous silica 149, *150*, *151*
carbazole derivatives 122, 123, *124*
carbon capsules 141, *142*, 143
carbon radical 17
cast films, bilayer structure within 166–168
catalytic activity, SAMs 34, 36
catechol 23
cathodic peak 13, 15
cationic polyelectrolytes *108*, 109, 114, *194*
cationic polymers 189
cell-surface display 196, 197
cellulose acetate 133
cerasome 112, 134, *136*, 189
CH stretching frequency region *18*, 19
charge-transfer interactions, LbL assembly 125
chemical etching 17, 18
chemical sensors 139–143
chemisorption 8
chiral amphiphiles 171
chiral recognition/discrimination 77, 78, 79
cholesterol 89, *90*, 91–93
circular dichroism spectroscopy 76
clay minerals 185, 186, 190
click chemistry 122, *123*
cloning 196, *197*
cobalt-porphyrin derivative 35
coenzymes, SAMs 35
collapse pressure 50, 52
colloidal crystals 129, 130, 191
colloidal layers 191–195
colloidosome 193
command surface 98
composites, solvophilic/solvophobic parts 175, *176*
compression speed 50, 52, *53*, 54
concanavalin 125
concave active site 36
concentrated polymer brush *179*, 180, 182–184

condensed phases
– BAM detection 56
– Gibbs monolayers 44, 45
– Langmuir monolayers 49, 52, *53*, 54
– liquid condensed phase 48–50, *51*, 52
controlled drug-release 149, 150, 152, 153
convex active site 36
cooperative binding, mesoporous 141
copper sulfate 122, *123*
covalent bonding, LbL assembly 121, 122, *123*
coverage, alkanethiol orientation 12, *13*
Creutzfeldt–Jakob disease (CJD) 176
critical micelle concentration (cmc) 44, 164
critical packing parameter 161
critical transition temperature (T_c) 163, *163*, 165, 166, 167, 170
crown ethers 80, *81*
crystalline complexes, formation *160*, 164, *166*, 195
crystallization, colloidal particles 192
Cu^{2+}-TBEA complex 23, *24*
cusp points 44
cyclic voltammetry (CV)
– Au(111) surface *13*
– gold electrode *28*
– LB films *95*
– LbL films *120*
cyclobis(paraquat-*p*-phenylene) 23, *25*
cyclodextrins (CDs) 25, 80, *81*, 98, 99
cyclohexane 141, *143*
cylinders, block copolymers 198, *199*

d

dangling bonds 17, 19
dark-field light microscopy *165*
Debye–Hückel lengths 192
decanethiol ($C_{10}SH$) 12, *13*
defect density 11
defect-healing process 9
dendrimer
– LbL assembly 110
– PAMAM *109*, 110
dendritic structure, Langmuir monolayer *53*, 54
deposition
– amphiphile turnover during 67–69, *71*
– horizontal lifting-up 69–71, *72*
– instruments for 65–67
– types *63*, *64*
deposition cycle *66*, 67, *70*, 71
deposition records 67
desorption, alkanethiols 13, *14*
diblock-copolymers 184, *198*, *199*

dichroic ratios 60, *61*
didodecyl dimethylammonium bromide 158
dielectric constant (ε) 74
dielectric mirror 139
differential scanning calorimetry (DSC) 163
diglycine derivatives 177
dihydrocholesterol 92
L-α-dimyristoylphosphatidic acid (DMPA) 91
dip pen nanolithography 23
direct methanol fuel cells (DMFCs) 137, *138*
DNA, LbL assembly 110, *148*, 150, *151*
DNA sequencing 196, *197*
N-dodecylgluconamide 77, *78*
domain boundary, SAMs 11, *12*
domain size, copolymers 198
double-chain amphiphiles 175, *176*
double-stripe patterns 14, 15
drug delivery systems 167
– controlled drug release 149, 150, 152, 153
dual-polarization interferometry (DPI) 120
dye 83, *84*, *91*, 114
dynamic host cavity 80, 82, *83*

e

elastomeric slider *27*
electrical conductivity 164
electrochemical coupling, LbL assembly 122, 123, *124*
electrochemical luminescence (ECL) 34, *148*, *149*
electrochromic devices 139
electron acceptor 29, *30*
– *see also* electron donor–acceptor films
electron acceptor–sensitizer–donor systems (A/S/D) 83, *84*, *85*
electron-density profiles, fatty acids *62*
electron donor–acceptor films
– fullerene C_{60} *85*, *86*, *87*
– optical logic gates *88*
electron relay 29, *30*
electron transfer
– photoinduced 29–34
– SAMs 28, 29
electron transfer rate constant 28, 29
electronic devices 198
electrostatic adsorption 4, 107, *108*
electrostatic interactions
– colloidal particles 192–194
– LbL assembly 107, 108, 111, 121

– lipid monolayers 92
– molecular recognition 73–77
elution 196, *197*
environmental uses, LbL films 139, *140*
enzymes
– LB films 97, 144, 145
– LbL assembly 110, 125, *126, 130*, 131, 144–148
– multienzyme reactor *146, 147*
– SAMs 35, *36*
– *see also specific enzymes*
epoxy polymers 190
erucic acid 52, *53*
Escherichia coli 196
etching
– chemical 17, 18
– LbL technique 132, 133
EXOR logic gates *88, 89*
expansivities, Langmuir monolayers 49, 50
external reflection spectroscopy 58

f

fatty acids, *see* long-chain acids
ferrocene 29, 30
ferrocenylalkanethiols 28, 29–32
ferrocenylundecanethiol (FcC$_{11}$SH) 9, *10*, 25
fibers, self-assembled 169–179
fibrous aggregates 169
film thickness, LB method 3
flavin adenine dinucleotide 77
flexible bilayer tube *166*
fluid mosaic model 157, *158*
fluorescence microscopy
– lipid monolayers *91*
– molecular recognition 79
Fourier Transform Infrared (FTIR) spectroscopy 57–61
– aggregate morphology *178, 179*
– molecular recognition 76
fragrance sensor 168
free-radical polymerization 181
friction coefficient 183
friction force microscopy (FFM) 27
frictional properties, SAMs 28
FTIR, *see* Fourier Transform Infrared (FTIR) spectroscopy
fuel cells 137, *138*
fullerene C$_{60}$ 31, 85–87, 126

g

galectin 25
ganglioside 89, *90*

gaseous phase 50, *51*
gauche-defect accumulation 26, 27
gel to liquid crystal phase transition 163, *163*, 167
gelators 173
Gemini amphiphiles 175, *176*
Gibbs free energy 52
Gibbs monolayers 43–45, *46*
glass transition temperature 184
glucoamylase *146, 147*, 148
gluconolactone 147, 148
glucose 147, 150
glucose oxidase (GOD) 97, 118, 147, 148
– reaction activity 144, *145*
– stability *145*
– thermostability *146*
glycerophospholipids 89, *90*
glycogen 125
gold, alkanethiol reaction 7, 8–15
gold electrode *13, 14*
– A/S/D LB films on 84
– catalysis at 35, *36*
– electron transfer 28
– photoinduced electron transfer 31–33
– sensor applications 23, 24
gold film 9, *10–12*
– nanotribology 26, 27
– surface coating 24
graft copolymers 198
graft density *179*, 180, 182
graft polymers 179, 181
grafting-from method 180
grafting-to method 180
graphene 143
graphene-sheet/ionic liquid 143
graphite 143
graphoepitaxy 200
grazing incidence X-ray diffractometry (GIXD) 61–63
growth of Myelin figure 164, 166
guanidinium 74, *77*
gyroid 198, *199*

h

H-bonding, *see* hydrogen bonding
helical superstructure 169–172, *172*
herringbone structure 14, 15
heterogeneous catalysis 36
hexanethiol (C$_6$SH) *13, 14*
hierarchic interaction 177–179
hierarchic structures, LbL assembly 134, 149
hollow capsules 129
hollow tubes 170, 172

homogeneous catalysis   36
horizontal lifting-up deposition   69–71, 72
horizontal scooping-up   71–73
host–guest intermolecular interaction   23, 25, *73, 74*, 76
– dynamic host cavity   80, 82, 83
– macrocyclic hosts   79, 80, *81*
hybrid ultrathin films
– self-standing   190
– see also organic–inorganic hybrids
hydrogen bonding
– biomimetic mineralization   95, 97
– chiral recognition   77, 78
– hierarchic interaction   177, *178*
– immobilization molecular arrangement   175–177, *178*
– LbL assembly   121, *122*
– molecular recognition   73–77
hydrogen-terminated Si surface   17, 19, 21
hydrophobic effect   175
12-hydroxyoctadecanoic acid   169
N-(γ-hydroxypropyl)-tridecanoic acid amide (HTRAA)   45, 46

i
imidazole ligand   87
immobilization
– glucose oxidase   144, *145*
– hydrogen bonding   175–177, *178*
indole   23
infrared (IR) vibrational bands   60, *61*
infrared reflection absorption spectroscopy (IRRAS)   60, *61*
– cadmium stearate   57, *58*
– molecular recognition   76
– PM-IRRAS   59, *60*
infrared spectrum, FcC$_{11}$SH   9, *10*
initiators, living-radical polymerizations   182
ink-jet printing   133
inorganic materials
– LbL assembly   111
– see also organic–inorganic hybrids
interdigit structures   12
intermolecular interactions   173
– colloidal particles   191, 192
– H-bonding   175–177
– organic–inorganic hybrids   185
– see also host–guest intermolecular interaction
interpenetrating network (IPN)-like structures   190

IRRAS, see infrared reflection absorption spectroscopy (IRRAS)
isobaric thermal treatment   56

k
Krafft point (K_p)   163

l
lamellae, block copolymers   198, 199
laminar flow, colloids   192
Langmuir–Blodgett (LB) films   2, 3, 7, 107
– amphiphile turnover during deposition   67–69, *71*
– colloidal particles   194
– concept   43
– enzyme activity   144, 145
– FTIR spectroscopy   57, 58
– functions and applications   73–99
– horizontal lifting-up deposition   69–71, *72*
– horizontal scooping-up   71–73
– instruments for deposition   65–67
– organic–inorganic hybrids   186
– preparation and characterization   43–73
– structure types   63, *64*
Langmuir–Blodgett technique   63–73
– and LbL process   114
Langmuir monolayers   43
– A–T isobars   47–50
– lipids   89–93
– measurement of properties   45–47
– π–A isotherms   50–52
– stability   52–56
– transfer to solid supports   63–73
Langmuir–Schaeffer technique   69
Langmuir trough   46, 47
– horizontal lifting-up deposition   70
– LB film deposition   65, 66
layer-by-layer (LbL) assembly/films   4
– analytical methods   119, 120, *121*
– automated machine   128
– colloidal particles   194
– concept and mechanism   107, 108
– functions and applications   136–153
– microshells   152
– nanoshells   153
– organic–inorganic hybrids   186
– preparation and characterization   109–136
– self-standing hybrid films   190
– three-dimensional   129–136
LB films, see Langmuir–Blodgett (LB) films
LbL assembly, see layer-by-layer (LbL) assembly/films
lead UPD monolayer   23, *24*
lecithin   157, 164, 165

lectin   25, 26
leucine fastener   179
light-driven electron pump   83, 84
line tension   92
lipid bilayers   4, 5
– LbL assembly   113, 114, 134, 136
– organic–inorganic hybrids   189
– see also bilayer structure
lipid layers   74
lipid monolayers   77, 89, 194
– electrostatic interactions   92
– phase diagrams   92, 93
– π–A isotherms and fluorescence microscopy   91
lipid rafts   93
lipid vesicles   112
lipids, Langmuir monolayers   89–93
liposomes   4, 5, 157
liquid condensed (LC) phase   48–50, 51, 52
liquid expanded (LE) phase   48–50, 51, 52
lithography
– dip pen nanolithography   23
– nanoimprint   200
– photolithography   1, 133, 184
living-radical polymerizations   180, 181, 182
long-chain acids
– electron-density profiles   62
– LB structure and deposition   64
– π–A isotherms   50, 51, 52, 55
luminescence, SAMs   34
lung surfactants   93, 94

m
M13 filamentous phage   196
macrocyclic hosts   79, 80, 81
macroscopic interface   74
mad cow disease (BSE)   176
maltoside-terminated alkanethiol (MalC$_{12}$SH)   25, 26
manganese carbonate   150, 151
material-binding peptides   195–197
materials-release profiles   149, 151
3-mercaptopropionic acid (MPA)   23
mesoporous carbon   140, 141, 142, 143
mesoporous silica
– capsules   149, 150, 151
– layers   187
– spheres   130, 131, 132, 133
metal alkoxide, LbL assembly   127
metal–ligand interactions, LbL assembly   121, 122
metal nanoparticles, colloidal layers   195
metal-oxide-semiconductor field effect transistors (MOSFETs)   139

metal oxides
– living-radical polymerizations   182
– organic–inorganic hybrids   189, 190
methyl methacrylate (MMA)   182, 186
methylviologen (MV^{2+})   29
micelles
– critical micelle concentration (cmc)   44, 164
– formation   160, 161, 162–165
microcantilever objects   133, 135
microcontact printing   22, 23
microscopic interface   74
mineralization, biomimetic   94, 95, 96, 97, 187
miscibility   173
molecular area–temperature (A–T) isobars   47–50
– calcium arachidate   55, 56
molecular interface   74
molecular photodiodes   83–85
molecular recognition   73–83
molecular shape   162
monolayers
– in situ characterization at subphase surface   56–63
– see also Gibbs monolayers; Langmuir monolayers; lipid monolayers; self-assembled monolayers
multiasperity contacts   27
multiblock-copolymers   198
multienzyme reactors   146, 147
multilamellae structure   167
multiple-angle incidence resolution spectroscopy (MAIRS)   58
mushroom conformation   179, 180
Myelin figure   164, 165, 166
myristic acid   50, 51, 52

n
N$^+$-C$_4$-Bph-Ala-OC$_{12}$ 171, 172
N$^+$-C$_6$-Azo-Ala-OC$_{12}$ 171, 172
N$^+$-C$_{10}$-Azo-Ala-OC$_{12}$ 172
N$^+$-C$_{11}$-L-Glu(OC$_{12}$)$_2$ 170, 170, 171
N-methyl octadecyl urea   49
Nafion membrane   137
nanoformulation, drugs   153
nanoimprint lithography   200
nanoparticles
– LbL assembly   111
– metal, colloidal layers   195
– semiconductor   32, 33, 195
nanoporous polyelectrolytes   131, 132, 133
nanosheets, LbL assembly   111
nanotribology   26–28
nanotubes, LbL assembly   132, 134

negative staining method   162
neonatal respiratory distress syndrome (NRDS)   93
nitrobenzene   25
nitroxide-mediated polymerization (NMP)   181
number of degrees of freedom (F)   46, 47

o

occupied molecular area (A)   47
– see also surface pressure–occupied molecular area (π–A) isotherms
octadecanoic acid   95, 97
octadecanol   95, 97
n-octadecyl mercaptan (OM)   23, 24
octadecyl urea   49, 50
octadecylsilicon substrate   19, 20
– orientation of chain   20
octadecylsiloxane   16
octanethiol monolayer   11, 24
oil-in-water emulsions   193
opals, formation   191
optical logic gates   87, 88, 89
organic–inorganic hybrids   112, 184–191
– preparation methods   185, 186
– types   185, 186
organic solvents, in gelation   173, 174
organoalkoxysilanes   15, 188, 188, 189
organogels   169, 171–173, 174
organosilanes, on silica surfaces   15–17
orotate   77
Ostwald ripening process   11
overadsorption   108
oxygen reduction reaction (ORR)   35

p

p-polarized radiation   60, 61
p/s band intensity ratios   61
palmitic acid   48, 51
patterned polymer brushes   183
Peltier elements   47
pentadecanoic acid   51, 52
peptides
– H-bonding   175, 177
– material-binding   195–197
permeability, LbL films   148, 150
permeation control   139
pH 2 subphase   51, 54
pH 3 subphase   53
phage-display (PD) method   196, 197
phase diagrams
– behenic acid   63
– block copolymers   198, 199
– lipid monolayers   92, 93

phase separation structures   185, 187
phase transitions
– Langmuir monolayers   50–52
– lipid monolayers   91
phenanthrene   87, 88
phenylboronic acid   150
phosphate bilayer membrane   175
phosphate–guanidinium interaction   74
phosphate–metal interactions   121, 122
phosphatidylcholine   89, 90, 92, 93
phosphatidylglycerol   89, 90
phosphoethanolamine   89, 90
phosphoinositol   89, 90
phosphoserine   89, 90
photoactive LbL films   137
photocurrent, logic gates   88, 89
photocurrent conversion device   123, 124
photodiodes, molecular   83–85
photoelectronics, LB films   83–88
photoinduced electron transfer   29–34
photoisomerization   98
photolithography   1, 133, 184
photopolymerization   68
photoresponsive films   98, 99
photosensitizer   29, 30
photoswitch   87, 88, 89
photovoltaic cells   137, 138
physicochemical applications, LbL films   121, 125, 137–143, 154
Pickering emulsion   193
pin-on-disk tribometry   27, 28
pin-stripe patterns   12
π–A isotherms, see surface pressure–occupied molecular area (π–A)
polarization modulation infrared reflection absorption spectroscopy (PM-IRRAS)   59, 60
polarized microscopy   164, 165
poly(acrylic acid) (PAA)   109, 122, 131, 132, 189
poly(allylamine hydrochloride) (PAH)   109
poly(amidoamine) dendrimer (PAMAM)   109, 110
poly(diallyldimethylammonium chloride) (PDDA)   109, 111, 114, 115, 117, 118, 119, 120, 134, 140
poly(dimethylsiloxane) (PDMS) stamp   22, 23, 133
polyelectrolytes
– bola-amphiphiles   112, 113
– cast films from   168
– clay hybrids   190
– colloidal particles   194
– LbL assembly   108, 109

– – dendrimer   *110*
– – inorganic   111
– – lipid bilayer   *113*, 114, 134, *136*
– – silica   114–120
– – 3D   *130*, 131, *132*, *133*
poly(ethyleneimine) (PEI)   *109*, 146, 147
poly(ferrocenylsilane)   150
polyion-complex   168
polymer brushes
– definition   179
– preparation   180–182
– properties and applications   182–184
poly(methyl methacrylate) (PMMA)   125, 183, 184
poly(p-phenylenevinylene) (PPV)   109
poly(sodium styrenesulfonate) (PSS)   *109*, *110*, 111, 117, 143, 147, 150
poly(sodium vinylsulfonate) (PVS)   109
poly(styrene-*b*-methyl methacrylate) (PS-*b*-PMMA)   199
polystyrene (PS)   184, 193
polyvinyl alcohol (PVA)   189, *190*
porphyrin (Por)   29–32
– CoPor derivative   35
– photoresponsive films   *98*, 99
porphyrin-fullerene dyads   *86*, 87
porphyrin-fullerene layers   126
potassium poly-styrenesulfonate   168
pressure–sliding-speed diagram   27
primary interactions   177, 178
propagation of monomer   181
proteins
– crystals   195
– LbL films   131, 132, *133*
– LbL shells   152
– surfactant proteins   93
– *see also* enzymes

q

quartz crystal microbalance (QCM)   9
– LbL assembly   110, 111, 114, 115, *116*, 127, *128*, 141, 149

r

readsorption, alkanethiols   13, 14
recognition components, LbL assembly   141, *142*, 143
redox reaction, SAMs   28, 29
relaxation time   54
reversed bilayer structure   169, 172
reversible addition-fragmentation chain transfer (RAFT)   181
ribbons, self-assembled   169–179
rosary-like structure   165

rotational anisotropy   19
rotational misfits   11
rugate filter   139
ruthenium tris-bipyridyl complex   34

s

s-polarized radiation   60, *61*
SAMs, *see* self-assembled monolayers (SAMs)
scanning electron microscopy (SEM)
– bilayer membrane   162
– LbL films   *117*
– SAMs   22
scanning probe microscopy (SPM)   23
scanning tunneling microscopy (STM)
– Au(111) surface   *10*, *11*, *12*, *14*, 24
– H-terminated Si(111) surface   20, *21*
– SAMs   9, 10
scrapie   176
secondary interactions   177, 179
sedimentation, colloidal particles   191, *192*
self-assembled fibers, tubes and ribbons   169–179
self-assembled monolayers (SAMs)   2, *3*, 7–37, 107
– advanced applications   28–36
– functions and applications   21–28
– future perspective   36, 37
– preparation and characterization   8–21
– on a solid substrate   8
self-assembly   7, 198
– bottom-up approach   1
– colloidal particles   191
– kinetics   9
self-standing LbL assembly   *135*
self-standing ultrathin films   190
SEM, *see* scanning electron microscopy (SEM)
semiconductor nanoparticles   32, 33, 195
semidilute polymer brush   *179*, 180, 182
sensor applications
– fragrance sensor   168
– LB films   95, 97, 98
– LbL films   139–143, 148, *149*
– SAMs   23, *24*, 25, 26
Si–C bonding   7, 17–21, 29, 31
sick-building syndrome   139
silica
– LbL assembly   111, 114–120
– mesoporous capsules   149, *150*, *151*
– mesoporous layers   187
– mesoporous spheres   130, 131, *132*, *133*
– organosilanes on surface   15–17

silicon substrate   8
– Si(111)-C18   18, *19*
– Si–C bonding   7, 17–21, 29, 31
single-asperity contacts   27
size-exclusion properties, polymer brushes   183, 184
small-angle X-ray scattering (SAXS)   114
soft contacts   27
sol-gel reaction
– LbL assembly   126, *127*
– organic–inorganic hybrids   188, 189
solar cells   137
solid phase   50, *51*, 52
solid substrate
– LB film fabrication   63–73
– SAM   8
solubility in water, amphiphiles   160, 161
soluble monolayers, *see* Gibbs monolayers
solvophilic/solvophobic parts   175, *176*
sonication, surfactants   157, *158*
spermidine   150, *151*
spheres, block copolymers   198, 199
sphingolipids   89, *90*, 93
sphingomyelin   89, *90*, 93
sphingosine chain   89, *90*
spin-coating
– colloidal particles   193
– hybrid films   190
– LbL assembly   127, *128*
spray LbL assembly   128, *129*
stacking misfits   11, *12*
star-shaped copolymers   198
stearic acid   50
stereocomplexation   125, *126*
steroid cyclophanes   80, *82*
stimuli-free controlled release   149, *151*
STM, *see* scanning tunneling microscopy (STM)
storage stability, LbL films   145
strain rate of compression   54
styrene   20, *21*
subphase surface
– heating and cooling   48
– *in situ* characterization of monolayers   56–63
substrate   8
– block copolymer films   199
– LB film fabrication   63–73
sugar guest recognition   97, *98*
sugars, H-bonding   176
sulfur adsorption sites   10
sum frequency generation (SFG) spectroscopy   18, *19*
supramolecular polymers   176

supramolecular recognition   *126*
surface balance   46, 48, 65
surface binding group   8
surface coating and patterning, SAMs   22, 23, *24*
surface compressibility   52
surface-enhanced Raman scattering (SERS)   120
surface force analysis   108
surface-initiated graft polymerization   181, *182*
surface plasmon resonance (SPR)   120
surface pressure
– Gibbs monolayer   44
– Langmuir monolayer   48–52
surface pressure–occupied molecular area (π–A) isotherms   45, 46, 47, 50–52
– arachidic acid   54
– behenic acid   63
– erucic acid   53
– lipid monolayers   91
– molecular recognition   76
– time of observation   54, 55
surface pressure–temperature (π–T) isochore   47
surface pressure–time (π–t) adsorption isotherm   44, 45, 46
surface tension   43, 44, 48
surfactant proteins   93
surfactants
– colloidal particles and   193
– crystalline complex   166
– formation   *160*, 161
– Gibbs monolayers   43–45, *46*
– lung surfactants   93, 94
– micelle formation from   164
– sonication   157, *158*

t

tannic acid   140, 141
Teflon barriers   66
– Langmuir monolayers   46, *47*
tellurium nanoclusters   33
TEM, *see* transmission electron microscopy (TEM)
terminal group   8, 10
N-tetradecyl-γ,δ-dihydroxypentanoic acid amide (TDHPA)   78
thermal expansivities   49, 50
thermostability
– alkanethiol SAMs   26, 27
– LbL films   145, *146*
2,2´-thiobisethyl acetoacetate (TBEA)   23, *24*

thiol (-SH) group 7, 8
three-dimensional LbL assembly 129–136
tilt angle 9
– misfits 11, 12
– octadecyl chain 20
time of observation 54, 55
titanium oxide 137
6-(*p*-toluidino)naphthalene-2-sulfonate (TNS) 82
top-down approach 1, 2, 197–199
transition moment 57, 59, 60
transition region 48, 49
transmission electron microscopy (TEM)
– bilayer membrane 158, 162
– cast films 168
– colloidal layers 195
– helical superstructure 172
– LbL assembly 134
transmission spectra, cadmium stearate 57, 58
triad molecules 84, 85
tribology
– nanotribology 26, 27
– polymer brushes 183
trichlorosilyl (-SiCl$_3$) group 7, 8, 15, 16
tri(ethylene glycol)-alkanethiol (EGC$_2$SH) 25, 26
trimethoxysilyl (-Si(OCH$_3$)$_3$) group 7, 8, 15
triple-chain amphiphiles 175, 176
truth table 88, 89
tubes
– flexible bilayer 165
– hollow 169, 172
– LbL assembly 132, 134
– self-assembled 169–179
turnover, amphiphiles 67–69, 71

u

ultrathin assemblies, preparation methodologies 1, 2
2,4-di(*n*-undecylamino)-6-amino-1,3,5-triazine (2C$_{11}$H$_{23}$-melamine) 76

underpotentially deposited (UPD) lead monolayer 23, 24
untilting 26, 27
uracil 76
uranyl acetate 162

v

vacancy islands (VIs) 11, 12, 15
van der Waals interaction 8
vaterite 94, 96
vertical dipping technique 63–73
vesicles 4, 5

w

wash step 196, 197
water contact angle 193
water evaporation 149
water-in-oil emulsions 193
wettability control 139
Wilhelmy-type surface balance 46, 48, 65
Wilhemy plate 60

x

X-ray reflectometry 61–63
X-type deposition 63, 64
X-type LB film structures 63, 64, 65
– from horizontal lifting-up deposition 69, 71

y

Y-type deposition 63, 64
– horizontal lifting-up 69, 71
Y-type LB film structures 64, 65, 67, 68
– from horizontal lifting-up deposition 71, 72
yttrium iron garnet 111

z

Z-type deposition 63, 64
– turnover of amphiphiles 67, 68
Z-type LB film structures 64, 65
zinc porphyrin 86, 87